Extreme Low-Power Mixed Signal IC Design

Armin Tajalli • Yusuf Leblebici

Extreme Low-Power Mixed Signal IC Design

Subthreshold Source-Coupled Circuits

Springer

Armin Tajalli
Ecole Polytechnique Fédérale
de Lausanne (EPFL)
Microelectronic Systems Lab. (LSM)
Station 11, 1015 Lausanne
Switzerland
armin.tajalli@epfl.ch

Yusuf Leblebici
Ecole Polytechnique Fédérale
de Lausanne (EPFL)
Microelectronic Systems Lab. (LSM)
Station 11, 1015 Lausanne
Switzerland
yusuf.leblebici@epfl.ch

ISBN 978-1-4419-6477-9 e-ISBN 978-1-4419-6478-6
DOI 10.1007/978-1-4419-6478-6
Springer New York Dordrecht Heidelberg London

Library of Congress Control Number: 2010934294

Printed on acid-free paper

Springer is part of Springer Science+Business Media (www.springer.com)

*To my father, Hossein, my mother, Maryam,
my wife, Paris, my little daughter, Ayrine and
my family: Azin, Ali, and Alaleh.*

–Armin Tajalli

Contents

Part I Scalable and Ultra-Low-Power Digital Integrated Circuits

List of Figures

List of Tables

Acknowledgments

Many people have helped us in preparing this book. Professor Eric Vittoz (EPFL & CSEM, Switzerland) has kindly supported this work by his valuable hints and feedbacks. His deep knowledge in the field of Microelectronics provided this opportunity for us to understand and go deeper into the subject. Some parts of this work are mainly devoted to close collaboration with Prof. Elizabeth J. Brauer (North Arizona University) and Prof. Massimo Alioto (University of Siena), and we would like to appreciate them for their very useful hints and helps.

We would also like to appreciate all the people who have helped us accomplish this work. Special thanks goes to Stéphane Badel for his very valuable help during physical design of test chips; Milos Stanisavljevic, Michele Mercaldi, and Bertrand Rey for their contribution in design of multiplier circuit; Mohammad Beikahmadi for design of ADC encoder and standard cell libraries; Nikola Katic for behavioral modeling of $\Delta\Sigma$ modulator; and Sylvain Hauser who provided the test setups for prototype measurements.

We would also like to appreciate Alain Vachoux and Alexandre Schmid for their kind technical support during this work. We are grateful to our colleagues in Microelectronic Systems Laboratory (LSM) for the very nice time and fruitful discussions and collaborations: Thomas Liechti, Vahid Majidzadeh Bafar, Torsten Mahne, Milos Stanisavljevic, Hossein Afshari, Yuksel Temiz, Niel Joye, Fengda Sun, and Alessandro Cevrero.

Acronyms

$\Delta\Sigma$	Delta-sigma modulator
ADC	Analog-to-digital data converter
Amp, AMP	Amplifier
AMS	Analog-mixed-signal
ASIC	Application specific integrated circuit
BJT	Bipolar junction transistors
BMS	Battery management system
BW	Bandwidth
CAD	Computer aided design
CCO	Current-controlled oscillator
CK, CLK	Clock signal
CML	Current-mode logic
CMOS	Complementary MOS
CPC	Charge-pump circuit
CT	Continuous-time
DAC	Digital-to-analog date converter
DEM	Dynamic element matching
DFF	D-type flip-flop
DIBL	Drain-induced barrier lowering
DPM	Dynamic power management
DT	Discrete-time
DR	Dynamic range
DRC	Design rule check
DVS	Dynamic voltage scaling
DWA	Dynamic weighted averaging
EDP	Energy-delay product
FAI	Folding and interpolating ADC
FoM	Figure of merit
FIR	Finite impulse response (digital filters)
FN	Fowler–Nordheim
FPAA	Field programmable analog array
FPGA	Field programmable gate array
g_m-C	transconductance-C filter

HDL	Hardware design language
HVT	High threshold voltage MOS device
IC	Integrated circuit
IC	Inversion coefficient
IIR	Infinite impulse response (digital filters)
LER	Line edge roughness
LSB	Least significant bit
LVS	Layout versus schematic check
LVT	Low threshold voltage MOS device
MCML	MOS current-mode logic
MI	Medium inversion
MOS	Metal-oxide-semiconductor solid-state device
MOSFET	Metal-oxide-semiconductor field-effect transistor
MOSFET-C	MOSFET-C filter continuous-time topology
MSB	Most significant bit
MTCMOS	Multi-threshold CMOS technology/topology
NM	Noise margin
NRZ	Nonreturn to zero
NTF	Noise transfer function
Op Amp	Operational amplifier
OSR	Over-sampling ratio
OTA	Operational transconductance amplifier
PAR	Place and rout
P_{diss}	Power dissipation
PDP	Power-delay product
PFD	Phase-frequency detector
PLL	Phase-locked loop
PVT	Process, voltage (supply), and temperature variation
$R\Delta\Sigma$	Ring oscillator based delta–sigma modulator
RB	Replica bias
RCX	Resistor/capacitor extractor
RD	Read signal in memory
RDF	Random dopant fluctuation
REF	Reference (voltage, current, frequency, etc.)
RMS	Root mean square
ROC	Ring oscillator based quantizer
RZ	Return to zero
SA	Sense amplifier
SCE	Short channel effect
SCL	Source-coupled logic
SFB	Source follower buffer
SI	Strong inversion
SNDR	Signal-to-noise and -distortion ratio
SNM	Static noise margin
SNR	Signal-to-noise ratio

SRAM	Static random access memory
STF	Signal transfer function
STSCL	Subthreshold source-coupled logic
UDSM	Ultra-deep-sub-micron technology
ULP	Ultra-low power
VCO	Voltage-controlled oscillator
VHDL	Versatile hardware design language
VLSI	Very large scale integration
V_T, V_{TH}	Threshold voltage of MOS devices
WI	Weak inversion
WR	Write signal in memory
WSN	Wireless sensor network
XOR	Exclusive-or logic gate

Chapter 1
Introduction

Design flexibility and *power consumption* in addition to the cost, have always been the most important issues in design of integrated circuits (ICs), and are the main concerns of this research, as well.

Energy Consumptions: Power dissipation (P_{diss}) and energy consumption are especially important when there is a limited amount of power budget or limited source of energy. Very common examples are portable systems where the battery life time depends on system power consumption. Many different techniques have been developed to reduce or manage the circuit power consumption in this type of systems. Ultra-low power (ULP) applications are another examples where power dissipation is the primary design issue. In such applications, the power budget is so restricted that very special circuit and system level design techniques are needed to satisfy the requirements. Circuits employed in applications such as wireless sensor networks (WSN), wearable battery powered systems [1], and implantable circuits for biological applications need to consume very low amount of power such that the entire system can survive for a very long time without the need for changing or recharging battery [2–4]. Using new power supply techniques such as energy harvesting [5] and printable batteries [6], is another reason for reducing power dissipation. Developing special design techniques for implementing low power circuits [7–9], as well as dynamic power management (DPM) schemes [10] are the two main approaches to control the system power consumption.

Design Flexibility: *Design flexibility* is the other important issue in modern integrated systems. There are many applications requiring integrated systems with a reconfigurable characteristics [11]. This property enables users to employ a system at different applications or at different situations without significant extra cost. Many new electronic products are designed to be used in different standards. Modern handheld devices, for example, are pocket sized computing equipments with capability of covering different applications or standards [12].

In some designs, reconfigurability is considered as the main specification of a system. For example, to optimize the power consumption versus frequency of operation (f_{op}), a system should bear a very wide tuning range. In such systems, power consumption is adjusted with respect to the operating frequency in a very wide

A. Tajalli and Y. Leblebici, *Extreme Low-Power Mixed Signal IC Design: Subthreshold Source-Coupled Circuits*, DOI 10.1007/978-1-4419-6478-6_1,

range [13]. The DPM concept in digital systems has been developed based on this property of the digital CMOS[1] circuits in which operating conditions of circuit can be adjusted over a very wide range.

Subthreshold MOS: Exponential I–V characteristics of subthreshold MOS devices [14] provides this opportunity to operate the circuit in a very wide bias current conditions with very small variation on the bias voltage levels. In other words, subthreshold MOS devices are suitable for implementing current-mode circuits with very wide tunability. The possibility to change the bias current in a wide range, especially provides appropriate bases to construct wide frequency tuning range circuits.

The other interesting property of the subthreshold MOS devices is that they are operating in a very low current density levels which is very convenient for ULP applications. Meanwhile, the devices in this regime exhibit maximum transconductance (g_m) to bias current (I_{DS}) ratio, i.e., g_m/I_{DS}, that means power efficiency of the MOS circuit can be maximized in this regime [7, 14].

Research Aspects: In this book, the main properties of subthreshold CMOS device for implementing flexible and ULP circuit will be exploited. As will be seen later, subthreshold MOS devices can be employed to implement very low power analog and digital integrated circuit with adjustable characteristics in a very wide range. Using subthreshold MOS devices, the main building blocks for constructing a mixed-signal integrated circuit with a very wide tuning range and very low power consumption will be developed. In the proposed circuits, power consumption scales proportional to the operating frequency. While the tunability of power consumption versus operating frequency is the main concern of this work, the possibility for changing the other parameters such as dynamic range (DR) in analog-to-digital converters (ADC), will also be investigated.

1.1 Applications of Widely Adjustable Circuits and Systems

Flexibility in adjusting the specifications of a circuit or system can be applied to different parameters such as operation frequency (f_{op}), dynamic range (DR), power consumption (P_{diss}), and even functionality. This concept is especially well developed in digital circuits where wide flexibility can be attained using CMOS logic topology [15]. The capability for reconfiguring the functionality and performance as well as possibility for changing the operation frequency in a very wide range make static digital CMOS circuits very suitable for implementing *flexible* or *reconfigurable* integrated systems. In addition, a top-level controlling systems can be employed to adjust the supply voltage of CMOS digital circuits with respect to the operation frequency or throughput, and hence optimize the system power consumption with respect to the work load [13]. Field programmable gate array (FPGA) circuits are good examples for reconfigurable digital integrated systems.

[1] Complementary metal-oxide-semiconductor (CMOS)

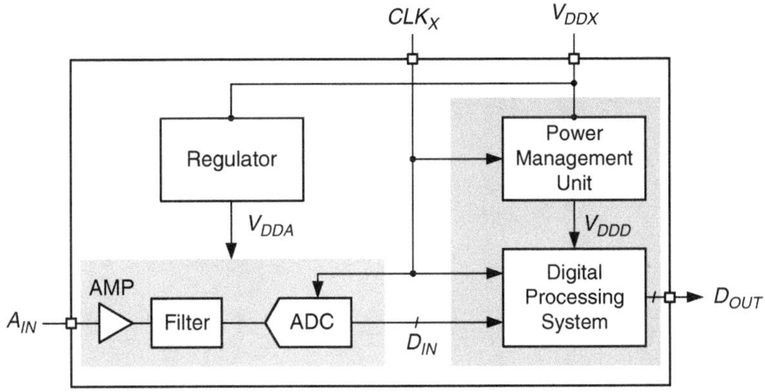

Fig. 1.1 Generic mixed-mode integrated system with a dynamic power management for digital part

In contrary, implementing flexible or reconfigurable analog integrated circuits is very challenging. Most of the conventional analog circuits can tolerate only a few percent of variation over their biasing condition. For example, the maximum acceptable variation on supply voltage of an analog circuit must of the time does not exceed 10–25%, depending on the design. This statement is also true for internal biasing condition of an analog circuit. Variation of a couple of tens of millivolts, can simply move a transistor from active region to linear region and hence reduce the circuit performance.

This limitation on scalability of analog circuits reduces the efficiency of power management of digital section employed inside a larger mixed-mode integrated system, such as the system depicted in Fig. 1.1. In this structure, a simple analog front-end is used to amplify the input signal, filter the noise and unnecessary signals, and then convert the analog signal to digital signal using an ADC. Digital part can do more precise and complicated processing on the signal and make it ready for the final usage. Due to the sensitivity of the analog circuits to supply variation, generally a precise regulator is needed to produce the appropriate supply voltage for this part of the circuit. This regulator can also reduce the noise injection from digital part to the sensitive analog front-end.

As illustrated in Fig. 1.1, the digital section benefits a dynamic supply voltage scaling (DVS) scheme for controlling the power consumption of this part with respect to the system clock frequency [16]. Whenever the input data rate is reducing, DVS will reduce the supply voltage in order to lower the power consumption of the digital part.

Figure 1.2 illustrates a more demanding configuration in which a central power management unit controls the power consumption of both digital and analog parts with respect to the input data rate. This unit generates proper supply voltage and internal clock frequency for digital part. For analog section, power consumption can be adjusted through a proper signal (here, a bias current I_{BA} has been used).

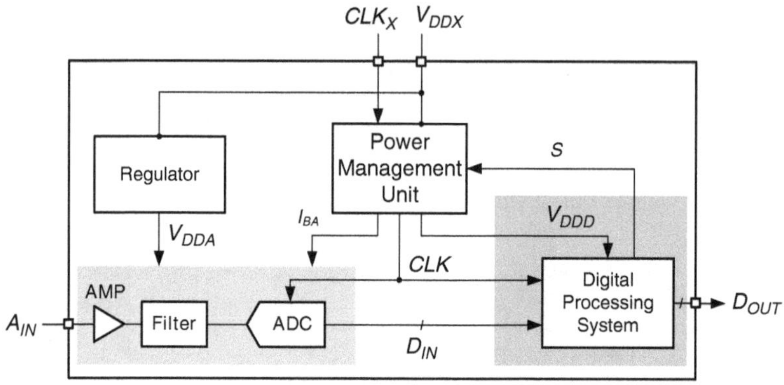

Fig. 1.2 A mixed-mode integrated system with dynamic power management for the entire system

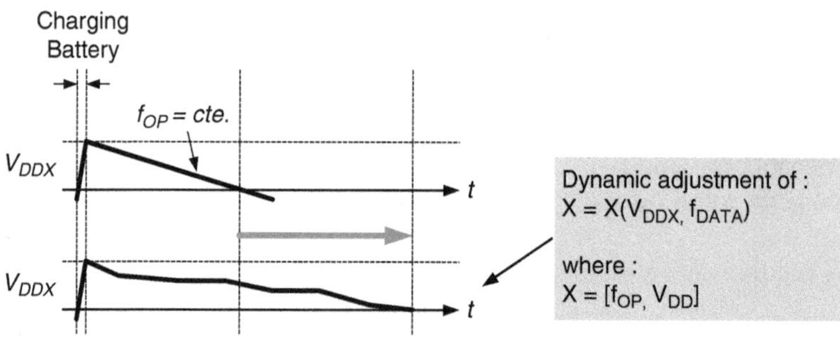

Fig. 1.3 Conceptual timing diagram for two systems, one without battery management system and the other one with a system controlling the power dissipation with respect to the battery voltage and data throughput

An internal phase-locked loop (PLL), for example, can generate the internal clock of the system (CLK). In this configuration, signal S generated by the digital section is used to indicate the required speed of operation. In a more general case, signal S can be generated by other parts of the system. For example, in a battery operating system, battery supply voltage (V_{DDX}) can be used as a measure for adjusting the system power consumption and hence controlling the battery life time[2], as illustrated in Fig. 1.3 [17, 18].

To implement such a system, it is necessary to design analog and digital circuits that can be operated in a wide frequency range with scalable power consumption. In addition to a wide tuning range for operating frequency (f_{op}) and power dissipation (P_{diss}), adjusting the dynamic range (DR) of the system can also help to implement a more power efficient system. In analog circuits, generally P_{diss} has a strong dependence on DR and hence a small reduction on DR (when the system can tolerate it)

[2] Battery management system (BMS)

can help to save considerable amount of power. It can be shown that the minimum power consumption of an class-A analog circuit is approximately [19]:

$$P = \frac{8\pi kT}{\eta_V \cdot \eta_I} \times SNR \times f_{op} \tag{1.1}$$

where $\eta_V = V_{in,pp}/V_{DD}$ is the ratio between the peak-to-peak signal swing and supply voltage, η_I is the efficiency of using supply current, k is Boltzmann's constant, T stands for temperature, and SNR is signal-to-noise ratio. From (1.1), it is clear that the circuit power consumption increases with SNR and operation frequency. Here, it is assumed that the integrated noise voltage and the required bias current of a class-A circuit are: $\overline{v_n^2} = kT/C$ and $I_{bias} = 2\pi f_{op} C \hat{V}$, where $\hat{V}$ is the signal voltage swing. Including distortion, the required power consumption can be even more. By technology scaling, η_V and η_I can change considerably and make the circuit less power efficient [19].

In many modern applications, such as biological products, implantable systems, and sensor networks, using a power management scheme similar to Fig. 1.2 is going to be unavoidable. In these applications, power consumption is extremely critical and it is necessary to develop more advance technique for controlling the system power dissipation.

1.1.1 Performance Scalability and Requirements

Most of the integrated circuits are designed to be operational with an acceptable performance even if the biasing or environmental conditions are changing. Having enough tuning range also makes it possible to adjust the circuit specification on desired conditions using some auxiliary circuits. However, generally the adjustability range of circuits are quite limited. Figure 1.4 describes the operation of a circuit when the biasing condition is changing. In this figure, B_0 represents the nominal

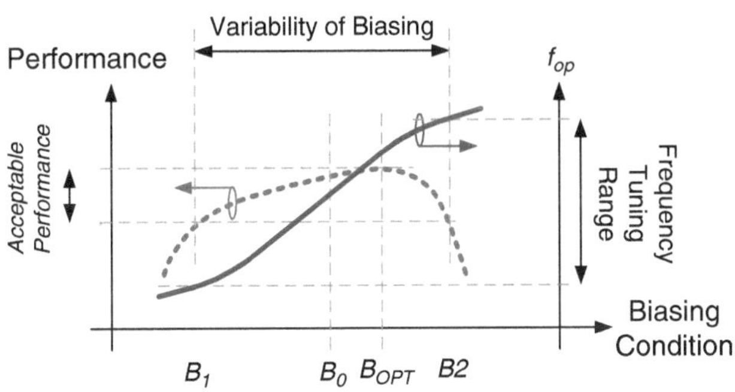

Fig. 1.4 Conceptual diagram to explain the acceptable frequency tuning range. Here, B_0 represents the nominal biasing condition and B_{opt} is the optimum bias point to maximize the performance

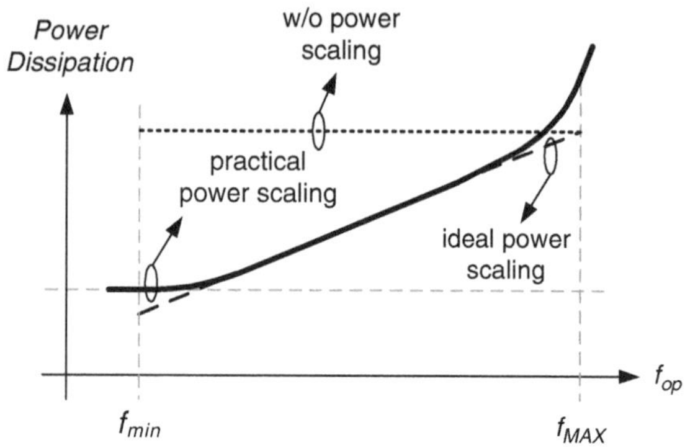

Fig. 1.5 Power-efficient frequency-scaling

biasing condition which is generally very close to the optimum operation condition, B_{OPT}. As long as the performance of the circuit remains within an acceptable range, the bias current can be changed (B_1–B_2) and corresponding to that it is possible to change the tunable parameter of circuit (which is operation frequency f_{op} in Fig. 1.4).

Power efficiency (η_P) is one of the main concerns in design of widely adjustable circuits. Scaling the operation frequency without scaling the power consumption will result in a design with very poor power efficiency. As shown in Fig. 1.5, to have a successful widely tunable circuit, it is necessary to scale the power, although in practice it might be impossible to keep the power efficiency constant for the entire tuning range. Close to lower frequency limit, generally the bias current of the periphery circuits, and also stand-by or leakage current become comparable to the circuit power consumption and hence the efficiency will drop in this region. Also, in very high frequencies, the effect of parasitic capacitances and other nonideality effects prevents having a linear power versus frequency scaling. Therefore, the power efficiency in high frequencies will not be as good as the power efficiency in the medium frequency ranges.

1.2 Prior Art

1.2.1 Digital Circuits

1.2.1.1 Static CMOS Logic

As mentioned before, the concept of power–frequency scalability has been extensively exploited in CMOS digital circuits and systems mainly for power minimization purpose. Illustrated in Fig. 1.6a, it is possible to change the maximum operation frequency of a CMOS digital circuit by adjusting the supply voltage. Hence, it is possible to adjust the operating frequency with respect to the work load.

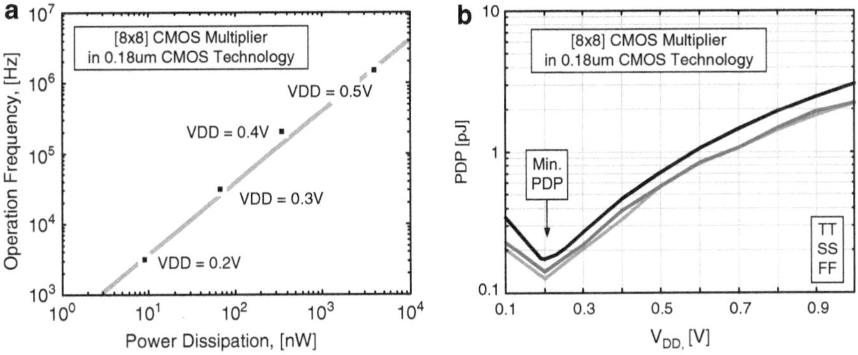

Fig. 1.6 (**a**) Simulated tuning range of a CMOS (8×8) Cary–Save multiplier achieved by adjusting the power supply designed in CMOS 0.18 μm. The tuning range can be extended even more by increasing the supply voltage (V_{DD}) above 0.5 V. (**b**) Simulated power-delay product this circuit versus supply voltage in different corner cases

This wide variability gives the possibility to optimize the system performance. As illustrated in Fig. 1.6b, it is possible to find a specific supply voltage (V_{DD}) to optimize the circuit performance. In this figure, power-delay product (PDP) or in other words power consumption per operation has been selected as a figure of merit (FOM), although other measures can be also utilized. It is noticeable that the optimum point is almost independent to the process corners. In high supply voltages, the main part of the power consumption is due to the dynamic power dissipation while in low supply voltages, the power consumption is dominated by the leakage current (mainly subthreshold leakage current). At very low supply voltages (or equivalently low operating frequencies), leakage currents construct the dominant part of the power consumption. Therefore, in this region of operation reducing the supply voltage does not help very much to reduce the dynamic energy consumption.

1.2.1.2 Other Logic Styles

Other types of digital circuits generally show a wide tuning range. Among them, source-coupled logic circuits (SCLs) or MOS current-mode logic (MCML) circuits are more popular for implementing mixed-mode circuits. Low sensitivity to the supply voltage in addition to low noise injection to the supply lines or substrate is mainly desirable for designing high performance circuits [20–22]. As will be shown in Chap. 3, in this topology there is a very good control on the circuit power consumption that makes it very attractive for ULP applications. Implemented in subthreshold region, this topology can also provide a very wide tuning range. The tuning range when the circuit is biased in strong inversion is limited.

1.2.2 Analog Circuits

1.2.2.1 Circuits Using Switchable (Programmable) Components

Achieving a wide tuning range in analog circuits, on the other hand, is very challenging. Conventional design techniques alow for less than 10–20% variations on the biasing condition of a circuit which is only sufficient to compensate for the process, supply voltage, and temperature (environmental) variations[3]. There are few circuits reported with a relatively wide tuning range without using switchable or programmable components [23]. Using switchable components and blocks is one possible approach that has been used to increase the tuning range of the circuits [24]. In this approach, passive or active switchable components can be utilized to increase the adjustability range.

Figure 1.7 shows an example in which a wide tuning range transconductor-C integrator has been implemented using switchable transconductors (G_m) and capacitors. In this way, it is possible to adjust the filter cutoff frequency linearly by changing the load capacitance or transconductance values as described by:

$$H(s) = \frac{G_m/C}{s}. \tag{1.2}$$

In this simple example, changing the value of transconductance by switching G_m cells will result in different cutoff frequencies, while power consumption will be also scaled proportional to the equivalent G_m value. On the other hand, switching the capacitors can show the same result in cutoff frequency without changing the power consumption. In the latter case, dynamic range or more precisely noise level changes by modifying the capacitance, and hence for high cutoff frequencies

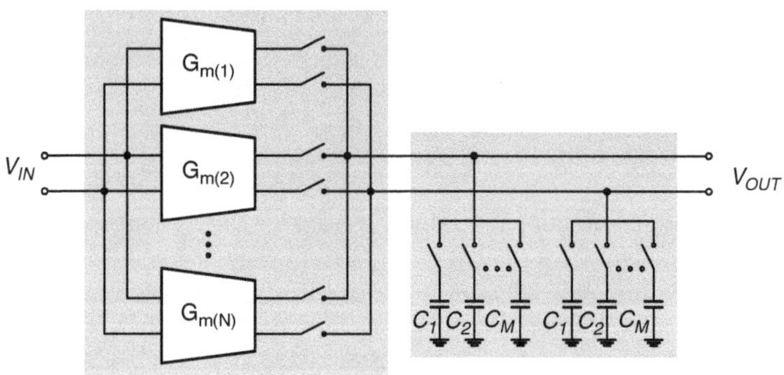

Fig. 1.7 Programmable continuous-time integrator uses switchable capacitors and transconductors to adjust the cutoff frequency

[3] PVT: process, supply voltage, and temperature (environmental) variations

where the size of the load capacitors is less, noise level can be very high. It is clear that both approaches for adjusting the cutoff frequency would need very large silicon area.

This approach has been used for implementing different analog building blocks such as transconductor-C filters [24] and MOSFET-C filters [25]. However, it becomes quickly difficult and inefficient to use this approach for implementing more complex analog blocks such as data converters.

1.2.2.2 Switched-Capacitor Circuits

The other possibility for implementing flexible analog integrated circuits is using discrete-time (switched-capacitor) analog circuits [26]. In this type of circuits, it is possible to adjust the frequency characteristics of circuits using an external clock. For example, Fig. 1.8 shows a low-pass switched-capacitor filter in which the cutoff frequency is:

$$f_c = f_{\mathrm{CLK}} \cdot \frac{C_S}{C_I}. \tag{1.3}$$

where f_{CLK} stands for the clock frequency. Therefore, it is possible to adjust the filter cutoff frequency precisely by adjusting the clock frequency in a relatively wide range [27]. Of course it is necessary to scale the power consumption of the amplifier used in this switched-capacitor filter to keep the amplifier non-ideality effects negligible. This needs to design an amplifier with a capability of changing the bias current in a very wide range. As will be explained in Chap. 7, implementing such an amplifier in subthreshold region is possible. Hence, switched-capacitor circuits can be successfully employed in design of widely tunable analog circuits.

Since the capacitors are constant in this scheme, DR of the circuit remains unchanged. Possibility for changing the DR by changing the size of capacitors (e.g., similar to the technique shown in Fig. 1.7), provides more flexibility. In this case, DR can be reduced by reducing the size of capacitors. Thereby, it is possible to reduce the power consumption of the amplifier proportional to the size of capacitors

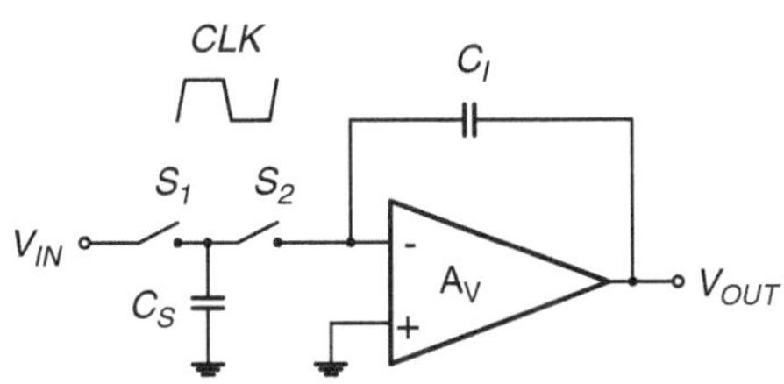

Fig. 1.8 A simplified switched-capacitor integrator. The capacitor C_S and the switches S_1 and S_2 are resembling a resistance. The charge transfer of this resistance depends on the clock frequency as well as the size of C_S (sampling capacitance). Therefore, the cutoff frequency of the entire circuit depends on clock frequency and the size of sampling capacitor as indicated in (1.3)

Fig. 1.9 Companding technique for implementing high DR circuits [29]

and hence have a more power efficient circuit. In this way, switched-capacitor topology can offer a power scalable circuit with respect to both operation frequency and also DR.

1.2.2.3 Log-Domain Circuits

This type of circuits are based on the logarithmic I–V characteristics of semiconductor devices. This property makes it possible to change the bias current in a very wide range while the bias voltages change slightly proportional to the logarithm of bias current. Therefore, it is possible to change the circuit bias current over a few decades, and hence have a relatively wide tuning range. Bipolar transistors as well as MOS devices biased in subthreshold regime are exhibiting logarithmic characteristics and can be utilized for this purpose. This technique has been used to implement log-domain filters with very wide tuning range while the cutoff frequency is adjusted by changing the bias current [28].

The logarithmic (exponential) I–V characteristics of semiconductor devices can also be exploited the *companding* technique for implementing high DR circuits [29]. Based on this approach, the input signal is first compacted by a nonlinear circuit, $z = f(x)$. Then, the required processing will be done on the companded signal using an appropriate nonlinear circuit. Finally, the signal is converted back using another nonlinear circuit with the inverse transfer function of the input stage, i.e., $y = f^{-1}(w)$. A simple block diagram of a companding architecture is shown in Fig. 1.9. This technique is especially attractive for low voltage designs where companding technique helps to reduce the required voltage headroom of the circuitry. The log-domain filter family is a specific example for companding systems.

1.3 Organization

Design and implementing *widely adjustable* integrated mixed-signal system with *very low power consumption* are the main concerns of this work. To achieve the required specifications, some new techniques are developed based on the intrinsic characteristics of the subthreshold MOS devices. Here is the organization of this report.

Before going into the details, Chap. 2 gives a short overview on the physics and the modeling concepts of MOS devices biased in subthreshold regime. This chapter also reviews very briefly the main leakage mechanisms in CMOS digital circuits. An analytical approach for studying the main issues in design of ULP digital CMOS circuits has been also described in this chapter. In Chap. 3, some new techniques for implementing ULP source-coupled logic (SCL) circuits will be explained. Using subthreshold SCL (STSCL) circuits instead of conventional static CMOS logic style, provides the possibility of reducing circuit power consumption well below the limitation of static CMOS circuits which is mainly due to the subthreshold leakage (residual channel) current.

To implement complicated STSCL digital systems, a library of standard cells is required. Implementing high performance and optimized standard cell libraries is briefly discussed in Chap. 4.

Chapter 5 will describe some techniques for improving the performance of STSCL circuits. Although STSCL circuits can be employed to reduce the power consumption, however, still conventional static CMOS circuits can exhibit better power-delay performance in some specific conditions. The techniques developed in this chapter will help to make the performance of the STSCL systems comparable or better than their CMOS counterparts. To complete the discussion, Chap. 6 deals with some techniques for implementing compact and low leakage memory elements. This chapter also studies the performance of STSCL circuits in very low activity rate conditions.

Continuous-time filters (CTFs) and ADCs are the two main analog building blocks for implementing a mixed-signal system. In Chap. 7, two different approaches are developed to implement CTFs with widely adjustable cutoff frequency. The continuous-time MOSFET-C and transconductor-C filters introduced in this chapter, both are exhibiting a very wide frequency tuning range while consuming proportional to their cutoff frequency.

In Chaps. 8 and 9, two different concepts for developingfolding-and-interpolating (FAI) and also $\Delta\Sigma$ ADCs have been proposed. Both ADCs exhibit a very wide tuning range and proportionally scalable power consumption. The proposed FAI ADC can be utilized in medium range resolution applications, while the $\Delta\Sigma$ ADC can be employed for high DR systems.

Chapter 10 brings some design techniques for implementing very wide tuning range phase-locked loop (PLL) circuits. As described in Sect. 1.1, PLLs can be utilized to adjust the operating conditions of the digital or the analog circuits in a mixed-signal system.

The work concludes with a summary on the main results achieved by the proposed approaches, and also the main contributions of this research in Chap. 11. This chapter also includes the perspectives offered by this research.

References

1. K. Ueno, T. Hirose, T. Asai, and Y. "CMOS smart sensor for monitoring the quality of perishables," *IEEE J. Solid-State Circuits*, vol. 42, no. 4, pp. 798–803, Apr. 2007
2. T.-H. Lin, W. J. Kaiser, and G. J. Pottie, "Integrated low-power communication systems design for wireless sensor networks," in *IEEE Communications Magazine*, pp. 142–150, Dec. 2004

3. D. Suvakovic and C.A.T. Salama, "A low V_t CMOS implantation of an LPLV digital filter core for portable audio applications," in *IEEE Transactions on Circuits and Systems-II: Analog and Digital Signal Processing*, vol. 47, no. 11, pp. 1297–1300, Nov. 2000
4. L. S. Wong, and et al., "A very low-power CMOS mixed-signal IC for implantable pacemaker applications," *IEEE J. Solid-State Circuits*, vol. 39, no. 12, pp. 2446–2456, Dec. 2004
5. D. Steingart, S. Roundy, P. Wright, and J. W. Evans, "Micropower ma terials development for wireless sensor networks," *MRS Bull.*, vol. 33, no. 4, pp. 408–409, Apr. 2008
6. D. Steingart, C. C. Ho, J. Salminen, J. W. Evans, and P. Wright, "Dispenser printing of solid polymer-ionic liquid electrolyte for lithium ion cells," in *IEEE International Conference on Polymers and Adhesives in 139 Microelectronics and Photonics (Polytronics 2007)*, pp. 261–264, Jan. 2007
7. E. Vittoz and J. Fellrath, "CMOS analog integrated circuits based on weak inversion operation," *IEEE J. Solid-State Circuits*, vol. 12, no. 3, pp. 224–231, Jun. 1977
8. K. Roy, A. Agrawal, and C. H. Kim, "Circuit techniques for leakage reduction," in *Low-Power Electronics Design*, Editor C. Piguet, CRC, 2005
9. E. Vittoz, "Weak inversion for ultimate low-power logic," in *Low-Power Electronics Design*, Editor C. Piguet, CRC, 2005
10. V. R. von Kaenel, M. D. Pardon, E. Dijkstra, and E. A. Vittoz, "Automatic adjustment of threshold and supply voltage for minimum power consumption in CMOS digital circuits," *IEEE Symp. Low Power Electron.*, pp. 78–79, Oct. 1994
11. C. D. Salthouse and R. Sarpeshkar, "A practical micropower programmable bandpass filter for use in bionic eras," *IEEE J. Solid-State Circuits*, vol. 38, no. 1, pp. 63–70, Jan. 2003
12. R. Bagheri, A. Mirzaei, S. Chehrazi, M. Heidari, M. Lee, M. Mikhemar, W. Tang, and A. Abidi, "An 800 MHz to 5 GHz software-defined radio receiver in 90 nm CMOS," *Int. Solid-State Circuits Conf. (ISSCC) Dig. Tech. Papers*, pp. 1932–1941, Feb. 2006
13. M. Horowitz, T. Indermaur, and R. Gonzalesz, "Low-power digital design," *IEEE Int. Symp. Low Power Electron. Design*, pp. 8–11, Oct. 2004
14. C. C. Enz and E. A. Vittoz, *Charge-based MOS Transistor Modeling*, Wiley, 2006
15. S.-M. Kang and Y. Leblebici, *CMOS Digital Integrated Circuits*, McGraw-Hill, 2003
16. S. G. Narendra and A. Chandrakasan, *Leakage in Nanometer CMOS Technologies*, Springer, 2006
17. A. Szumanowski and Y. Chang, "Battery management systems based on battery nonlinear dynamics modeling," *IEEE Trans. Vehicular Tech.*, vol. 57, no. 13, pp. 1425–1432, May 2008
18. H. J. Bergveld, W. S. Krujt, and P. H. L. Notten, *Battery Management Systems - Design by Modeling*, Kluwer, 2002
19. A.-J. Annema, B. Nauta, R. van Langevelde, and H. Tuinhout, "Analog circuits in ultra-deep-submicron CMOS," *IEEE J. Solid-State Circuits*, vol. 40, no. 1, pp. 132–143, Jan. 2005
20. J. M. Musicer and J. Rabaey, "MOS current mode logic for low power, low noise CORDIC computation in mixed-signal environment," *Proc. Int. Symp. Low Power Electron. Dessign (ISLPED)*, pp.102–107, 2000
21. P. Heydari and R. Mohanavelu, "Design of ultrahigh-speed low-voltage CMOS CML buffers and latches," *IEEE Tran. Very Large Scale Integration (VLSI) Syst.*, vol. 12, no. 10, pp. 1081–1093, Oct. 2004
22. A. Tajalli, E. J. Brauer, Y. Leblebici, and E. Vittoz, "Subthreshold source-coupled logic circuit design for ultra low power applications," *IEEE J. Solid-State Circuits*, vol. 43, no. 7, pp. 1699–1710, Jul. 2008
23. A. Tajalli and A. Adibi, "A 1.5-V supply, video range frequency, Gm-C filter," *Proc. IEEE Symp. Circ. Syst. (ISCAS)*, vol. 2, pp. 148–151, Geneva, Switzerland, May 2000
24. B. Pankiewicz, M. Wojcikowski, S. Szczepanski, and Y. Sun, "A field programmable analog array for CMOS continuous-time OTA-C filter applications," *IEEE J. Solid-State Circuits*, vol. 37, no. 2, pp. 125–136, Feb. 2002
25. T. Hollman, S. Lindfors, M. Lansirinne, J. Jussila, and K. A. I. Halonen, "A 2.7-V CMOS dual-mode baseband filter for PDC and WCDMA," *IEEE J. Solid-State Circuits*, vol. 36, no. 7, pp. 1148–1153, Jul. 2002

26. R. Gregorian and G. C. Temes, *Analog MOS Integrated Circuits for Signal Processing*, Wiley, 1986
27. U.-K. Moon, "CMOS high-frequency switched-capacitor filters for telecommunication applications," *IEEE J. Solid-State Circuits*, vol. 35, no. 2, pp. 212–220, Feb. 2000
28. C. Enz, M. Punzenberger, and D. Python, "Low-voltage log-domain signal processing in CMOS and BiCMOS," in *IEEE Transactions on Circuits and Systems-II: Analog and Digital Signal Processing*, vol. 46, no. 3, pp. 279–289, Mar. 1999
29. Y. Tsividis, "Externally linear, time-invariant systems and their application to companding signal processing," in *IEEE Transactions on Circuits and Systems-II: Analog and Digital Signal Processing*, vol. 44, no. 2, pp. 65–85, Feb. 1997

Chapter 2
Subthreshold MOS for Ultra-Low Power

This chapter provides a brief review on modeling of MOSFET devices especially for weak-inversion (WI) devices.[1] The main issues associated with WI design such as variation due to PVT, mismatch effects, and device noise are briefly addressed. Meanwhile, a review on the main problems for implementing ULP CMOS circuits is provided. At the end of this Chapter, an analytical approach for systematic design of digital CMOS circuits operating in WI region with optimum energy consumption and acceptable reliability is proposed.

2.1 MOS Technology

The first proposal for implementing metal-oxide-semiconductor field-effect transistors (MOSFETs) can be traced back to 1930, when Lilienfeld and Heil patented the initial concept independently [1–3]. However, successful implementation was demonstrated after a few decades in 1960. Simple topology of MOSFETs in addition to their small area, makes it possible to implement very large-scale integrated (VLSI) circuits. This property is especially demanding for implementing digital systems with very powerful processing capabilities. Commercial requirements have pushed the need for fabricating ICs with more powerful processing capabilities or more number of devices per chip area for the past couple of decades as depicted in Fig. 2.1. These properties have made the MOSFET technology to be the mainstream in design of high performance integrated circuits.

MOSFET transistors are generally used as switching devices in digital circuits with close to zero off current and very large turn on current. In static CMOS topology, the steady state current of a logic gate is very small [4]. In analog applications MOSFET devices are employed as active devices generally biased in strong inversion (SI) to be able to operate at high frequencies and at the same time keep the

[1] MOS device operates in weak inversion (WI) when the channel underneath the gate is weakly inverted by absorbing carriers. When the channel is completely inverted, the device will be in strong inversion (SI). The region in between is usually called medium inversion (MI) [1].

A. Tajalli and Y. Leblebici, *Extreme Low-Power Mixed Signal IC Design: Subthreshold Source-Coupled Circuits*, DOI 10.1007/978-1-4419-6478-6_2,

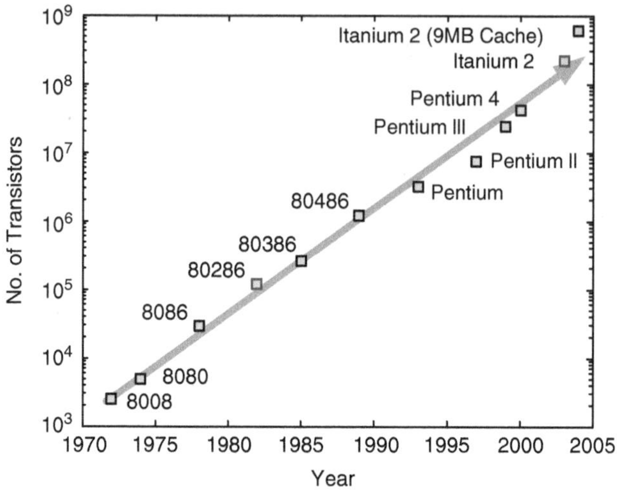

Fig. 2.1 Exponential increase of number of transistors on a single chip thanks to the CMOS technology scaling and comparison to the prediction made in [8]

noise level very low. On the other hand, subthreshold (or WI) MOSFET devices are suitable for ULP applications where the device current density is very low [5].

Since most of the circuit topologies that are developed in this work are based on subthreshold MOS devices, in the rest of this chapter a very brief review on the subthreshold MOS devices and their modeling techniques will be presented.

2.2 Device Modeling

A profound background on device modeling is essential to design high performance circuits. This section provides the necessary background for design and analysis that will be carried out throughout the rest of this work. Figure 2.2 shows the structure of NMOS and PMOS transistors which are the main building blocks for implementing CMOS integrated circuits.

2.2.1 I–V Characteristics

The issue of MOSFET modeling in subthreshold regime has been extensively addressed in [1, 4, 6], and [7]. The EKV model[2], first presented in [7], is based on an interpolation approach which can be used for all different regions of oper-

[2] Enz–Krummenacher–Vittoz (EKV).

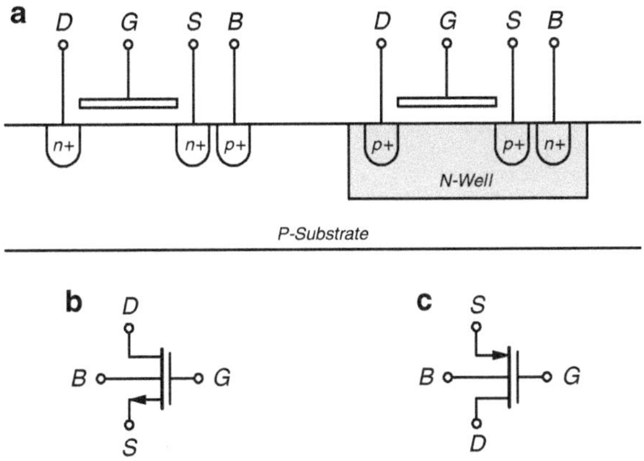

Fig. 2.2 (**a**) Structure of NMOS and PMOS devices. Symbol for (**b**) NMOS and (**c**) PMOS devices

ation of an MOS device. In this model, all the voltage levels are referred to the local substrate voltage (not to the source voltage of a MOSFET device as it is usual in BSIM model [1]). This property is especially interesting in this work where the bulk of transistor is used frequently as the second gate (or back gate [1]) of a device to provide more design flexibility. Based on EKV model, the drain current of an NMOS transistor can be calculated by [7]:

$$I_{\rm DS} = 2n\mu_e C_{\rm ox} U_T^2 \frac{W}{L_e}\left(\ln^2\left(1+{\rm e}^{\left(\frac{V_P-V_S}{2U_T}\right)}\right) - \ln^2\left(1+{\rm e}^{\left(\frac{V_P-V_D}{2U_T}\right)}\right)\right) \tag{2.1}$$

where:

- n is the subthreshold slope factor which depends on process parameters as well as biasing condition, and is usually between 1 and 1.5,
- μ_e in $(m^2/(V \cdot s))$, is the effective carrier mobility in the channel and is different for electrons and holes:

  $$\mu_0 = A \cdot {\rm e}^{B\sqrt{N_{\rm ch}}} \tag{2.2}$$

  where $N_{\rm ch}$ represents the channel doping density. For NMOS devices: $A = 1150$ and $B = -5.34 \times 10^{-10}$, and for PMOS devices: $A = 317$ and $B = -1.25 \times 10^{-9}$. Carrier mobility, also depends on Electric field.
- $C_{\rm ox} = \epsilon_{\rm SiO_2}/t_{\rm ox}$ is gate oxide capacitance per unit area, $\epsilon_{\rm SiO_2} = K_{\rm SiO2}\epsilon_0$ is dielectric constant of SiO_2, $\epsilon_0 = 8.8541878176 \times 10^{-12}\,{\rm Fm}^{-1}$, $k_{\rm SiO2} = 3.9$, and $t_{\rm ox}$ is oxide thickness,
- $U_T = kT/q$ is the thremodynamic voltage, $k = 1.3806504 \times 10^{-23}\,{\rm J{\cdot}k}^{-1}$ is Boltzmann's constant[3], T is the absolute junction temperature, and $q = 1.602 \times 10^{-19}$ C[4] is the elementary electron charge,

[3] Although this coefficient is called by the name of Austrian physicist, Ludwig Boltzmann, it has been first introduced by German scientist Max Planck, in his derivation of the law of the black body radiation in December 1900 (see: [9], and also *http://www.wikipedia.org*).

[4] [C] ≡ [A·s].

- W and L_e are the effective channel width and length of the device,
- V_P is device pinch off voltage.

The first term in (2.1) is called forward channel current, I_F, and the second term is called reverse channel current, I_R. Also, specific current of the device is defined as: $I_S = 2n\mu_e C_{ox} U_T^2$.

To complete the calculations using (2.1), it is necessary to calculate the values of V_P and n. The pinch off voltage depends on the gate voltage (V_G) as:

$$V_P = V_G - V_{T0} - \gamma \cdot \left(\sqrt{V_G - V_{T0} + \left(\sqrt{\Psi_0} + \frac{\gamma}{2}\right)^2} - \left(\sqrt{\Psi_0} + \frac{\gamma}{2}\right) \right) \quad (2.3)$$

where V_{T0} stands for the device threshold voltage and is equal to the gate voltage when the mobile inversion charge density in the channel (Q_{inv}) is zero, or [4]

$$V_{T0} = V_{\text{FB}} + \Psi_0 + \gamma\sqrt{\Psi_0} \quad (2.4)$$

where V_{FB} is the flat band voltage, and

$$\gamma = \frac{\sqrt{2q\epsilon_s N_{\text{ch}}}}{C_{\text{ox}}} \quad (2.5)$$

is the substrate factor or body effect, $\epsilon_s = K_{\text{Si}}\epsilon_0$ is the Si dielectric constant ($K_{\text{Si}} = 11.7$), N_{ch} is the doping concentration in the substrate, $\Psi_0 = 2\Phi_F + mU_T$ is the surface potential[5], $\Phi_F = U_T \ln(N_{\text{ch}}/n_i)$ is the substrate Fermi potential, and n_i stands for the intrinsic carrier concentration of Si[6]. The derivation of the gate voltage with respect to the pinch off voltage is defined as the device subthreshold slope factor given by[7]:

$$n \equiv \frac{\mathrm{d}V_G}{\mathrm{d}V_P} = 1 + \frac{\gamma}{2\sqrt{\Psi_0 + V_P}} \quad (2.6)$$

which can be simplified to:

$$\frac{1}{n} = 1 - \frac{\gamma}{2\sqrt{V_G - V_{T0} + \left(\gamma/2 + \sqrt{\Psi_0}\right)^2}}. \quad (2.7)$$

It can also be shown that:

$$V_P \simeq \frac{V_G - V_{T0}}{n}. \quad (2.8)$$

[5] In this equation, m depends on the region of operation [7].

[6] $n_i = 3.1 \times 10^{16} T^{3/2} \exp\left(-\frac{7000}{T}\right)$.

[7] Experimental results in this work show that when one of the junctions in the MOS device becomes forward bias, this equation will not be precise enough. Using a modified substrate doping concentration can solve the problem. The other possibility is adding a bipolar device to the proposed MOS device in a proper configuration (see Chap. 3).

In SI, (2.1) can be simplified to:

$$I_{\mathrm{DS}} \approx \frac{n\mu_e C_{\mathrm{ox}}}{2}\frac{W}{L_e}\left((V_P - V_S)^2 - (V_P - V_D)^2\right)$$
$$\approx \frac{\mu_e C_{\mathrm{ox}}}{2n}\frac{W}{L_e}(V_G - nV_S - V_{T0})^2 . \tag{2.9}$$

It is noticeable that the current sensitivity to the source voltage is n times more than to the gate voltage. In other words: $g_{\mathrm{ms}} = n \times g_m$. Assuming that n is equal to one, then (2.9) can be simplified to the conventional equation. In linear (triode) region where: $n \cdot V_D \leq V_G - V_{T0}$, the drain current is given by

$$I_{\mathrm{DS}} = n \cdot \beta \cdot \left(\frac{V_G - V_{T0}}{n} - \frac{V_D + V_S}{2}\right) \cdot (V_D - V_S) \tag{2.10}$$

In WI:

$$I_{\mathrm{DS}} \approx 2n\mu_e C_{\mathrm{ox}}\frac{W}{L_e}U_T^2\left(\mathrm{e}^{\frac{V_P - V_S}{U_T}} - \mathrm{e}^{\frac{V_P - V_D}{U_T}}\right).$$
$$\approx 2n\mu_e C_{\mathrm{ox}}\frac{W}{L_e}U_T^2\mathrm{e}^{\frac{V_G - V_{T0}}{nU_T}}\left(\mathrm{e}^{\frac{-V_S}{U_T}} - \mathrm{e}^{\frac{-V_D}{U_T}}\right). \tag{2.11}$$

where all the voltages are referred to the substrate and V_{T0} is independent to the V_{SB}. In this work, we are frequently using (2.1), (2.9), and (2.11) for analysis purposes.

2.2.2 *Second Order Effects*

2.2.2.1 Mobility Reduction Due to Vertical Field

By increasing the vertical electric field, the carriers tend to flow closer to the silicon–oxide interface which causes more carrier scattering and mobility degradation as a result. To include the effect of mobility degradation due to the vertical electric field, mobility can be replaced by the following value:

$$\mu_e = \frac{\mu}{1 + \theta \cdot V_P} \tag{2.12}$$

where θ is a constant coefficient between 0.1 and 1 V^{-1} [7,10]. A very approximate value for θ is: $\theta \approx 2 \times 10^{-9}/t_{\mathrm{ox}}$ which shows more degradation in thinner gate oxides [10].

2.2.2.2 Velocity Saturation

Carrier mobility is proportional to the electric field by $v = \mu E$ or more precisely [4]:

$$v = \frac{\mu_e E}{\sqrt[n]{1 + \left(\frac{E}{E_C}\right)^n}} \tag{2.13}$$

where $n = 1$ for electrons and $n = 2$ for holes, and $E_C = v_{\text{sat}}/\mu_e$ is critical electric field. In high electric field values when it becomes comparable to E_C, the carrier velocity saturates due to the scattering phenomena. The scattering of carriers by high-energy phonons is the main reason for this speed limitation. In silicon, the carrier speed saturates at about $v_{\text{sat}} = 10^5$ (m/s) when the electric field approaches to about $E_{\text{sat}} \approx 10^6$ (V/m) [10].

As the device current depends on carrier velocity, this effect is generally modeled as the following [7]:

$$I_{\text{DS}} = \frac{I_{\text{DS0}}}{1 + \frac{V_D^* - V_S}{L \cdot E_{\text{sat}}}}. \tag{2.14}$$

Here, V_D^* is equal to V_D for triode MOS ($V_D < V_{\text{Dsat}}$), and equal to V_{Dsat} for saturated MOS ($V_D > V_{\text{Dsat}}$). Also, I_{DS0} is the current calculated without velocity saturation effect.

One of the main issues with the velocity saturation is that in short channel devices where V_{Dsat} is becoming larger than LE_{sat}, then the device current approaches:

$$I_{\text{DS}} \approx \frac{\mu C_{\text{ox}}}{2} W(V_G - nV_S - V_{T0}) E_{\text{sat}} \tag{2.15}$$

which does not depend on channel length. In this case, the saturation voltage can be approximated by [4]

$$V_{\text{DSsat}} = \sqrt{\frac{2 v_{\text{sat}} L_e (V_G - V_{T0})}{n \mu_e}}. \tag{2.16}$$

As can be seen, based on (2.15), the quadratic relationship between current and voltage is modified to a first order linear equation. Generally, the relationship between current and voltage in strong inversion is modeled with an equation with the order of $1 < \alpha < 2$ [11, 12].

2.2.2.3 Channel Length Modulation

When drain voltage is larger than the pinch off voltage, pinch off point starts to move towards the source and reduce the channel length as a result by ΔL. Therefore, the drain current will be increased proportional to the channel length reduction as [4]:

$$I_{\text{DS}} = \frac{I_{\text{DS0}}}{1 - \frac{\Delta L}{L}} \tag{2.17}$$

The channel length reduction can be calculated by [7]:

$$\Delta L \approx \zeta \cdot \sqrt{V_D - V_P} \tag{2.18}$$

where

$$\zeta = \frac{2\epsilon_S}{\gamma C_{\text{ox}}} = \sqrt{\frac{2\epsilon_S}{q N_{\text{ch}}}}. \tag{2.19}$$

Generally, a simplified model for channel length modulation is used. In this approach, a resistance (output resistance) is put in parallel to the drain-source of a MOS device. The value of this resistance can be calculated using $g_{\text{ds}} \approx I_{\text{DS}}/(\lambda \cdot L)$. This approach is similar to introducing Early voltage in bipolar transistors where in MOS devices, the Early voltage can be defined as: $V_A = \lambda \cdot L$. By increasing the channel length or reducing the bias current, the parasitic effect of the channel length modulation can be reduced.

2.3 Design Considerations in Subthreshold

In this section, some of the main issues associated with MOS devices biased in subthreshold regime, such as variability, noise, and matching are addressed very briefly. As will be seen later, these nonideality effects can increase the design cost in terms of area, energy consumption, and reliability.

2.3.1 PVT Variation

Rewriting (2.11) in the form of:

$$I_{\text{DS}} \approx I_0 \mathrm{e}^{\frac{V_G}{nU_T}} \left(\mathrm{e}^{\frac{-V_S}{U_T}} - \mathrm{e}^{\frac{-V_D}{U_T}} \right) \tag{2.20}$$

it clearly illustrates the exponential I–V characteristics of a MOS device biased in WI (subthreshold) regime.[8] This characteristic is on one hand useful for implementing widely tunable circuits, while on the other hand, it represents the high sensitivity of circuit to PVT variations. For example, any small variation on the device threshold voltage (V_{T0}) will be translated to exponential variation on the bias current.

[8] $I_0 = 2n\mu_e C_{\text{ox}} \frac{W}{L_e} U_T^2 \mathrm{e}^{\frac{-V_{T0}}{nU_T}}$.

It is also instructive to calculate the temperature dependence of the bias current in subthreshold regime. Assuming $\mu = \mu_0(T/T_0)^{\alpha}$:[9]

$$\frac{\partial I_{DS}}{\partial T} \approx I_{DS} \cdot \left(\frac{\alpha+2}{T} - \frac{\partial V_{T0}/\partial T - V_{T0}/T}{nU_T} \right). \tag{2.21}$$

To derive this equation, the temperature dependence of subthreshold slope factor has been ignored. Meanwhile, it is assumed that $V_S << U_T$ and $V_D >> U_T$ which is not the case for all the possible configurations. Based on (2.21), it is possible to show that in WI:

$$I_{DS} = I_{DS0} \cdot e^{\frac{F}{T_0}} \cdot \left(\frac{T}{T_0} \right)^G \cdot e^{-\frac{F}{T}} \tag{2.22}$$

where $G = \alpha+2-\theta q/(nk)$, $F = qV_{T0}/(nk)$ which is independent of temperature, and $V_{T0} \approx V_{T00} + \theta(T - T_0)$.[10] On the other hand in SI, the temperature variation of the device current can be calculated by:

$$I_{DS} = I_{DS0} \cdot \left(\frac{T}{T_0} \right)^{\alpha} \cdot \left(\frac{V_G - nV_S - V_{T0} - \theta(T - T_0)}{V_G - nV_S - V_{T00}} \right)^2. \tag{2.23}$$

The thermal variation of the bias current is depicted in Fig. 2.3. As illustrated in this figure and can be concluded from (2.22) and (2.23), by moving toward subthreshold region, the variations due to the temperature increases very rapidly.

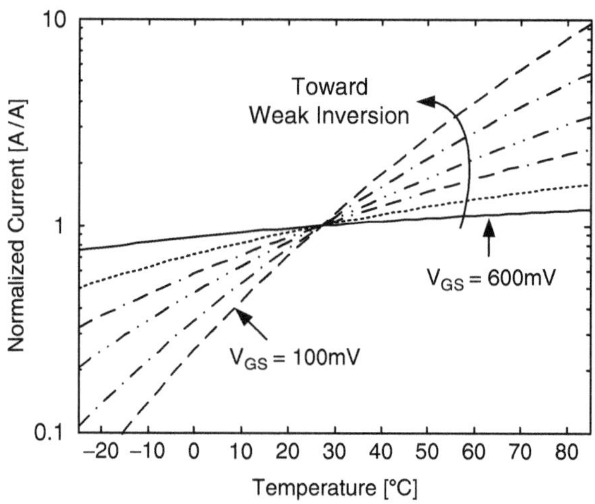

Fig. 2.3 Bias current dependence on temperature variations. In this figure, the bias current is normalized to the nominal bias current at $T = 27°C$

[9] Here, T_0 is the temperature in which μ_0 has been measured. Meanwhile, α is equal to −2.4 for electron and −2.2 for hole in Si [1].

[10] Here, it is assumed that the threshold voltage linearly depends on temperature and the proportionality factor is θ and threshold voltage at T_0 is equal to V_{T00} [4].

2.3.2 Matching

Device mismatch is one of the most important design issues especially in design of high performance analog and digital systems in modern ultra-deep-submicron (UDSM) technologies. Experiments show that the two main sources of introducing mismatch among devices are difference in threshold voltage (ΔV_T) and current factor ($\Delta\beta$, where $\beta = \mu C_{ox} W/L_e$). The difference among devices raised from difference in V_T and β have random nature with a normal distribution where their mean values are V_{T0} and β_0 [13]. The variance of these parameters can be presented by

$$\sigma^2(\Delta V_T) = \frac{A_{\mathrm{VT}}^2}{W \cdot L} \tag{2.24}$$

$$\left(\frac{\sigma(\Delta\beta)}{\beta}\right)^2 = \frac{A_\beta^2}{W \cdot L} \tag{2.25}$$

where proportionality constants A_{VT} and A_β are technology dependent parameters.

For simple current mirrors and differential pair configurations, it can be shown that the mismatch between current values and input referred voltage offset are, respectively:

$$\left(\frac{\sigma(\Delta I_{\mathrm{DS}})}{I_{\mathrm{DS}}}\right)^2 = \left(\frac{\sigma(\Delta\beta)}{\beta}\right)^2 + \left(\frac{g_m}{I}\right)^2 \sigma^2(\Delta V_T) \tag{2.26}$$

$$\sigma^2(\Delta V_{\mathrm{GS}}) = \sigma^2(\Delta V_T) + \left(\frac{I}{g_m}\right)^2 \left(\frac{\sigma(\Delta\beta)}{\beta}\right)^2 \tag{2.27}$$

Since the value of g_m/I has its maximum value in WI (Fig. 2.5), and regarding (2.26) and (2.27), it is expected that the voltage matching improves slightly by moving towards WI,[11] while the current matching degrades. This implies that implementing current mirrors with acceptable level of matching will be much more difficult in WI region compared to the current mirrors implemented in SI region.

Figure 2.4 shows the expected value of the input referred offset of an NMOS differential pair circuit by technology scaling. Although the value of A_{VT} and A_β are improving by technology scaling, however, the size of devices are reducing as well, and consequently the expected offset value increases considerably. Depicted in Fig. 2.4, the input referred offset increases by a factor of about 12 mV/decade by technology scaling.

[11] Generally, the term which depends on V_T variation is dominant over the term depending on the variations due to β. Therefore, the expected reduction on the input referred offset voltage is not considerable.

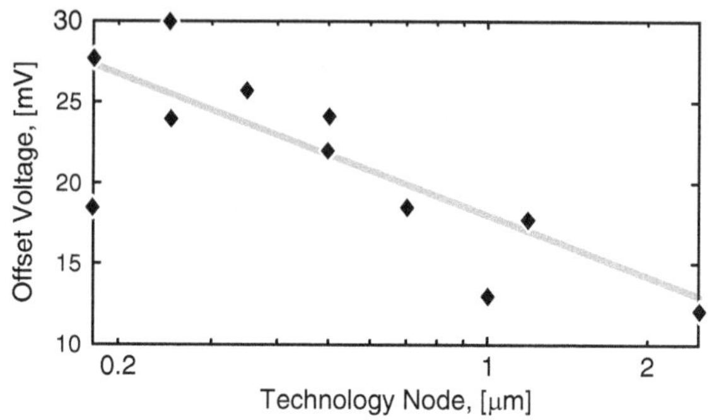

Fig. 2.4 Expected offset voltage at the input of a differential pair circuit by technology scaling when minimum size devices are utilized. Data values are extracted from [13]

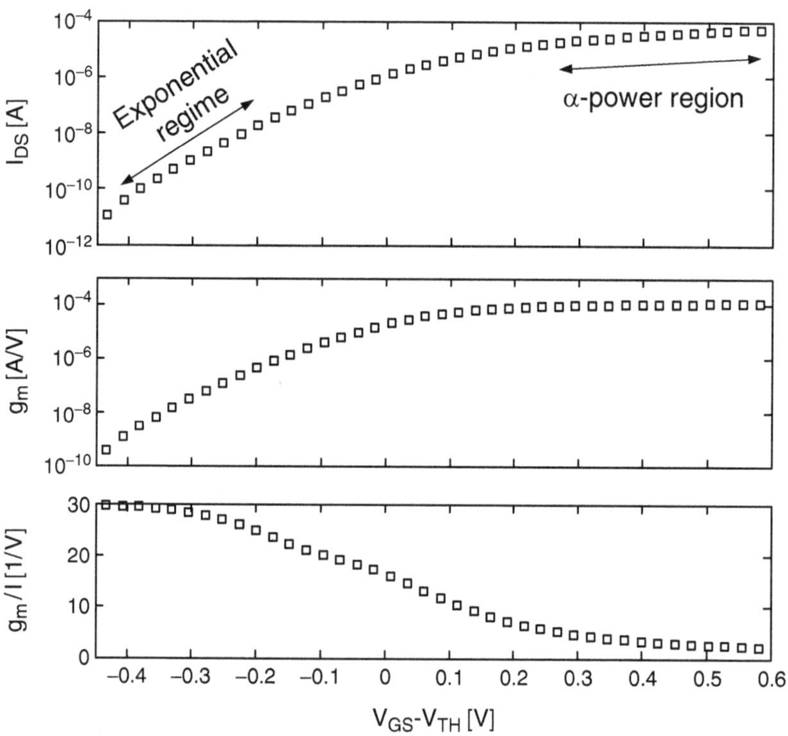

Fig. 2.5 Dependence of bias current, transconductance, and g_m/I on gate overdrive voltage: $V_{GS} - V_T$

2.3.2.1 Physical Mechanism of V_T Fluctuation

Threshold voltage of an MOSFET device can be expressed by:

$$V_T = V_{FB} + 2\phi_B + \frac{Q_d}{C_{ox}} \tag{2.28}$$

where Q_d is the depletion layer charge and ϕ_B is the surface potential. Based on this, any variation on channel doping concentration, surface state charge density (Q_{ss}), and gate oxide thickness can result in variation on the device threshold voltage. The variation on surface potential, $\delta\phi_B$, can be estimated by $\delta\phi_B \approx U_T \cdot \delta N_A / N_A$, where δN_A is the fluctuation on substrate doping [14]. It can be shown that threshold voltage fluctuation due to the random dopant fluctuation (RDF) can be estimated by [14]:

$$\sigma_{V_T} = \frac{\sqrt[4]{q^3 \epsilon_{\mathrm{Si}} \phi_B}}{\sqrt{2}\epsilon_{\mathrm{ox}}} \cdot t_{\mathrm{ox}} \cdot \sqrt[4]{N_A} \cdot \frac{1}{\sqrt{W_{\mathrm{eff}}(L_{\mathrm{eff}} - W_d)}} \tag{2.29}$$

where W_d is the average of the maximum PN junction depletion layer width of the drain n^+ region. This expression indicates that threshold voltage fluctuation increases approximately by a factor of $\sqrt[4]{\kappa}$ by technology scaling where $\kappa > 1$ is the scaling factor under constant field scaling rule [14]. Some more recently published reports are proposing the following expression for the standard deviation of the device threshold voltage [15]:

$$\sigma_{V_T} = 3.19 \times 10^{-8} \cdot t_{\mathrm{ox}} \cdot \sqrt[2.5]{N_A} \cdot \frac{1}{\sqrt{W_{\mathrm{eff}} L_{\mathrm{eff}}}} \tag{2.30}$$

which indicates a stronger dependence on channel doping concentration compared to (2.29).

To prevent the increase on threshold voltage variation with technology scaling, V_T adjustment method needs to be modified. For example, instead of controlling the depletion layer charge, new gate materials could be used to avoid increasing substrate doping concentration. There are other sources for increasing the threshold voltage variability, such as line edge roughness and oxide thickness variation. While the effect of line edge roughness can be neglected, the variation of threshold voltage due to the oxide thickness is about half of the variation due to the RDF [16].

2.3.2.2 Mismatch due to Gate Leakage

The gate leakage current adds a new source of device mismatch which should be included in the calculations specially in thin oxide devices. The variation on drain current including the gate leakage mismatch is [17]:

$$\frac{\sigma^2_{I_{\mathrm{DS}}}}{I^2_{\mathrm{DS}}} \approx \left(\frac{A_{\mathrm{VT}}}{\sqrt{\mathrm{WL}}} \cdot \frac{g_m}{I_{\mathrm{DS}}}\right)^2 + \left(\frac{X_{\mathrm{IGS}}}{\sqrt{\mathrm{WL}}} \cdot \frac{I_G}{I_{\mathrm{DS}}}\right)^2 \tag{2.31}$$

where $X_{\mathrm{IGS}} \approx 0.03$.

2.3.3 Noise

The model that is used generally to estimate the noise of MOS device, the drain thermal noise and the gate voltage flicker noise are

$$i_{n,d}^2 = 4kT\gamma g_m \tag{2.32}$$

$$v_{n,f}^2 = \frac{4kT\rho}{\mathrm{WL}} \cdot \frac{1}{f^\alpha} = \frac{k_f}{\mathrm{WLC_{ox}}} \cdot \frac{1}{f^\alpha} \tag{2.33}$$

where γ and ρ represent excess noise factor. Flicker noise is inversely proportional to the frequency f, and $k_f = 4kTC_{\mathrm{ox}}\rho$ [6]. The empirical coefficient k_f for NMOS devices is essentially independent of bias, fabricator and technology ($k_{f,\mathrm{NMOS}} = 10^{-24}$), while for PMOS devices, this coefficient is smaller[12] and depends on biasing condition [18]. To reduce the effect of flicker noise, the most effective way is to increase the device dimensions [19].

To have a unified thermal noise model for SI and WI regions, γ (excess noise factor) has been defined as the following in [6]:

$$\gamma = \frac{2n}{3 + \left(\frac{g_m}{I_{\mathrm{DS}}} n U_T\right)^2} \tag{2.34}$$

which results in $\gamma = n/2$ in WI and $\gamma = 2n/3$ in SI. The thermal noise power-spectral density can also be interpolated from WI to SI using the following function:

$$\frac{1}{R_N} = g_m \times \frac{1}{1+i_f} \cdot \left(\frac{1+\alpha}{2} + \frac{2}{3} \cdot i_f \cdot \frac{1+\sqrt{\alpha}+\alpha}{1+\sqrt{\alpha}}\right) \tag{2.35}$$

where $\alpha = i_r/i_f$, i_f and i_r are the forward and reverse currents in the channel normalized to specific current $I_S \equiv 2n\beta U_T^2$,[13] and R_N is the equivalent noise resistance of the channel ($v_n^2 = 4kTR_N$). It is interesting to notice that the channel noise increases when device moves from saturation ($\alpha = 0$) to conduction ($\alpha = 1$). Also, the channel noise increases slightly when device moves from WI (low i_f values) toward SI (high i_f values).

In [1], the channel thermal noise has been calculated as

$$i_{n,d}^2 = 2qI^* \left(1 + \mathrm{e}^{-\frac{V_{\mathrm{DS}}}{U_T}}\right) \tag{2.36}$$

where I^* is the current in flat part of the I_{DS}–V_{DS} curve (or in other words: $V_{\mathrm{DS}} > 5U_T$). Although, this expression has been derived assuming the presence of

---

12 $k_{f,\mathrm{PMOS}}$ can be 50 times smaller than $k_{f,\mathrm{PMOS}}$ [19].

13 Forward and reverse currents can be calculated from (2.1) where the first term stands for forward current and the second term stands for reverse current.

thermal noise in channel, it is corresponding to shot noise associated with the dc flow produced by carriers crossing the source-channel barrier [1]. It is also noticeable that the current noise increases with reducing the V_{DS}.

In very high frequencies, where the transient time of carriers between source and drain becomes important, a new source of noise should be added to the MOS device model. The finite carrier transition time in the channel adds a positive term or equivalently a resistive part to the input impedance of a MOS device. The noise associated with this effect can be modeled by a noise current source at the gate with mean-square power of [10]:

$$i_{n,g}^2 = 4kT\delta g_g \Delta f = 4kT\delta \cdot \frac{\omega^2 C_{gs}^2}{5g_{d0}} \cdot \Delta f \tag{2.37}$$

where δ is typically 4/3. This noise is correlated with the drain thermal noise with correlation factor of $c \equiv i_{n,g} \cdot i_{n,d}^* \Big/ \sqrt{i_{n,g}^2 \cdot i_{n,d}^2} = j0.395$.

2.3.3.1 Noise Efficiency Factor

To be able to compare the noise performance of a specific design with other designs, noise efficiency factor (NEF) has been defined in [19]. For this purpose, the total equivalent input noise of an ideal bipolar transistor (including only thermal noise without considering the base resistance noise) has been defined to be the reference noise level:

$$v_{\text{rms,in,bip}} = \sqrt{\text{BW} \cdot \frac{\pi}{2} \cdot \frac{4kT}{g_m}} \tag{2.38}$$

where $g_m = I_C/U_T$ in a bipolar transistor (I_C is the collector current). Also, BW represents the circuit bandwidth. In case of a simple bipolar transistor, the bandwidth is f_t (transient frequency of a bipolar transistor or the frequency at which the current gain of transistor becomes one) [19]. To calculate the NEF for a circuit with equivalent input referred noise of $v_{\text{rms,in}}$:

$$\text{NEF} = \frac{v_{\text{rms,in}}}{v_{\text{rms,in,bip}}}. \tag{2.39}$$

For example, for a simple MOS transistor in SI:

$$v_{\text{rms,in,MOS}}^2 = \text{BW} \cdot \frac{\pi}{2} \cdot \frac{4kT}{\frac{2}{3}g_m} = \text{BW}\frac{\pi}{2} \cdot \frac{3kT(V_{\text{GS}} - V_T)}{I_{\text{DS}}} \tag{2.40}$$

Assuming that the device is operating on the boundary of SI (i.e., $V_{\text{GS}} - V_T = 2\sqrt{10}nU_T$), then NEF = 2.43 [19]. Therefore, equivalent noise of a CMOS design in SI with the same amount of power dissipation and bandwidth is about five times more than a bipolar design. In [20–24], some techniques for implementing amplifiers with very low NEF values have been reported. Chopper stabilization have been used in [22] to reduce the flicker noise and the offset voltage. In [23], careful current partitioning technique has been used to improve the NEF to 3.81 in a folded-cascode operational transconductance amplifier (OTA). To reduce the NEF to 1.8, partial OTA sharing technique has been introduced in [24]. In this design, large size devices have been used to make the flicker noise effect negligible.

2.3.3.2 Noise Due to the Gate Leakage

The noise of gate leakage current is a shot noise as the noise in other types of PN junctions. The noise current density can be expressed by [17]

$$S_{\mathrm{IG}} = 2qI_{\mathrm{GS}} \tag{2.41}$$

that should be included in the estimation of circuit noise. To calculate the gate leakage current the following expression can be used [17]:

$$I_{\mathrm{GS}} = A \cdot V_{\mathrm{INV}} \cdot V_{\mathrm{GS}} \cdot \mathrm{e}^{B \cdot V_{\mathrm{GS}}} \tag{2.42}$$

which represents exponential dependance of the gate current on the gate-source voltage of a device. In this equation:

$$A = \frac{I_{\mathrm{GINV}}}{2} \cdot \mathrm{e}^{-\frac{3}{2} \cdot \frac{B_{\mathrm{INV}}}{X_B}} \tag{2.43}$$

and

$$B = \frac{3}{8} \cdot \frac{B_{\mathrm{INV}}}{X_B^2} \tag{2.44}$$

and

$$V_{\mathrm{INV}} = nU_T \cdot \ln\left(1 + \mathrm{e}^{\frac{V_{\mathrm{GS}} - V_T}{nU_T}}\right) \tag{2.45}$$

Here, X_B is the oxide potential barrier which is 3.1 V for electrons and 4.5 V for holes. I_{GINV} and B_{INV} are physical parameters depending on t_{ox}, L, and W. For electrons:

$$I_{\mathrm{GINV}} = 1.6 \cdot 10^{-4} \cdot \frac{\mathrm{WL}}{t_{\mathrm{ox}}^2} \tag{2.46}$$

and

$$B_{\mathrm{INV}} = 2.9 \cdot 10^{10} \cdot t_{\mathrm{ox}}. \tag{2.47}$$

These values can be replaced in (2.42) to estimate the gate leakage current.

2.4 Ultra-Low-Power Design Using Subthreshold MOS

Using subthreshold MOS devices for implementing low-voltage and very low-power analog and digital circuits can be traced back to the 1970s [5, 25]. While in most of the applications at that time MOS devices were employed in strong-inversion, the need for reducing the power consumption and supply voltage encouraged the designers to develop special design techniques for using subthreshold MOS devices. Some industrial applications such as low-power quartz wristwatches [26] promoted even more the researchers to establish the required bases to simplify and increase the reliability of using subthreshold MOS devices. For this purpose, many different design and device modeling techniques have been proposed [5, 7].

In [25], in 1970, it was shown that it is possible to reduce the supply voltage of a CMOS inverter down to $V_{DD} \approx 4U_T$ with sufficient gain for logic operation. Therefore, it is possible to use CMOS logic circuits deeply biased in subthreshold regime. This means that when the speed of operation is not the premier design issue, it is possible to reduce the supply voltage and hence reduce the power dissipation of a system which is mostly proportional to the dynamic power consumption.

Afterwards, the concept of low-power design using reduced supply voltage has been developed even further to construct more complex integrated circuits with possibility of dynamic power management [27]. In this type of system, supply voltage can be scaled in a very wide range to minimize the power dissipation with respect to the operation frequency or work load [28].

Figure 2.6 shows the trends in semiconductor industry based on the data points and predictions made in 2001 [29]. All the parameters in these two graphs are normalized to their nominal values in the year 2001. While device channel length (L) has been scaled down progressively, the scaling for supply voltage, V_{DD}, and gate oxide thickness have not been as aggressive as scaling of channel length. As illustrated in these graphs, there is a very rapid increase in the static power

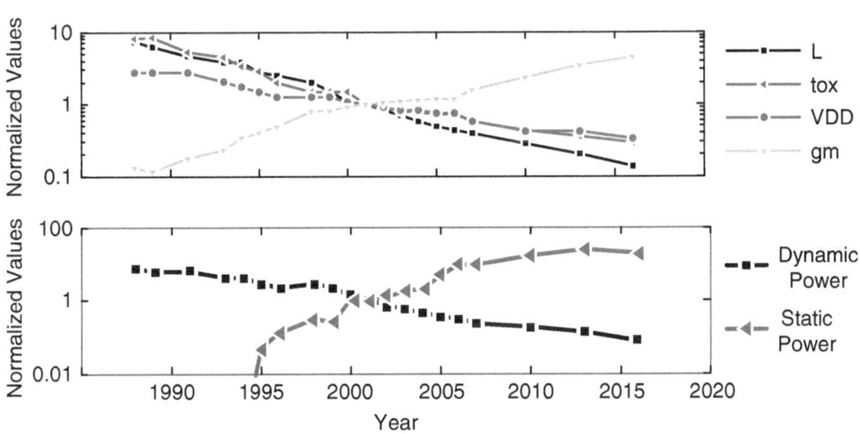

Fig. 2.6 ITRS predictions for device scaling and power dissipation at 2001 [29]

consumption that becomes more and more pronounced in more advanced technology nodes. Therefore, to design ULP systems in modern technologies, special care is required to overcome this problem.

Emerging new applications that require very low power consumption, has made subthreshold circuits very popular. In these type of applications, energy consumption and cost are the most important parameters with medium (1 Mspe – 10 Msps) or low (10 ksps – 100 ksps) data throughput systems [16]. Lowering the supply voltage even below threshold voltage of devices leads to quadratic reduction of the circuit dynamic power. This technique is also helpful to reduce the leakage or static power consumption of conventional CMOS circuit topologies implemented in modern nano-scale technologies.

In the following sections, the two main issues in design of ultra-low power digital circuits, i.e., static power dissipation and variability, will be reviewed. In more advanced deep-submicron MOS technologies, these two problems are more pronounced. Therefore, if not necessary, generally older technologies can be used for implementing energy-constrained circuits that does not require a high performance, such as in RFIDs, bio-implants, and sensor network. In some applications, the energy-constrained circuit needs to have a high performance while occasionally is operational [30]. In such cases, dynamic voltage scaling can be employed to scale the circuit power consumption and performance by moving from subthreshold region to superthreshold (above threshold) region. In such bursty applications, an advanced CMOS technology needs to be used to support the required specifications during high performance mode of the operation [30]. Advanced MOS technologies also have been used for implementing energy-constrained circuits which are supporting a high-performance application. In these cases, special design techniques are required to implement subthreshold circuits which suffer from high leakage current and very wide parameter variability [30].

2.4.1 MOS Transistor Leakage Mechanisms

While the static power consumption of static CMOS circuits have been ignored in early CMOS technologies [31], it has become a major challenge in UDSM technologies. Figure 2.7 describes the main leakage mechanisms in a deep sub-micron MOS device. Among different types of leakage, subthreshold residual channel leakage current and gate tunneling currents are more essential. The main sources of static power consumption in CMOS logic circuits that are more pronounced in modern technologies are briefly explained in this section (see also: [32–34]).

2.4.1.1 Scaling Rules

To keep the transistor performance on an acceptable level, in addition to scaling the device length, L, it is necessary also to scale gate oxide thickness, t_{ox}, junction

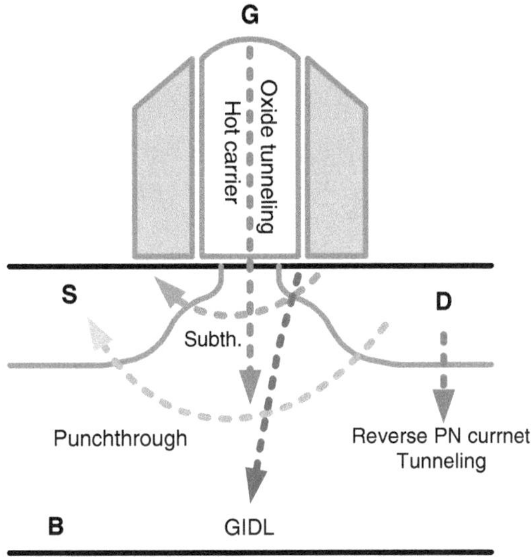

Fig. 2.7 Leakage current sources in a MOS device

depth, X_j, and depletion depth, D. This proportional scaling results in an acceptable device aspect ratio defined by

$$K_{\mathrm{AR}} = \frac{L}{\sqrt[3]{t_{\mathrm{ox}} X_j D \frac{\epsilon_{\mathrm{Si}}}{\epsilon_{\mathrm{ox}}}}}. \tag{2.48}$$

Unfortunately, it is difficult to keep the device K_{AR} on an acceptable level in very deep sub-micron technologies. Specially, maintaining the vertical sizes on desired value is very difficult. As will be seen in the next section, when gate oxide approaches scaling limits, there is a rapid increase in gate oxide leakage. Therefore, it is difficult to scale down the gate oxide thickness as device channel length. This constraint prevents having appropriate device K_{AR}.

2.4.1.2 Gate Tunneling

Oxide leakage is due to tunneling of carriers through the gate oxide. In more advanced technologies where oxide thickness, t_{ox}, is reducing and hence the field across the oxide is increasing, the tunneling phenomena becomes more significant. The gate tunneling current is due to the two different mechanisms: Fowler–Nordheim (FN) tunneling, and direct tunneling. The FN tunneling current density is given by [4]

$$J_{\mathrm{FN}} = \frac{q^3 E_{\mathrm{ox}}^2}{16\pi^2 \hbar \phi_{\mathrm{ox}}} \exp\left(-\frac{4\sqrt{2m^* \phi_{\mathrm{ox}}^3}}{3\hbar q E_{\mathrm{ox}}}\right) \tag{2.49}$$

where E_{ox} is the field across the oxide, ϕ_{ox} is the effective height for electron in the conduction band, and m^* is the effective mass of an electron in the conduction band of silicon. On the other hand, the current density of the direct tunneling is [4, 35]

$$J_{DT} = \frac{q^3 E_{ox}^2}{16\pi^2 \hbar \phi_{ox}} \exp\left(-\frac{4\sqrt{2m^* \phi^3}}{3\hbar q E_{ox}} \cdot \left(1 - \sqrt{\left(1 - \frac{V_{ox}}{\phi_{ox}}\right)^3}\right)\right) \tag{2.50}$$

By reducing the gate oxide thickness, the direct tunneling current increases rapidly.

In analog applications, it is possible to model the gate leakage current by a conductance (g_{tun}) in parallel to the gate capacitance (C_g) [17]. In frequencies higher than $f_g = g_{tun}/(2\pi C_g)$, the input impedance is capacitive while for frequencies lower than f_g, it is resistive. As shown in [17], the gate cutoff frequency can be calculated by

$$f_g = \frac{g_{tun}}{2\pi C_g} \approx A \cdot V_{GS}^2 \cdot e^{t_{ox}(V_{GS}-13.6)} \tag{2.51}$$

where t_{ox} is in (nm) unit, and A is a constant number (1.5×10^{16} for NMOS transistors and 0.5×10^{16} for PMOS devices). When f_g is about 0.1 Hz for 0.18-μm CMOS, it increases to about 1 MHz in 65-nm CMOS [13].

2.4.1.3 Subthreshold Conducting

Subthreshold (weak inversion) conduction current is due to the drift of minority carriers at $V_{GS} < V_{TH}$. The minority carrier concentration in this region of operation is very low but not zero. The weak inversion current can be estimated using (2.11) where:

$$n = 1 + \frac{C_{dm}}{C_{ox}} = 1 + \frac{\epsilon_{Si}}{\epsilon_{ox}} \cdot \frac{t_{ox}}{W_{dm}}. \tag{2.52}$$

Here, W_{dm} is the maximum depletion region width, and C_{dm} is the capacitance of the depletion region [4]. The leakage current due to the subthreshold current is generally characterized by the subthreshold slope:

$$S = \left(\frac{d(\log_{10} I_{DS})}{dV_{GS}}\right)^{-1} = 2.3nU_T = 2.3U_T\left(1 + \frac{t_{ox}}{W_{dm}} \cdot \frac{\epsilon_{Si}}{\epsilon_{ox}}\right). \tag{2.53}$$

Subthreshold slope indeed represents how effectively the transistor can be turned off when V_{GS} is decreased below threshold voltage. As illustrated in Fig. 2.8, a lower subthreshold slope results in smaller off current, I_{OFF}. Higher value for V_T helps to reduce the off current. However, using high V_T devices (HVT) results in lower on current, I_{ON}, and hence increased gate delay.

2.4.1.4 PN Junction

Reverse biased PN junction leakage has two main components: the first one is due to the minority carrier diffusion and drift near the edge of the depletion region and the

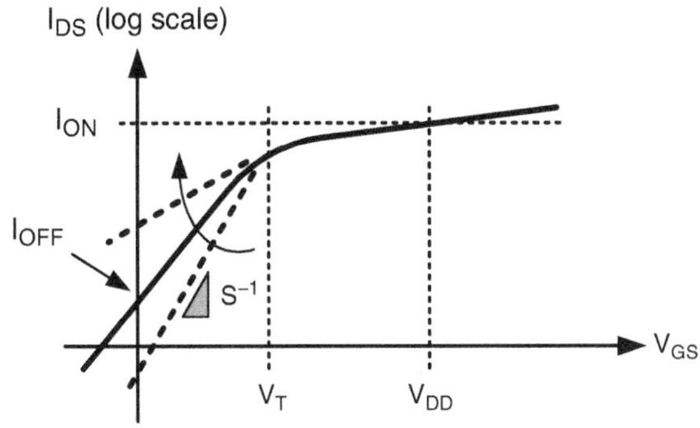

Fig. 2.8 I–V characteristics of an NMOS transistor and effect of subthreshold slope factor on off current of the device

second one is due to the electron–hole pair generation inside the depletion region of reverse-biased pn junction. When the p-side and n-side of the junction are heavily doped, which is the case in MOSFET devices, then the band-to-band tunneling current should be added to the estimations. The tunneling current density is given by [4]

$$J_{\mathrm{BB}} = \frac{\mathrm{AEV}_R}{\sqrt{E_g}} \exp \frac{-B\sqrt{E_g^3}}{E} \tag{2.54}$$

where $A = \sqrt{2m^*}q^3/(4\pi^3\hbar^2)$ and $B = 4\sqrt{2m^*}/(3q\hbar)$, m^* is effective mass of electron, E_g is the energy bandgap, V_R is the applied reverse biased voltage, E is the electric field at the junction, and $\hbar = h/(2\pi)$ and $h = 6.62606896 \times 10^{-34}$ (J.s) is Planck's constant. Assuming a step junction, the electric field can be calculated by

$$E = \sqrt{\frac{2qN_aN_d(V_R + V_{\mathrm{bi}})}{\epsilon_{\mathrm{Si}}(N_a + N_d)}} \tag{2.55}$$

where N_a and N_d are the doping concentration in P and N side of the junction.

2.4.1.5 DIBL

Drain voltage can affect the channel charge like gate voltage, especially in very short-channel devices. In short-channel devices because of proximity of the source and drain, drain voltage can influence the depletion region beneath the channel and hence change the channel potential. Drain-induced barrier lowering (DIBL) affects the leakage current by reducing the effective device threshold voltage [4]. In short-channel devices, the source-drain potential have a considerable effect on band bending over the channel. Therefore, the threshold voltage and consequently the subthreshold current of device can vary with this voltage. Indeed, in short-channel devices the depletion region of source and drain junctions interact to each other near the channel surface and will reduce the potential barrier between the two. Higher drain voltage or shorter channel length with enhance the DIBL effect. DIBL generally happens before the pinchthrough via the bulk occurs [33].

DIBL does not change the subthreshold slope. To reduce the effect of DIBL, higher surface and channel doping and shallow source and drain junction depths are required. The DIBL coefficient, η, can be expressed as [36]

$$\eta = \frac{1}{2\cosh\frac{L_{\text{eff}}}{2L_t}} \tag{2.56}$$

in which L_t is a characteristics length:

$$L_t = \sqrt{\frac{\epsilon_{\text{Si}} t_{\text{ox}} W_{\text{dm}}}{\epsilon_{\text{ox}} K}} \tag{2.57}$$

and K is a fitting parameter. Based on this expression, by scaling the transistor length, DIBL coefficient is increasing.

The bias current of a MOS device biased in subthreshold regime including DIBL and body effect can be modeled by [33]

$$I_{\text{DS}} = I_{\text{DSO}} \times \mathrm{e}^{\frac{V_{\text{GS}} - V_{T0} - \gamma V_S + \eta V_{\text{DS}}}{nU_T}} \times \left(1 - \mathrm{e}^{-\frac{V_{\text{DS}}}{U_T}}\right) \tag{2.58}$$

where:

$$I_{\text{DSO}} = \mu_0 C_{\text{ox}} \frac{W}{L_e} U_T^2 \mathrm{e}^{-\frac{\Delta V_T}{nU_T}}. \tag{2.59}$$

Here, ΔV_T is added to consider the threshold voltage variation from one transistor to the other one. The exponential dependence of I_{DSO} on ΔV_T shows the high sensitivity of the subthreshold current on process variation.

Regarding (2.58), the subthreshold leakage current could be calculated by

$$I_{\text{sub}} \approx I_{\text{DSO}} \times \mathrm{e}^{\frac{-V_{T0} - \gamma V_S + \eta V_{\text{DD}}}{nU_T}} \times \left(1 - \mathrm{e}^{-\frac{V_{\text{DD}}}{U_T}}\right) \tag{2.60}$$

which is very sensitive to DIBL effect.

2.4.1.6 GIDL

Gate-induced drain leakage (GIDL) is due to the high electric field near the Si–SiO_2 interface. The high gate-drain electric field can give sufficient energy to the electrons or holes to cross the interface potential barrier and pass through the oxide. This phenomena creates a current flow between drain and substrate. To reduce the GIDL effect, very high and abrupt drain doping concentration with very low series resistance should be used [37].

2.4.1.7 Hot Carrier

Hot carrier injection is due to the high electric field near the Si–SiO_2 interface [4]. High electric field can give sufficient energy to the carriers to cross the interface potential barrier and enter into the oxide layer [38].

2.4.1.8 Punchthrough

Due to the proximity of drain and source in short-channel devices punchthrough can happen [37]. In this case, the depletion region at the drain-substrate and source-substrate junctions extend into the channel. This phenomena will reduce the effective channel length. Therefore, increasing the reverse bias voltage across the junctions by increasing V_{DS} pushes the junction closer to each other. Punchthrough happens when the depletion regions merge together [37].

2.4.1.9 Channel Length Effect

The threshold voltage reduction of an MOS device when the device length is reducing is called threshold voltage rolloff [4]. The reduction of threshold voltage can be worsen in higher drain-source voltages due to DIBL effect. A nonuniform HALO doping can be used to mitigate this problem by reducing the depletion width and hence reducing the DIBL effect [39]. As a result, reverse SCE (RSCE) occurs and threshold voltage decreases by increasing the length of device [40].

2.4.1.10 Narrow-Width Effect

The threshold voltage of an MOS device also depends on the width of transistor [4,33,41,42]. Depending on isolation technologies, threshold voltage can be reduced or increased by reducing the channel width. With a less abrupt transition between the channel and the isolation, such as in local oxidation of silicon (LOCOS), the device threshold voltage increases with reducing the channel width. This effect is mainly because of extra depletion charge beneath the field oxide that should be added to the channel charge [34]. This effect is inverse for abrupt isolations such as in sealed interface local oxidation (SILO), and shallow trench isolation (STI).

2.4.1.11 Thermal Effect

The stand-by current of a transistor can change considerably by temperature. This variation is mainly due to carrier mobility (μ), thermal voltage (U_T), subthreshold slope factor (n), and threshold voltage [34]. Subthreshold slope (S) increases with temperature almost linearly, while threshold voltage decreases with temperature (the coefficient is about $-0.8\,mV/°C$) [4].

2.4.1.12 Short Circuit Current

Because of finite transition time at the input of a static CMOS gate, during a very short period of time both PMOS and NMOS devices are on and hence there is a short circuit current between V_{DD} and ground. This current can be considerable when V_{DD} is high and both PMOS and NMOS devices conducting in SI. When the logic circuits are biased in subthreshold regime, most of the time this current can be ignored [34].

2.4.2 *Leakage Reduction Techniques*

The total power consumption of a digital system is the sum of dynamic (P_D) and leakage (or static) power consumption (P_{leak}) can be approximated by [32]

$$P_{diss} \approx P_D + P_{leak} \tag{2.61}$$

where

$$P_D = \alpha f_{op} C V_{DD}^2 \tag{2.62}$$

and

$$P_{leak} = I_{leak} \cdot V_{DD} \tag{2.63}$$

where α stands for the average switching activity rate. To control the static power consumption of CMOS logic circuits which is going to be more and more pronounced in advance technologies, special techniques are needed to be used [4, 33]. Some of these techniques are briefly explained in the following.

2.4.2.1 Device Level Engineering

The leakage current, as explained before, depends on different physical phenomena and can be reduced by controlling the device dimensions (such as length, L, oxide thickness, t_{ox}, junction depth, X_j), and doping profile of the transistor.

In device engineering level, it is very important to control the short-channel effects (SCEs) by scaling down the device dimensions and choosing proper channel doping profile. Generally, it is very desirable to scale the device dimensions under constant field principle [4]. Using retrograde doping and halo doping are two possible approaches to control the SCEs [4].

2.4.2.2 Circuit Level Techniques

At the circuit level, it is possible to reduce the leakage current contribution through careful selecting voltage levels in different terminals of devices, and choosing proper devices with appropriate threshold voltages. Careful device sizing is the

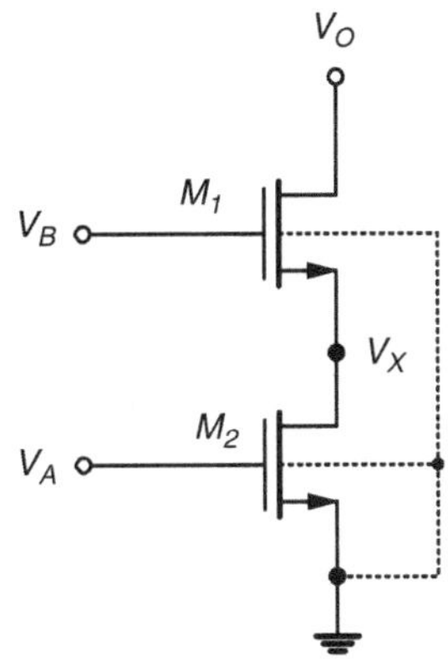

Fig. 2.9 Stacking technique to reduce the leakage current

other possibility to reduce the leakage current. In many ultra-low power designs, the length of MOS devices is selected slightly larger than the minimum size to reduce the leakage current and have less variability [16, 43], and [44]. It is also possible to use special circuit topologies to control the static current [34].

A common circuit technique that can be used for reducing the leakage current, as an example, is using stacked transistors (stacking effect). This technique, depicted in Fig. 2.9, can reduce the leakage current by one order of magnitude compared to a single transistor configuration [45, 46]. The main issue associated with this technique is the dependence of leakage current on input data vector [47].

Multiple threshold voltage CMOS technologies (MTCMOS) provide this possibility to use different types of devices for different purposes. In other words, one can use HVT devices for reducing the leakage current and use LVT devices in critical paths where the speed of operation is important. To fabricate multiple threshold devices in a technology, it is possible to change the channel doping, oxide thickness, or using transistors with different length, or body bias. There are some advanced techniques that are changing the threshold voltage of devices with respect to the operating condition through controlling the body voltage [48].

The leakage power consumption can also be controlled by supply voltage scaling. The dynamic power consumption, as shown in (2.62), is proportional to the square value of V_{DD}. Therefore, it is possible to control the dynamic power consumption by adjusting the supply voltage very effectively. It has been also shown that supply voltage scaling can help to reduce the static power consumption of digital circuits by decreasing the DIBL effect [49].

2.5 Impacts of Variation on Subthreshold CMOS Operation

Variability and static leakage current are the two main concerns in design of digital systems in advanced nano-scale CMOS technologies [4, 50]. Both of these issues are more pronounced in ultra-low power (ULP) systems, where the transistors are mostly biased in subthreshold regime in order to reduce the static and dynamic power consumption [16]. The exponential I–V characteristics of MOS devices in this regime of operation exacerbates the circuit sensitivity to the variation of device parameters. Circuit reliability, delay, and energy consumption are among the most important issues that are affected by process variation [16, 30].

In the field of digital design, gate delay variation due to the process variation has been always an important concern. This effect is more pronounced in subthreshold logic circuits where current of MOS devices exponentially depends on gate voltage and threshold voltage. Therefore, any small variation in the device parameters can change considerably the peak current (device on current) and the off current of the device, and hence change the gate delay and also the static and the dynamic current consumption of the circuit [16,54]. Figure 2.10 shows the effect of process variation on different device parameters in CMOS 65-nm technology. As can be seen, by moving towards subthreshold regime (lower V_{DD} values), the amount of variation on the cell turn on current, I_{ON}, and delay, t_d, increases rapidly. The variation on turn off current, I_{OFF}, is always high because this current is always determined by the subthreshold current. It is also noticeable that the ratio of turn on to turn off current, γ, degrades considerably by reducing the supply voltage.

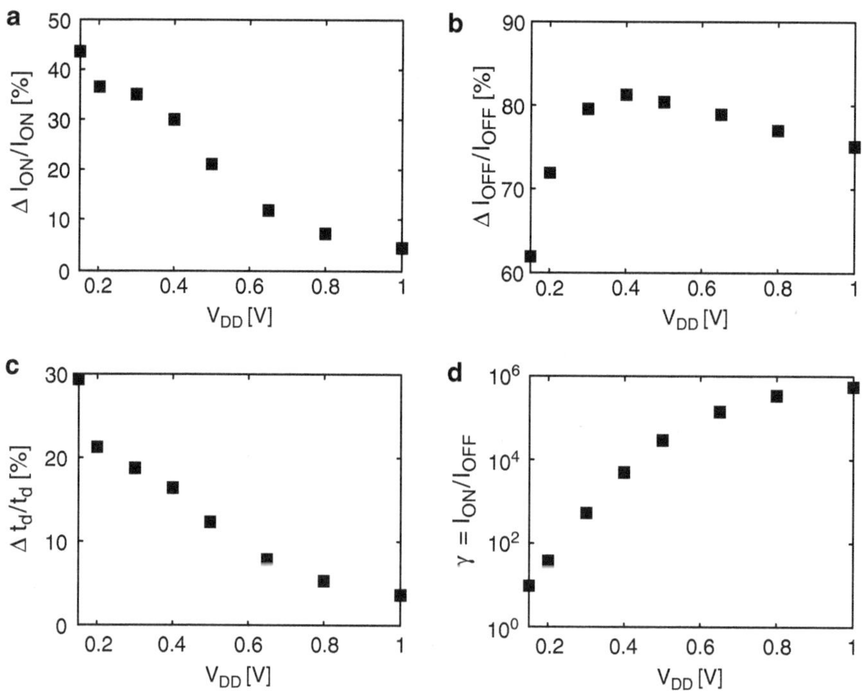

Fig. 2.10 Variation on: (**a**) I_{ON} current, (**b**) I_{OFF} current, and (**c**) delay of a NAND gate implemented in 65 nm CMOS technology. (**d**) Typical value of $\gamma = I_{ON}/I_{OFF}$

In addition, process variation and device mismatch can degrade the circuit reliability. For example, device parameter variation can degrade the static noise margin (SNM) of memory or logic cells considerably [16, 30]. To compensate the effect of process or environmental variations, many different techniques have been proposed. A common approach for mitigating this effect is to use up-sized channel length devices which helps to reduce the variability and improve subthreshold factor, simultaneously [16]. The other possibility is to increase the circuit supply voltage to a high enough value to make sure that the circuit will remain operational even in presence of variation [30].

As described in [30], not all the ultra-low power systems are required to be integrated in a modern nano-scale CMOS technology. However, still there are several very important ultra-low-power applications that are needed to be implementing in such advanced technology nodes. In such cases, special techniques are required to cope with the device variability and the leakage current and yet keep the performance high. Some recent studies show that using devices larger that the minimum feature size or increasing the supply voltage can help to compensate the effect of variability [16, 30]. The price that comes with the up-sizing of devices or increasing the supply voltage is augmentation in system energy consumption. Preliminary analysis show that the benefit of technology scaling in terms of energy consumption starts to diminish for 45/32-nm technology nodes and below [16].

Here, the goal is to provide a more methodological approach for proper device sizing and choosing the supply voltage of a digital CMOS circuit in order to maximize the benefit of technology scaling. In this methodology, the effect of circuit activity (duty) rate and also interconnects can be involved in the analysis to have a more precise estimation of the system performance.

This section provides an analytical approach for estimation the impact of variability on the main design parameters, namely noise margin, energy consumption, and gate delay. The results of this analysis will be used in Sect. 2.5.3 to explore the behavior of a digital system in course of technology scaling and exploring the optimal approach for choosing circuits parameters, such as size of devices and supply voltage.

2.5.1 Noise Margin

Noise margin, NM, is a measure of robustness of a logic gate again external or internal perturbations such as noise, and variation [51, 52]. Generally, a nonnegative noise margin for combinational logic cells ($NM \geq 0$) and a positive noise margin for sequential circuits is necessary ($NM > 0$).

To explore the effect of process variation on logic cell operation, in this section, the NM of an inverter will be analyzed. Since ULP applications are the main concern of this work, we are assuming that all the devices are biased in subthreshold regime. In other words, we are assuming that the circuit supply voltage is not more than threshold voltage of MOS devices.

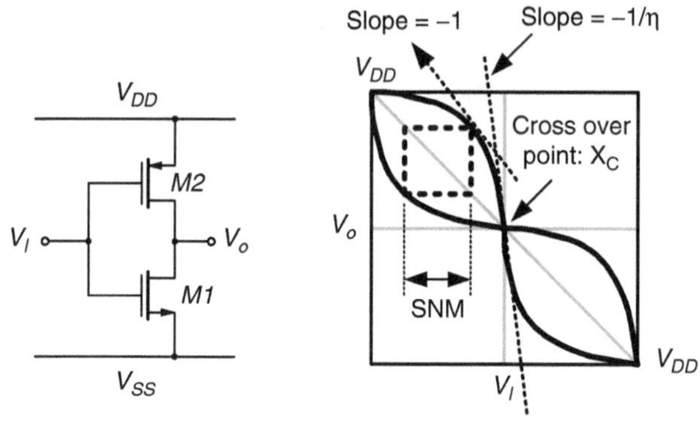

Fig. 2.11 A sample CMOS inverter and the corresponding Butterfly curve used for estimating NM

Figure 2.11 shows a CMOS inverter and the corresponding Butterfly curve that is generally used to measure NM. Using the EKV model [7], the bias current of an NMOS device biased in subthreshold could be estimated by:

$$I_{\mathrm{DS}} = I_0 \cdot \mathrm{e}^{\frac{V_G + \eta V_D}{nU_T}} \cdot \left(1 - \mathrm{e}^{\frac{-V_D}{U_T}}\right) \cdot \left(1 + \frac{V_{\mathrm{DS}}}{V_A}\right) \tag{2.64}$$

where η is used to model the drain-induced barrier lowering (DIBL) effect [4, 33], V_A represents the effect of finite output resistance, and I_0 is defined by

$$I_0 = 2n\mu_e C_{\mathrm{ox}} \frac{W}{L_e} U_T^2 \mathrm{e}^{-\frac{V_{T0}}{nU_T}}. \tag{2.65}$$

All voltages are referred to the bulk of the device [7]. Although not necessary, in the rest of this section it is assumed that the subthreshold slope factor, n, and V_A values are equal for NMOS and PMOS devices in order to simplify the analysis. For real estimations made in Sect. 2.5.3, the precise values of n and V_A for NMOS and PMOS devices have been used.

To maximize the typical NM of the gate, the relative size of PMOS and NMOS devices should be selected such that satisfy this requirement:

$$I_{\mathrm{DS,NMOS,0}}\,|_{V_{\mathrm{IN}}=V_{\mathrm{DD}}/2} = I_{\mathrm{SD,PMOS,0}}\,|_{V_{\mathrm{IN}}=V_{\mathrm{DD}}/2}\,. \tag{2.66}$$

With this constraint, the crossover voltage will be as close as possible to $V_{\mathrm{DD}}/2$ and logic cell will have a relatively symmetric rising and falling transitions. The zero index in (2.66) stands for nominal conditions without including the device mismatches or process variation.

Now, to calculate the voltage transfer characteristic (VTC) of a CMOS inverter, the following equation should be solved [43]:

$$I_{\mathrm{DS,NMOS}} = I_{\mathrm{SD,PMOS}} \tag{2.67}$$

or

$$I_{\text{DS,NMOS},0} \times (1 + \Delta I_{\text{DS,NMOS}}) = I_{\text{SD,PMOS},0} \times (1 + \Delta I_{\text{SD,PMOS}}) \tag{2.68}$$

where ΔI_{DS} and ΔI_{SD} are used to include the deviations on transistor current respect to the nominal value in presence of process variations. This results in:

$$K_{\Delta} \cdot \mathrm{e}^{-\frac{V_{\text{DD}}(1+\eta)}{nU_T}} \cdot \mathrm{e}^{\frac{2V_I}{nU_T}} = \mathrm{e}^{-\frac{2\eta V_O}{nU_T}} \cdot \frac{1 - \mathrm{e}^{\frac{-V_{\text{DD}}+V_O}{U_T}}}{1 - \mathrm{e}^{\frac{-V_O}{U_T}}} \cdot \frac{1 + (V_{\text{DD}} - V_O)/V_A}{1 + V_O/V_A}. \tag{2.69}$$

which represents the DC VTC of the inverter. In this equation, the effect of all parameters related to process variations is summarized in K_{Δ} which can be estimated by:

$$K_{\Delta} = \frac{1 + \Delta I_{\text{DS,NMOS}}}{1 + \Delta I_{\text{SD,PMOS}}} = \mathrm{e}^{\frac{\Delta V_T}{nU_T}} \cdot \frac{1 + \Delta\beta_N/\beta_N}{1 + \Delta\beta_P/\beta_P} \tag{2.70}$$

in which:

$$\delta V_T = \mid V_{T0,P} + \Delta V_{T0,P} \mid - (V_{T0,N} + \Delta V_{T0,N}). \tag{2.71}$$

The term K_{Δ} includes threshold voltage variation and also variation on transistor $\beta = \mu C_{\text{ox}} W/L_{\text{eff}}$ value. The nominal value of K_{Δ} when there is no parameter variation is one. It is also interesting to notice that based on (2.71), VTC variation due to threshold voltage depends only on relative variation on threshold voltage of NMOS and PMOS devices.

Figure 2.12a depicts the calculated VTC of an inverter using (2.69) whereas process variation has been included in the equation. Figure 2.12b and c show the static noise margin and input–output VTC crossover point (X_C) calculated using (2.69) in comparison to the transistor level simulation results. As can be seen, there is a very good agreement between (2.69) and the transistor level simulation results.

Excluding Process Variation: In the first step, a simplified model for NM of an inverter operating in subthreshold regime will be derived using (2.69). This simplified model can be especially interesting to predict the circuit reliability in course of technology scaling.

Regarding (2.64), in the presence of DIBL effect the small signal output conductance of a MOS device will change to:

$$g_{\text{out}} \approx g_{\text{ds}} + g_{\text{DIBL}} = \frac{I_{\text{DS}}}{V_A} + \frac{I_{\text{DS}}}{nU_T/\eta} \tag{2.72}$$

which means DIBL reduces the output impedance of MOS devices, which results in circuit gain reduction. As the gain of a CMOS gate directly affect the noise margin of a cell, therefore, it is expected that DIBL effect causes noise margin degradation.

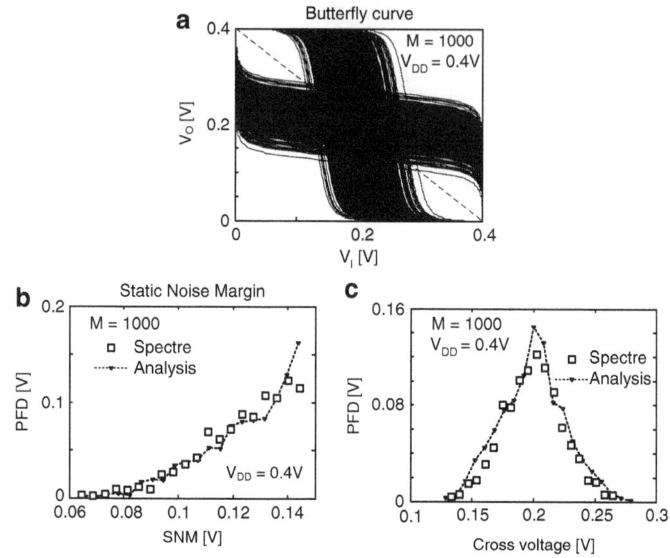

Fig. 2.12 Comparing the estimated static noise margin based on (2.69) and transistor level simulation results. (**a**) The calculated VTC based on (2.69) including process variations. (**b**) Static noise margin in comparison to the transistor level simulations (**c**) Input–output crossover point, X_C

Ignoring the finite output resistance of the MOS devices for simplicity and using (2.69), it can be seen that the slope of VTC close to the transient point is:

$$\frac{\partial V_O}{\partial V_I} \approx \frac{-1}{\eta}. \tag{2.73}$$

which means that the gain of an inverter will be limited by the DIBL factor in advanced CMOS technologies. Also, it is clear that η needs to be much smaller than unity to have enough gain for reliable logic operation. To estimate the static noise margin, based on definition, the points in which the slope of VTC becomes -1 should be calculated:

$$\frac{\partial V_O}{\partial V_I} = -1. \tag{2.74}$$

The slope of VTC can be calculated using (2.69). Based on this analysis, the static noise margin of an inverter which is biased in subthreshold without including process variations can be estimated by:

$$\mathrm{NM}_0 = \left(\frac{V_{\mathrm{DD}}}{2} - U_T \ln\left(\frac{1}{D \cdot (1-D)}\right)\right) - \eta \cdot \left(\frac{V_{\mathrm{DD}}}{2} + 2U_T \ln(1-D)\right) \tag{2.75}$$

where:

$$D = \frac{n}{n + 2(1-\eta)}. \tag{2.76}$$

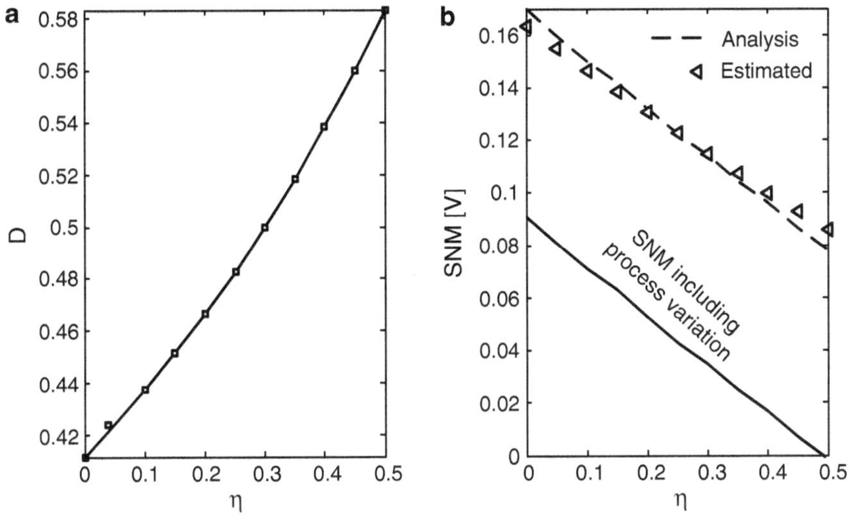

Fig. 2.13 (**a**) Parameter D versus η. (**b**) NM_0 based on analysis in comparison to the NM_0 value calculated using (2.75). This graph also shows the lower limit on NM when process variation is included. Here, $V_{\mathrm{DD}} = 0.4\,\mathrm{V}$ and $V_T = 0.5\,\mathrm{V}$

Again, index zero means that there is no parameter variation in this estimation. Parameter D depends on subthreshold slope factor, n, and DIBL coefficient, η. It is important to notice that based on (2.75), NM depends on DIBL coefficient and by increase of η, NM reduces. Therefore, to have a positive NM value, DIBL coefficient needs to be much smaller than one. It is also noticeable that NM degrades when n, or equivalently subthreshold slope (S) increases. Figure 2.13a shows the value of D versus η. Figure 2.13b compares the estimated value for NM based on (2.75) and the precise value of NM calculated from (2.69) which is showing a very good agreement.

It is also possible to derive a very crude approximation for NM just for having a better understanding of the effect of DIBL:

$$\mathrm{NM}_0 \approx \frac{V_{\mathrm{DD}}}{2} \cdot (1 - \eta) \tag{2.77}$$

which indicates that NM reduces almost linearly with increase of η value, and the only way to compensate this effect is to increase the supply voltage.

Including Process Variation: To derive (2.75), the effect of device parameter variations considered in K_Δ has been ignored. Including the device variations and after some analysis, it can be shown that NM is sensitive to process variation and the reduction on NM can be modeled by:

$$\mathrm{NM} = \mathrm{NM}_0 - \frac{nU_T}{2} \cdot \ln K_\Delta \tag{2.78}$$

By replacing K_Δ:

$$\mathrm{NM} = \mathrm{NM}_0 - \left| \frac{\delta V_T}{2} - \frac{nU_T}{2} \cdot \ln\left(\frac{1+\Delta\beta_N/\beta_N}{1+\Delta\beta_P/\beta_P}\right) \right| \tag{2.79}$$

It is important to notice that any variation on threshold voltage difference degrades the NM value regardless of the sign of this variation. Indeed, the maximum NM can be achieved by setting the crossover point to $V_{\mathrm{DD}}/2$ and since variations on the threshold voltage difference will move this point to the left or to the right, it will degrade the NM value regardless of sign of variations. As the variation on β, especially logarithm of β as appears in (2.79), is negligible in comparison to the variation on threshold voltage,[14] although not necessary, this equation can be simplified to

$$\mathrm{NM} = \mathrm{NM}_0 - \frac{\delta V_T}{2} \tag{2.80}$$

As the crossover point (X_C shown in Fig. 2.11) depends on δV_T, any variation on difference of threshold voltage of PMOS and NMOS devices will be reflected on NM. In Fig. 2.13b, the degradation on NM due to the process variation has been shown. It can be seen that in high DIBL coefficient values, noise margin degrades and in presence of variations it will be really difficult to design a gate with sufficient NM.

Using (2.80), it is possible to estimate the minimum acceptable size of transistors to have a positive noise margin, i.e., NM>0. Using $\sigma_{V_T}^2 = A_{\mathrm{VT}}^2/(W \cdot L_{\mathrm{eff}})$ [13], and assuming that variation on threshold voltage of PMOS and NMOS devices is uncorrelated and the width of PMOS device is R times larger than NMOS transistors, the effective width and length of NMOS device should be larger than:[15]

$$\sqrt{W_N L_N} > \frac{3}{2} \cdot \frac{A_{\mathrm{VT}}}{\mathrm{NM}_0} \cdot \sqrt{\frac{R+1}{R}} \tag{2.83}$$

to have a positive noise margin. In (2.83), NM_0 is the nominal noise margin without parameter variation and can be estimated by (2.75). To simplify the analysis, it is assumed that $A_{\mathrm{VT}} = \max\{A_{\mathrm{VT},P}, A_{\mathrm{VT},N}\}$. In (2.83), a coefficient of three has been added to the nominator to include the 3σ variation effect. Indeed, (2.83) puts a lower limit on device area which depends on supply voltage through NM_0. A larger transistor size means larger area, and more importantly more parasitic capacitance

[14] Based on ITRS suggestion, standard deviation on device length, L, needs to be less than 20% of its nominal value [29].

[15] Based on this estimation, there is a lower limit on effective physical length and width of transistors. Based on BSIM model, the effective length and width of transistors are [58]:

$$L_e = L + XL - 2 \times \mathrm{d}L \approx L + XL - 2 \times \mathrm{DLC} \tag{2.81}$$
$$W_e = W + XW - 2 \times \mathrm{d}W \approx W + XW - 2 \times \mathrm{DWC} \tag{2.82}$$

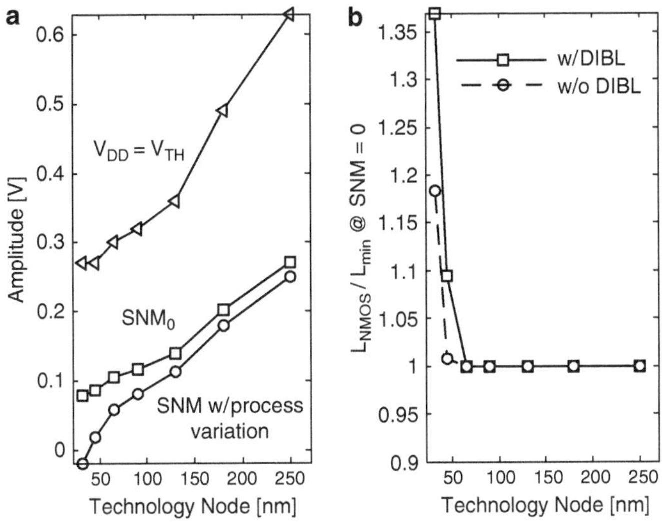

Fig. 2.14 (**a**) Noise margin of a subthreshold inverter biased with $V_{DD} = V_{T0}$ in course of technology scaling. The degradation of noise margin due to process variation has been also shown. (**b**) Minimum NMOS transistor length to have a positive noise margin in presence of process variation. The results have been shown with and without including the DIBL effect

which results in larger gate delay. Therefore, it is really important to keep the size of devices as small as possible. The implication of (2.83) is that by technology scaling, NM_0 of subthreshold CMOS circuits degrades due to more DIBL effect. Hence, the size of transistors could not be scaled down with the same ratio as the gate length scaling. Even if we ignore the DIBL effect, based on (2.83) the size of transistors could be scaled down only proportional to the improvement on A_{VT}.

Figure 2.14a shows the estimated noise margin for a CMOS inverter with technology scaling. For this estimation, minimum size devices have been used. By technology scaling, supply voltage is reducing and at the same time DIBL effect becomes more and more evident. This explains the drop of NM_0 in Fig. 2.14a. Including the process variation, NM starts to become negative for technology nodes below 65 nm. Figure 2.14b shows the minimum acceptable device length with respect to the device minimum feature size to keep the noise margin positive. In very deep technology nodes such as 16 nm, this ratio can be as high as 1.35.

2.5.2 Energy Consumption

Deriving a closed form equation for estimating the power dissipation of a CMOS systems is very complicated. Here, we are trying to calculate the power dissipation of a fundamental structure as a basis for more complicated topologies [53].

Figure 2.15a illustrates the proposed test structure and Fig. 2.15b depicts the simplified waveform of the current drawn from supply source by a single gate. The peak

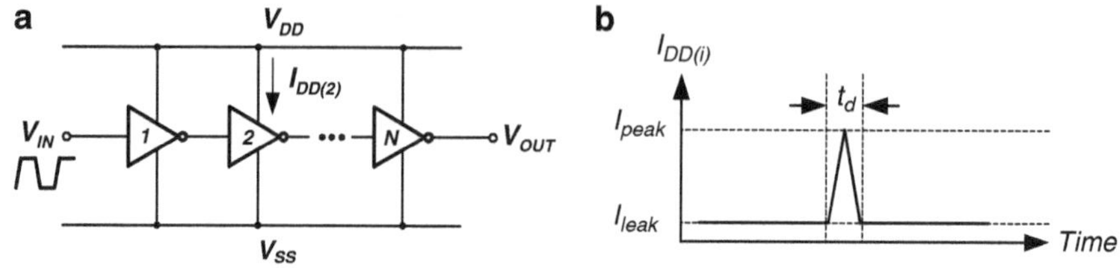

Fig. 2.15 (**a**) A chain of N identical CMOS gates. Note that the type of logic gate used in the chain is arbitrary. (**b**) Modeling the current waveform

current (I_{peak}) and the leakage current (I_{leak}) drawn form supply by each logic cell, both depend on V_{DD}, size, and aspect ratio of the devices. Meanwhile, I_{peak} depends on transition time at the input of the corresponding gate. To simplify the calculations, we are assuming that the transition time at the input of each gate is comparable to the intrinsic transition time at the output of that gate when it drives C_L. This assumption is very close to reality when the logic depth is high. With this constraint, I_{peak} will depend only on V_{DD}.

The rms (root mean square) power consumption of this circuit shown in Fig. 2.15a can be calculated by [53][16]

$$P_{\text{diss,CMOS},N} = V_{\text{DD}} \cdot \sqrt{\frac{1}{T}\int_0^T i_{\text{DD}}^2(t)\text{d}t}. \tag{2.84}$$

Considering the simplified waveform of Fig. 2.15b for supply current, the total rms power consumption of the circuit will be:

$$P_{\text{diss},N} \approx N I_{\text{leak}} V_{\text{DD}} \sqrt{1 + \frac{\alpha \cdot \lambda}{3} \cdot \left(\frac{\gamma^2}{N^2} + \frac{\gamma}{N} - 2\right)} \tag{2.85}$$

where, $\alpha = f_{\text{op}}/f_{\text{max}}$ represents the activity rate, $f_{\text{max}} = 1/(2t_d)$ is the maximum operation frequency of a single gate, $\gamma = I_{\text{peak}}/I_{\text{leak}}$, $f_{\text{op}} = 1/T$, and $\lambda = [N/2]$. Here, λ is used to take into account that supply current depends only on the current that is used for charging the load capacitances. As expected, the minimum power consumption of the circuit is determined by the leakage current when activity rate is very low ($\alpha \approx 0$). At higher operating frequencies where the dynamic power consumption becomes dominant, the power dissipation is proportional to the square root of the operating frequency. By increasing the logic depth, the total power consumption scales up proportionally while the maximum speed of operation reduces by the same factor. Based on (2.85), it can be found that for activity rates smaller than a "critical activity rate" (α_C) given by:

$$\alpha_C \approx \frac{3N^2}{\lambda \cdot \gamma^2} \approx \frac{6N}{\gamma^2} \tag{2.86}$$

[16] Please note that the derivation given here is based on the conventional definition of root-mean square (rms) power. Similar conclusions can also be derived using the average power definition.

the subthreshold leakage power consumption will be dominant, while for higher activity rates, the dynamic power consumption comprises the main part of the power. Since α_C is proportional to $1/\gamma^2 = (I_{\text{leak}}/I_{\text{peak}})^2$, α_C increases quadratically with reducing γ. This means that in more advanced CMOS technologies, the contribution of leakage current will be more evident, and α_C will be higher. On the other hand, when logic depth increases, α_C also increases which means the effect of leakage current becomes more dominant in structures with deeper average logic depth [53].

Based on Fig. 2.15b, the maximum operating frequency of a CMOS gate ($f_{\max}$) can be estimated by:

$$f_{\max} = \frac{1}{t_d} \approx \frac{I_{\text{peak}}}{2V_{\text{DD}}C_L}. \tag{2.87}$$

Sometimes a constant coefficient is added to this expression to take into account different sources of nonideality that has not been included in our simplified estimation [54, 55].

Having (2.85) and (2.87), and using EKV model one can estimate the energy consumption of a chain of N CMOS gates in a specific operating frequency ($f_{\text{op}} < f_{\max}/N$), and supply voltage:

$$E_{\text{diss},N} \approx \frac{2N^2V_{\text{DD}}^2C_L}{\gamma} \cdot \sqrt{1 + \frac{\alpha \cdot \lambda}{3} \cdot \left(\frac{\gamma^2}{N^2} + \frac{\gamma}{N} - 2\right)}. \tag{2.88}$$

This expression represents the dependence of energy consumption on logic depth, N, interconnect parasitic effects, C_L, and the activity rate, α. To complete the calculation, γ can be estimated by:

$$\gamma = \frac{I_{\text{peak}}}{I_{\text{leak}}} = \frac{I\ |_{V_D=V_{\text{DD}}/2, V_G=V_{\text{DD}}/2}}{I\ |_{V_D=0, V_G=V_{\text{DD}}}} \tag{2.89}$$

Using (2.64) and after simplifying the relationship:

$$\gamma \approx \mathrm{e}^{\frac{V_{\text{DD}}\cdot(1-\eta)}{2nU_T}} \tag{2.90}$$

It is clear from (2.90) that DIBL and subthreshold slope factor both could reduce γ value. Combining (2.90) with (2.77), one can show that:

$$\text{NM}_0 \approx nU_T \cdot \ln \gamma \tag{2.91}$$

Figure 2.16 compares the predicted noise margin from (2.91) and transistor level simulations in 65 nm technology. Although a very rough estimation, (2.91) indicates this very important result that γ directly affects the circuit reliability. It is noticeable that based on transistor level simulations, (2.91) is valid in all regions of operations including subthreshold and strong inversion.

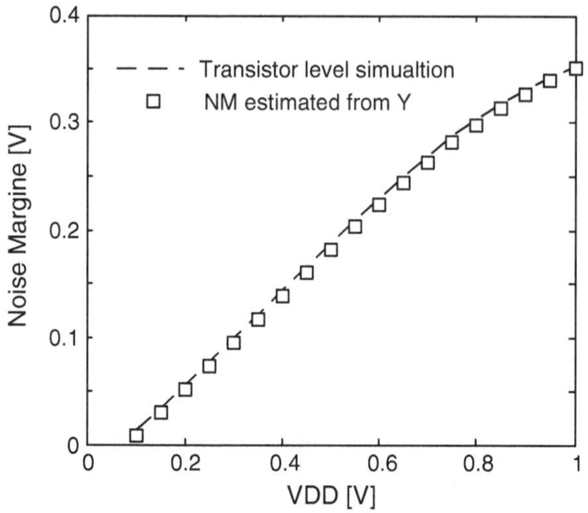

Fig. 2.16 Comparing noise margin resulted from transistor level simulations with the results from (2.91) in 65 nm technology

To have a reliable operation, nominal γ value should be large enough to overcome the effect of process variation on NM as presented before in (2.80):

$$\gamma > \exp\left(3 \times \frac{1}{nU_T} \times \frac{A_{\mathrm{VT}}}{\sqrt{W_N L_N}} \cdot \sqrt{\frac{R+1}{R}}\right) \tag{2.92}$$

or equivalently supply voltage needs to be larger than:

$$V_{\mathrm{DD}} > 3 \times \frac{A_{\mathrm{VT}}}{\sqrt{W_N L_N}} \cdot \sqrt{\frac{R+1}{R}} \cdot \frac{2}{1-\eta}. \tag{2.93}$$

This relationship represents a direct tradeoff between transistor area and supply voltage. A more precise lower limit on supply voltage can be extracted from (2.75).

Using High Threshold Voltage Devices: To reduce leakage current and hence power dissipation of an ULP digital system, there are two possibilities: either reducing the supply voltage or using high-V_T devices. Both approaches result in more gate delay values. However, most of the time in ULP circuits delay is not the primary issue and the delay increase can be tolerated. The main issue with supply voltage reduction is the reduction of noise margin as predicted by (2.75). Therefore, supply voltage reduction can be employed in subthreshold circuits only in the range that is allowed by (2.93). On the other hand, (2.92) implies that using high-V_T transistors could be a better choice than reducing supply voltage. The reason is that γ and NM are not affected by threshold voltage in the first order approximations. Hence, unlike down scaling the supply voltage, noise margin will not be degraded by increasing the threshold voltage.

2.5.3 Optimal Design with Technology Scaling

Having estimated the main circuit parameters such as noise margin, energy consumption, delay, and also having the relationship among these parameters, now we are ready to tackle this problem that what are the optimal design parameters to maximize the benefit from technology scaling.

In ultra-low power systems where energy consumption is the most critical parameter, the circuit operating condition is generally determined such that minimizes this parameter, i.e., [56]

$$\frac{\partial E_{\text{diss}}}{\partial V_{\text{DD}}} = 0 \tag{2.94}$$

Depending on system characteristics such as activity rate, interconnection parasitic effects, etc., the optimum supply voltage, $V_{\text{DD,opt}}$, in which energy consumption becomes minimum, is most of the time smaller than the device threshold voltage. Operating in subthreshold regime, it is necessary to make sure that variability will not affect the circuit performance. In other words, $V_{\text{DD,opt}}$ needs to be larger than the lower limit indicated in (2.93). Otherwise, either supply voltage, or the area of transistors should be increased.

Now we can use (2.88) to estimate the energy consumption of a digital system in different technology nodes. For this purpose, we use predictive technology model parameters to estimate the power consumption of a system in different CMOS technologies [16, 57].

2.5.3.1 A Low Activity Rate System Example

As an example, assume that the average system logic depth is $N = 20$, the activity rate is $\alpha = 0.1/(N)$, and the average load capacitance is $C_{L0} = 5\,\text{fF}$. A small fan-out of two has been considered for each gate, as well. To have a fair estimation, "low power" option in which devices have higher threshold voltage and less gate leakage current has been selected for this analysis.

The results for this estimation are shown in Fig. 2.17. Figure 2.17a depicts the minimum achievable energy consumption based on different strategies. The corresponding operating frequency and the supply voltage for minimum energy consumption are shown in Fig. 2.17b and c, while Fig. 2.17d shows this supply voltage normalized to the device threshold voltage at the corresponding technology node. As can be seen in Fig. 2.17a (grey line), by scaling the technology from 0.25 μm to around 65 nm, the energy consumption can be reduced. However, as technology continues scaling down, the minimum achievable energy consumption increases. In other words, technology scaling below 65 nm does not help to reduce the energy consumption of the circuit with the aforementioned conditions for activity rate and load capacitance. Based on Fig. 2.17d, for optimized energy consumption, the supply voltage needs be selected more and more close to the threshold voltage when the device feature sizes are decreasing. This is mainly due to the leakage current enhancement in more advanced technologies.

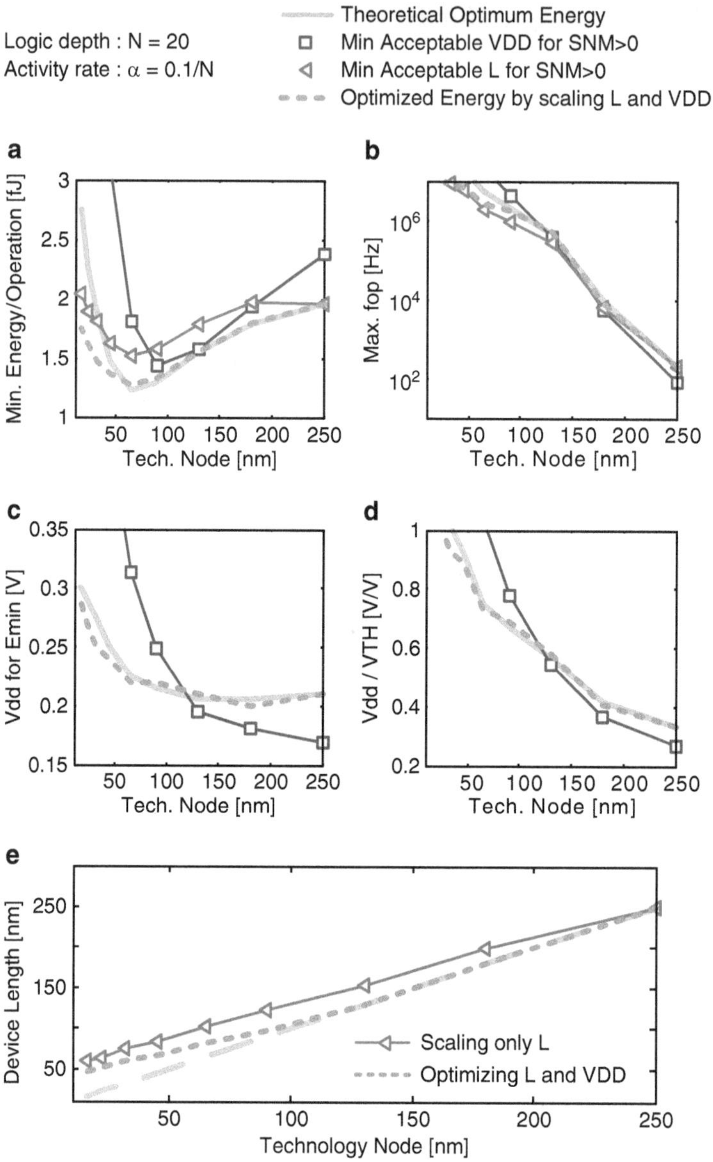

Fig. 2.17 (**a**) Optimum energy consumption by technology scaling ($\alpha = 0.1/N$, $N = 20$, $C_{L0} = 5\,\text{fF}$). (**b**) Corresponding operating frequency for optimum energy consumption. (**c**) Supply voltage in which energy consumption can be minimized. This figure also shows the minimum acceptable supply voltage to keep the noise margin positive. (**d**) Ratio of the optimum supply voltage to device threshold voltage by technology scaling. (**e**) Scaled device length to have a positive NM

However, to have a more practical estimation of energy consumption, we have to consider the process variation as well. In other words, not always the minimum energy consumption predicted by the grey line in Fig. 2.17d is achievable mainly because there are cases in which noise margin becomes unacceptably small due to the variations. As illustrated in Fig. 2.17c, the supply voltage for minimizing energy consumption is well below the acceptable level of V_{DD} for having a positive NM at technology nodes below 0.13 μm. This means that either the supply voltage or the device sizes are needed to be increase to improve the NM value to an acceptable level in these technology nodes.

Figure 2.17a depicts the energy consumption for three other cases as well: (a) scaling up the supply voltage to have a positive noise margin, (b) scaling up the size of device to improve the noise margin, and (c) using a combination of supply voltage and device size scaling to have the desired noise margin and at the same time keep the energy consumption as close as possible to the minimum achievable value. As depicted in Fig. 2.17a, a combination of supply voltage and device size scaling can result in the best performance in terms of energy consumption. Figure 2.17b compares the operating frequency for different design approaches. As depicted in this figure, the combinational approach does not give the best result in terms of delay, but still very close to the value expected by the initial optimized design resulted from $\partial E/\partial V_{DD} = 0$. Figure 2.17e shows the selected device length to have the desired noise margin based on different approaches. As depicted in this figure, the scaling of transistor size slows down below 90 nm node mainly because of compensating the effect of process variation.

Even using a combination of supply voltage and device size scaling, as illustrated in Fig. 2.17a, the energy consumption increases by moving to technologies below 65 nm node. In very deep technology nodes (below 65 nm), the proposed combinational approach gives a better result compared to the ideal estimations for the minimum energy consumption. The main reason for this improvement is that the size of transistors are slightly larger than the minimum value in the resulted circuit which can reduce considerably the leakage current as well as the DIBL effect.

2.5.3.2 A High Activity Rate System Example

Of course the result of the analysis depends on system specifications such as activity rate or loading effect. In any case, the relationships derived in this section can give a clear insight about the main design tradeoffs for implementing ULP systems in advanced CMOS technologies. For example, Fig. 2.18a–e shows the same graphs for a different condition in which activity rate is very high. In this case, V_{DD} scaling is more efficient than the device up-sizing for technology nodes above 0.13 μm and below this point, device up-sizing will result in less energy consumption. The optimized design combined of scaling both of these two parameters offers much better result, yet slightly higher than ideal energy consumption for all different technology nodes.

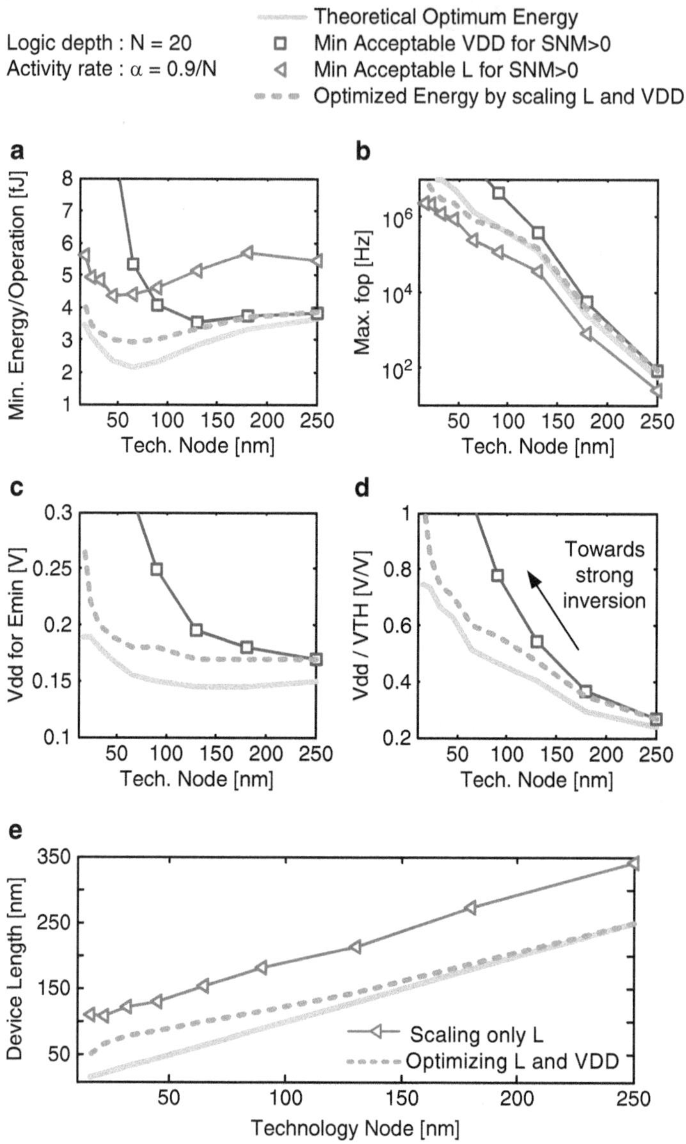

Fig. 2.18 (**a**) Optimum energy consumption by technology scaling ($\alpha = 0.9/N$, $N = 20$, $C_{L0} = 5\,\text{fF}$). (**b**) Corresponding operating frequency for optimum energy consumption. (**c**) Supply voltage in which energy consumption can be minimized. This figure also shows the minimum acceptable supply voltage to keep the noise margin positive. (**d**) Ratio of the optimum supply voltage to device threshold voltage by technology scaling. (**e**) Scaled device length to have a positive NM

2.5.3.3 Discussion

As a conclusion, a very careful design strategy for selecting optimum supply voltage or choosing proper device sizes is required to maximize the benefit from technology scaling for ultra-low power systems. Even in very deep technology nodes, still there is this possibility to minimize the energy increase, and hence control the energy loss. Of course this statement depends highly on high-level system specifications such as logic depth, activity rate, interconnections, and etc.

The other important result of this study is that the size of CMOS circuits biased in subthreshold regime can not be scaled as fast as the technology scaling permits. Indeed, because of effect of variation on circuit performance, the size of devices can only be scaled down proportional to the improvement in matching properties of MOS devices which can be represented by A_{VT}. The optimum device length which is shown in Figs. 2.17e and 2.18e depicts that the device sizes do not track the same path that technology scaling traverses. Depending on system specifications, the optimum device length is more than minimum technology feature size for technologies below 90 nm/0.13 μm.

2.5.4 Supply Voltage and Threshold Voltage Scaling for Optimal Design

The results of Sect. 2.5 provides the necessary basis for high-level analysis of digital CMOS circuits only by knowing few main process parameters in addition to the system specifications. Using these results, this section provides a more close look at the issue of performance optimization. In Sect. 2.5.3, we assumed that the circuit threshold voltage is given by technology and the only parameters that can be varied to reduce the energy consumption (or other convenient figures of merit), are supply voltage, V_{DD}, and device sizes. Now let us assume that there is this possibility to vary the device threshold voltage to reduce even more the circuit consumption. Indeed, (2.85) and (2.87) can provide the necessary analytical tools for this purpose. In addition, (2.75) and (2.79) can be used to limit our design space to the cases in which circuit reliability remains acceptable even in presence of process variation and hence make sure that the results of this study will be practically acceptable.

To generalize the study and find the optimum point, the design space should not be limited to only subthreshold region. Since in deriving (2.85) and (2.87) there has been no assumption regarding the region of operation, they can be used in our general analysis. However, the analysis has been carried out for estimating the noise margin are based on this assumption that the devices are biased in subthreshold regime. To avoid this problem, one can use (2.91) which is valid in all regions of operation, as it is depicted in Fig. 2.16.

Let us take the second example in previous section, where $\alpha = 0.9/N$, and try to minimize the system energy consumption by varying both supply voltage and threshold voltage. The result of this optimization is shown in Fig. 2.19. Comparing this figure with Fig. 2.18 reveals that it is possible to reduce the system energy consumption by adding the extra parameter of threshold voltage in process of optimization. To have minimum energy dissipation in different technology nodes, the analysis shows that the threshold voltage should be set to its maximum possible

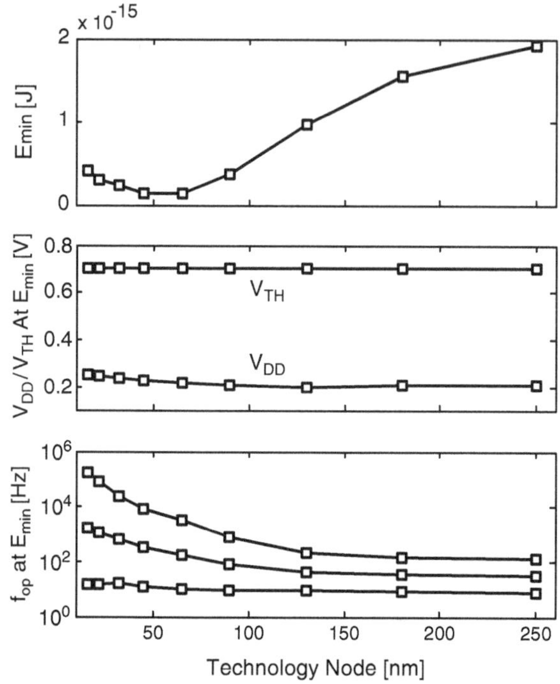

Fig. 2.19 Minimum energy consumption in different technology nodes when both supply voltage and threshold voltage are optimized. The optimum values for supply voltage and threshold voltage are also shown. Here, $\alpha = 0.9/N$. The bottom figure shows the nominal, the best, and the worst case operating frequency of the circuits in minimum energy consumption point

value (which in this example is set to be 0.7 V), while on the other side supply voltage tends to be very small, just enough to satisfy the noise margin requirement. In all the technology nodes, the devices are needed to be operated in week inversion. The achievable reduction in energy consumption can be as high as 30% in deep sub-micron technology nodes. Still, it can be seen that there is no clear benefit from energy consumption point of view to use technologies deeper than 45/65 nm for ultra-low power purposes. The other important achievement is that while Fig. 2.18a shows a very sharp increase in energy consumption at technology nodes below 65 nm, the new results in Fig. 2.19 exhibit a much slower slope for the mentioned technologies.

The expected operating frequency for the proposed system is also plotted in Fig. 2.19 including the maximum and minimum expected value due to the process variations (3σ variation). As the devices are operating in weak inversion, the variation is very high.

If we consider a different figure of merit in which delay or speed are playing a more important role, such as energy-delay product (EDP), then the results of optimization will change. Figure 2.20 illustrates the results of EDP optimization in different technology nodes. Again in each node appropriate V_{DD} and V_{TH} has been

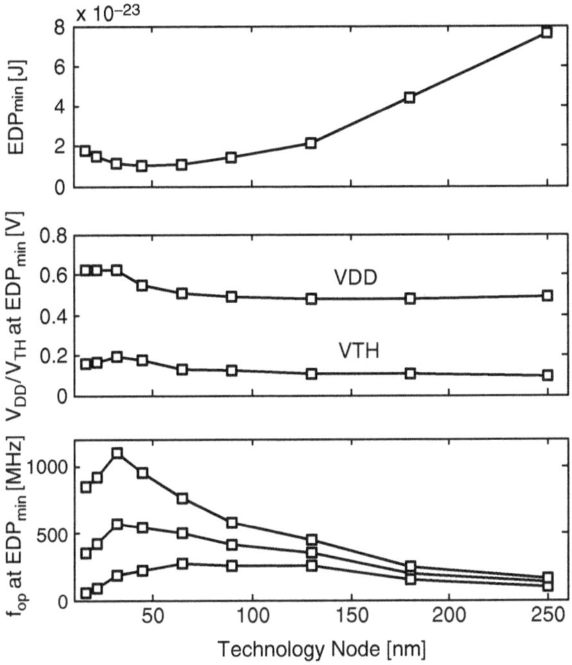

Fig. 2.20 Minimum energy-delay product in different technology nodes when both supply voltage and threshold voltage are optimized. The optimum values for supply voltage and threshold voltage are also shown. Here, $\alpha = 0.9/N$. The bottom figure shows the nominal, best, and worst case operating frequency of the circuits in minimum EDP point

determined to achieve the minimum possible EDP. Meanwhile, the size of devices and supply voltage in each node have been chosen such that satisfy the noise margin requirement.

As in this case delay has considered to have more importance, the resulted optimized values for threshold voltage are no more equal to the maximum allowed value (as it was the case in Fig. 2.19). To have a small delay, the optimization has resulted in circuits which are biased in above threshold (superthreshold) regime.

As indicated before, moving to deep sub-micron technology seems to be not the best choice always to reduce the energy consumption or EDP, especially bellow 45/32 nm nodes. On the other side, looking from a different perspective, and considering the examples shown in Figs. 2.19 and 2.20, one can see that the price should be paid in terms of PDP or EDP for going to deeper technology nodes below 65/45/32 nm can be minimized by a careful design. For example, in Fig. 2.20, the price for going from 45 to 16 nm is about 30% increase in energy consumption. In Fig. 2.20, EDP increases by moving into technology nodes deeper than 45 nm; however, the amount of increase is very small.

In the rest of this book, some techniques for implementing ULP digital and analog circuits based on subthreshold MOS devices will be described. The emphasis here is to address the main existing design issues such as leakage (static) current reduction and implementing reliable circuits in very low current densities.

References

1. Y. Tsividis, *Operation and Modeling of the MOS Transistors*, McGraw-Hill, 1999
2. R. G. Arns, "The other transistors: early history of the metal-oxide semiconductor field-effect transistor," in *IEE Eng. Sci. Educ. J.*, vol. 7, no. 5, pp. 233–240, Oct. 1998
3. J. E. Lilienfeld, "Method and apparatus for controlling electric current," US Patent no. 1745175, Jan. 1930
4. Y. Taur and T. H. Ning, *Fundamentals of Modern VLSI Devices*, Cambridge University Press, 1998
5. E. Vittoz and J. Fellrath, "CMOS analog integrated circuits based on weak inversion operation," *IEEE J. Solid-State Circuits*, vol. 12, no. 3, pp. 224–231, Jun. 1977
6. C. C. Enz and E. A. Vittoz, *Charge-Based MOS Transistor Modeling*, Wiley, 2006
7. C. C. Enz, F. Krummenacher, and E. A. Vittoz, "An analytical MOS transistor model valid in all regions of operation and dedicated to low-voltage and low-current applications," in *Analog Integrated Circuits and Signal Processing*, vol. 8, pp. 83–114, Jul. 1995
8. G. E. Moore, "Cramming more components onto integrated circuits," in *Electronics Magzine*, vol. 38, no. 8, Apr. 1965
9. M. Plank, "The Genesis and Present State of Development of the Quantum Theory (Nobel Lecture)," Jun. 1920
10. T. H. Lee, *The Design of CMOS Radio-Frequency Integrated Circuits*, Second Ed., Cambridge University Press, 2002
11. T. Sakurai and A. R. Newton, "Alpha-power law MOSFET model and its applications to CMOS inverter delay and other formulas," *IEEE J. Solid-State Circuits*, vol. 25, pp. 584–594, Apr. 1990
12. T. Sakurai and A. R. Newton, "A simple MOSFET model for circuit analysis," in *IEEE Transactions on Electron Devices*, vol. 38, pp. 887-894, Apr. 1991
13. P. Kinget, "Device mismatch and tradeoffs in the design of analog circuits," *IEEE J. Solid-State Circuits*, vol. 40, no. 6, pp. 1212–1224, Jun. 2005
14. T. Mizuno, J.-I. Okamura, and A. Toriumi, "Experimental study of threshold voltage fluctuation due to statistical variation of channel dopant number in MOSFET's," in *IEEE Transactions on Electron Devices*, vol. 41, no. 11, pp. 2216–2221, Nov. 1994
15. A. Asenov, A. R. Brown, J. H. Davies, S. Kaya, and G. Slavcheva, "Simulation of intrinsic parameter fluctuations in decananometer and nanometer-scale MOSFETs," in *IEEE Transactions on Electron Devices*, vol. 50, no. 9, pp. 1837–1852, Sep. 2003
16. D. Bol, R. Ambroise, D. Flander, and J. D. Legat, "Interests and limitations of technology scaling for subthreshold logic," in *Transactions on Very Large Scale Integration (VLSI) Systems*, vol. 17, no. 10, pp. 1508–1519, Oct. 2009
17. A.-J. Annema, B. Nauta, R. van Langevelde, and H. Tuinhout, "Analog circuits in ultra-deep-submicron CMOS," *IEEE J. Solid-State Circuits*, vol. 40, no. 1, pp. 132–143, Jan. 2005
18. A. A. Abidi, "Phase noise and jitter in CMOS ring oscillators," *IEEE J. Solid-State Circuits*, vol. 41, no. 8, pp. 1803–1816, Aug. 2006
19. M. S. J. Steyaert, W. M. C. Sansen, and C. Zhongyuan, "A micropower low-noise monolithic instrumnetation amplifier for medical purposes," *IEEE J. Solid-State Circuits*, vol. 22, no. 6, pp. 1163–1168, Dec. 1987
20. R. R. Harrison and C. Charles, "A low-power low-noise CMOS amplifier for neural recording application," *IEEE J. Solid-State Circuits*, vol. 38, no. 6, pp. 958–965, Jun. 2003
21. H. Wu and Y. P. Xu, "A 1V 2.3 μW biomedical signal acquisition IC," *IEEE Solid-State Circuit Conf. (ISSCC)*, pp. 119–120, Feb. 2006

22. T. Denison, K. Consoer, A. Kelly, A. Hachenburg, and W. Santa, "A 2.2 μW 94 nV/$\sqrt{Hz}$, chopper-stabilized instrumentation amplifier for EEG detection in chronic implants," *IEEE Solid-State Circuit Conf. (ISSCC)*, pp. 162–163, Feb. 2007
23. W. Wattanapanitch, M. Fee, and R. Sarpeshkar, "An energy-efficient micropower nerual recording amplifier," *IEEE Trans. Biomedical Circ. Syst.*, vol. 1, no. 2, pp. 136–147, Jun. 2007
24. V. Majidzadeh Bafar, A. Schmid, and Y. Leblebici, "A micropower neural recording amplifier with improved noise efficiency factor," to appear in *European Conference on Circuits Theory and Design (ECCTD)*, Antalya, Turkey, Aug. 2009
25. R. M. Swanson an dJ. D. Meindl, "Ion-implanted complementary MOS transistors in low-voltage circuits," *IEEE J. Solid-State Circuits*, vol. 7, pp. 146–153, Apr. 1972
26. E. Vittoz, B. Gerber, and F. Leuenberger, "Silicon-gate CMOS frequency divider for the electronicd wirst watch," *IEEE J. Solid-State Circuits*, vol. 7, no. 2, pp. 100–104, Apr. 1972
27. A. P. Chandrakasan and R. W. Broderson, "Minimizing power consumption in digital CMOS circuits," in *Proceedings of the IEEE*, vol. 83, no. 4, pp. 498–523, Apr. 1995
28. Z. T. Deniz, Y. Leblebici, and E. A. Vittoz, "On-line global energy optimization in multi-core systems using priciples of analog computation," *IEEE J. Solid-State Circuits*, vol. 42, no. 7, pp. 1593–1596, Jul. 2007
29. "International Technology Road Map for Semiconductors," 2001, [online], Available: *http://public.itrs.net*
30. B. H. Calhoun, S. Khanna, R. Mann, and J. Wang, "Sub-threshold circuit design with shrinking CMOS devices," in *IEEE International Symposium on Circuits and Systems*, pp. 2541–2544, May 2009
31. F. M. Wanlass and C. T. San, "Nanowatt logic using field-effect metal-oxide semiconductor triodes," *IEEE Solid-State Circuit Conf. (ISSCC)*, pp. 32–33, Feb. 1963
32. M. Anis and M. Elmasry, *Multi-Threshold CMOS Digital Circuits, Managing Leakage Power*, Kluwer, 2003
33. K. Roy, S. Mukhopadhyay, and H. Mahmoodi-Meimandi, "Leakage current mechanisems and leakage reduction techniques in deep-submicrometer CMOS circuits," in *Proceeding of the IEEE*, vol. 91, no. 2, pp. 305–327, Feb. 2003
34. P. R. van der Meer, A. van Staveren, and A. H. M. van Roermund, *Low-Power Deep Sub-Micron CMOS Logic*, Springer, 2004
35. K. Schuegraf and C. Hu, "Hole injection Sio2 breakdown model for very low voltage lifetime extrapolation," in *IEEE Transactions Electron Devices*, vol. 41, pp. 761–767, May 1994
36. Z.-H. Liu, C. Hu, J.-H. Huang, T.-Y. Chan, M.-C. Jeng, P. K. Ko, and Y. C. Cheng, "Threshold voltage model for deep-submicrometer MOSFETs," in *IEEE Transactions on Electron Devices*, vol. 40, no. 1, pp. 8695, Jan. 1993
37. K. Roy and S. C. Prasad, *Low-Power CMOS VLSI Circuit Design*, New York: Wiley, 2000
38. Y. Leblebici and S.-M. Kang, *Hot-carrier reliability of MOS VLSI circuits*, Kluwer, 1993
39. B.C. Paul, Raychowdhury, and K. Roy, "Device optimization for digital subthreshold logic operation," in *IEEE Transactions on Electron Devices*, vol. 52, no. 2, pp. 237–247, Feb. 2005
40. K. Tae-Hyoung, J. Kaene, E. Hanyong, and C. H. Kim, "Utilizing reverse short-channel effect for optimal subthreshold circuit design," in *IEEE Transactions on Very Large Scale Integration (VLSI) Systems*, vol. 15, no. 7, pp. 821–829, Jul. 2007
41. S. Chung and C.-T Li, "An analytical threshold-voltage model of trench-isolated MOS devices with nonuniformly doped substrates," in *IEEE Transactions on Electron Devices*, vol. 39, pp. 614–622, Mar. 1992
42. D. Fotty, *MOSFET Modeling with SPICE*, Englewood Cliffs, NJ: Prentice-Hall, 1997
43. S. Hanson, M. Seok, D. Sylvester, and D. Blauw, "Nanometer device scaling in subthreshold logic and SRAM," in *IEEE Transactions on Electron Devices*, vol. 55, no. 1, pp. 175–185, Jan. 2008
44. T.-H. Kim, J. Jeane, H. Eom, and C. H. Kim, "Utilizing reverse shortchannel effect for optimal subthreshold circuit design," in *IEEE Transactions on Very Large Scale Integration (VLSI) Systems*, vol. 15, no. 7, pp. 821–829, Jul. 2007
45. Y. Ye, S. Borkar, and V. De, "New technique for standby leakage reduction in high-performance circuits," *Dig. Tech. Papers Symp. VLSI Circuits*, pp. 40–41, Jun. 1998

46. Z. Chen, M. Johnson, L. Wei, and K. Roy, "Estimation of standby leakage power in CMOS circuits considering accurate modeling of transistor stacks," in *Proceedings of the International Symposium on Low Power Electronics and Design*, pp. 239–244, Aug. 1998
47. Z. Chen, L. Wei, A. Keshavarzi, and K. Roy, "IDDQ testing for deep submicron ICs: challenges and solutions," *IEEE Des. Test Comput.*, pp. 24–33, Mar.-Apr. 2002
48. C. Wann, F. Assaderaghi, R. Dennard, C. Hu, G. Shahidi, and Y. Taur, "Channel profile optimization and device design for low-power high-performance dynamic-threshold MOSFET," *Dig. Tech. Papers IEEE Int. Electron Devices Meeting*, pp. 113–116, Dec. 1996
49. A. J. Bhavnagarwala, B. L. Austin, K. A. Bowman, and J. D. Meindl, "A minimum total power methodology for projecting limits on CMOS GSI," *IEEE Trans. VLSI Syst.*, vol. 8, pp. 235–251, Jun. 2000
50. S. Mukhopadhyay, K. Keunwoo; C. Ching-Te, "Device design and optimization methodology for leakage and variability reduction in sub-45-nm FD/SOI SRAM," in *IEEE Transactions on Electron Devices*, vol. 55, no. 1, pp. 152–162, Jan. 2008
51. J. Lohstroh, E. Seevinck, and J. De Groot, "Worst-case static noise margin criteria for logic circuits and their mathematical equivalence," *IEEE J. Solid-State Circuits*, vol. 18, Dec. 1983
52. J. R. Hauser, "Noise margin criteria for digital logic circuits," *IEEE Transactions on Education*, vol. 36, Nov. 1993
53. A. Tajalli and Y. Leblebici, "Leakage current reduction using subthreshold source-coupled logic," in *IEEE Transactions on Circuits and Systems-II: Express Briefs (Special Issue on Nanocircuits)*, vol. 56, no. 5, pp. 347–351, May 2009
54. B. Zhai, S. Hanson, D. Blauw, and D. Sylvester, "Analysis and mitigation of variability in subthreshold design," in *Proceedings IEEE/ACM International Symposium Low-Power Electronics Design*, pp. 20–25, 2005
55. R. Gonzalez, B. M. Gordon, and M. A. Horowitz, "Supply and threshold voltage scaling for low power CMOS," *IEEE J. Solid-State Circuits*, vol. 32, no. 8, pp. 1210–1216, Aug. 1997
56. N. Verma, J. Kwong, and A. P. Chandrakasan, "Nanometer MOSFET variation in minimum energy subthrehsold circuits," in *IEEE Transactions on Electron Devices*, vol. 55, no. 1, pp. 163–174, Jan. 2008
57. Predictive Technology Model, [online], *http://www.eas.asu.edu/ ptm/*
58. X. Xi, and et al., *BSIM4.3.0 MOSFET Model - Users Manual*, University of California, Berkeley, 2003

Part I
Scalable and Ultra-Low-Power Digital Integrated Circuits

Chapter 3
Subthreshold Source-Coupled Logic

3.1 Introduction

Power and cost efficiency, flexibility, performance, and reliability of signal processing in digital domain have promoted designers to gradually replace the traditional analog domain signal processing with the signal processing in digital domain. The digital domain signal processing[1] has been proven to be a very powerful tool in many different applications such as in telecommunications, controlling systems, measurement equipments, etc., and hence plays a very important role in modern industrial products.

The demand for high-performance digital signal processing, calls for very powerful digital signal processors with low cost and low power consumption. For a long time, conventional CMOS topology has been very widely used for implementing high performance digital integrated circuits [1]. These type of circuits occupy a very small area, while their static power consumption is negligible and due to these properties, it is possible to implement very complex and hence high performance systems.

To improve the speed and implement more complex digital systems, CMOS technology has been continuously scaled down for the past few decades. Technology scaling, however, has made some of the secondary non-ideality effects in CMOS devices more pronounced. Among them, increase of device leakage current is a very important issue for digital circuits [2]. While the static power dissipation of digital CMOS integrated circuits implemented in conventional technologies has been negligible, device leakage current in deep-sub-micron CMOS technologies increases the static power considerably and hence reducing power efficiency.

As explained in the previous chapter, there are different sources for leakage current in a device. Subthreshold residual (leakage) current ($I_{L,\mathrm{STH}}$) and gate leakage current ($I_{L,G}$) are generally constructing the main part of device leakage current [2]. Reducing the device threshold voltage (V_T) to have enough current driving

[1] Digital signal processing, DSP.

A. Tajalli and Y. Leblebici, *Extreme Low-Power Mixed Signal IC Design: Subthreshold Source-Coupled Circuits*, DOI 10.1007/978-1-4419-6478-6_3,

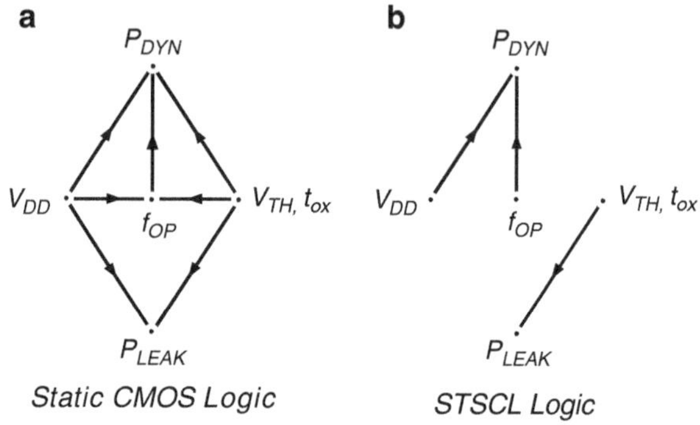

Fig. 3.1 Design space for (**a**) static CMOS and (**b**) STSCL logic styles

capability when the supply voltage has been continuously reduced by the technology scaling, is one of the main reasons for increasing $I_{L,\mathrm{STH}}$. On the other hand, reducing the gate oxide thickness (t_{ox}) for keeping the control of gate on channel charge on an acceptable level, increases the gate leakage current, $I_{L,G}$.

Figure 3.1a depicts the relationship among different design and process parameters in CMOS topology. Illustrated in this figure, the tight tradeoff among speed of operation (f_{op}), power consumption (P_{diss}), supply voltage (V_{DD}), and device parameters (such as t_{ox} and V_T) in conventional CMOS technology creates many challenges for implementing high performance systems especially for low-power applications. This design space can be compared with the design space of source-coupled logic (SCL) topology[2] depicted in Fig. 3.1b where the design tradeoffs are more relaxed.

In this chapter, a new topology for implementing digital circuits for ultra-low-power applications will be presented. For this purpose, a novel approach for implementing source-coupled logic (SCL) circuits biased in subthreshold regime will be described. In this topology, the speed of operation does not depend on supply voltage and threshold voltage of devices. This property, as illustrated in Fig. 3.1b, relaxes the design tradeoffs in ULP implementations. In addition, the current consumption of each cell can be controlled very precisely down to few pico-Amperes. Therefore, it is possible to reduce the system power consumption well below the subthreshold leakage current of conventional CMOS circuits.

In the rest of this chapter, the proposed ULP logic style will be introduced. The conditions for stable operation of the subthreshold SCL (STSCL) circuits and also performance of this type of circuits are analyzed. Experimental results have been provided to show the performance of the circuits in practice.

[2] Also called current-mode logic (CML) or MOS CML (MCML).

3.2 Conventional SCL Topology

In the following section, a brief review on conventional SCL circuits is provided. Meanwhile, an analytical approach for optimal design of a chain of SCL circuits is proposed [3]. This analysis can be used for optimized implementation of complicated SCL digital systems.

3.2.1 Circuit Topology

Background: The basic ideas of source-coupled logic circuits was mainly developed during the 1960s [4] for implementing high speed digital integrated circuits using bipolar [5–8]. The idea was afterwards used for designing GHz range SCL circuits in CMOS technology [9]. Nowadays, MOS SCL circuits are widely used in various demanding applications as high speed signal generators and signal processing units [10, 11]. Recently, standard libraries for implementing more complex digital systems using SCL topology has been developed to make the design and implementation of complex systems automatic [12].

Operation: The core of an SCL circuit is constructed based on NMOS differential pairs. The logic operation in SCL topology takes place in current domain and hence this type of logic circuits can inherently be very fast. Input and output voltages as well as the steered current, all are differential signals which is a key characteristic for reducing switching noise [12].

As illustrated in Fig. 3.2, for a simple inverter (buffer) circuit, the constant tail bias current (I_{SS}) is steered to one of the two output branches based on the

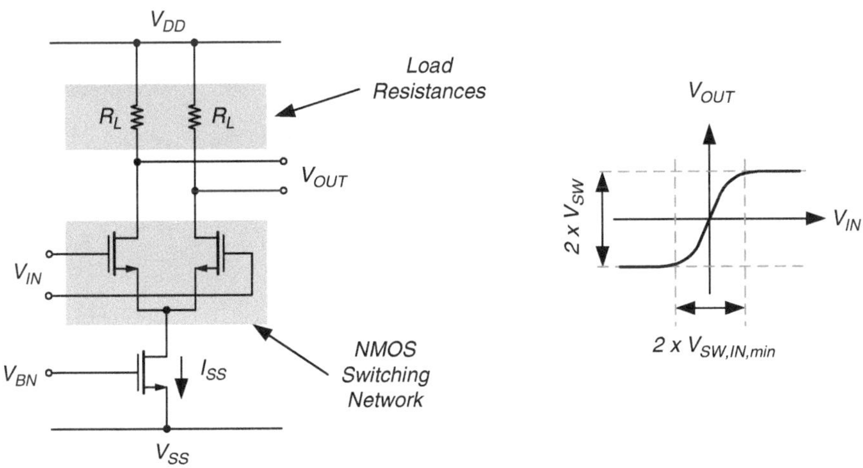

Fig. 3.2 A conventional SCL-based inverter/buffer circuit. The switching part can be composed of a complex network of NMOS source-coupled pairs to implement more complex logic functions [7, 13]. The load resistances, R_L, can be implemented using PMOS devices biased in triode region

desired logic operation. NMOS differential pairs (NMOS switching network) can be arranged in a proper way to implement the required logic operation. It is possible to implement more complex logic operations using appropriate NMOS differential pairs [7, 13]. Finally, the output logic current is converted to voltage by load resistances, R_L.

Strong Inversion Operation: Assuming that the devices are in SI and using EKV model, the differential output current, ΔI, can be calculated versus differential input voltage, ΔV_{IN}, by [12]

$$\frac{\Delta I}{I_{\mathrm{SS}}} = \sqrt{2} \cdot \frac{\Delta V_{\mathrm{IN}}}{V_t} \cdot \sqrt{1 - \frac{\Delta V_{\mathrm{IN}}^2}{2V_t^2}} \tag{3.1}$$

where $V_t = \sqrt{2n_n I_{SS}/\beta}$ denotes the voltage threshold for current switching in differential pair devices. The transconductance of the NMOS differential pair can be estimated by

$$G_m = \frac{2g_m}{\left(1 + \frac{\Delta I}{I_{\mathrm{SS}}}\right)^{-\frac{1}{2}} + \left(1 - \frac{\Delta I}{I_{\mathrm{SS}}}\right)^{-\frac{1}{2}}} \tag{3.2}$$

where:

$$g_m = \sqrt{\frac{\beta I_{\mathrm{SS}}}{n_n}} = \sqrt{2} \cdot \frac{I_{\mathrm{SS}}}{V_t}. \tag{3.3}$$

Here, it is assumed that the devices are in SI and there is no short-channel effect (SCE). When the channel length reduces close to the minimum technology feature size, velocity saturation will impact the device behavior. As explained in Chap. 2, this effect can be modeled by dividing the saturation current by $1 + V_{\mathrm{DSsat}}/(E_C L_e)$, where E_C is the critical electric field. Assuming $V_{\mathrm{DSsat}}/(E_C L_e) >> 1$, then [14]

$$I_{\mathrm{DS}} \approx \frac{n_n \beta}{2} E_C L_e \left(\frac{V_G - V_T}{n_n} - V_S\right). \tag{3.4}$$

Based on this:

$$\frac{\Delta I}{I_{\mathrm{SS}}} = \frac{\beta E_C L_e}{2 I_{\mathrm{SS}}} \cdot \Delta V_{\mathrm{IN}}. \tag{3.5}$$

which represents a linear input–output relationship. In this case, the transconductance of the device is given by

$$g_m = \frac{n_n \beta}{2} E_C L_e. \tag{3.6}$$

For a real case, when the device behaves between square-law and velocity saturation case, α-model for MOS devices can be used [15, 16]

$$I_{\mathrm{DS}} = \frac{\beta}{2} \cdot \left(\frac{V_G - V_T}{n_n} - V_S\right)^{\alpha} \tag{3.7}$$

As shown in [12], in this general case, the transconductance of a differential pair is

$$G_m = \frac{2g_m}{\left(1+\frac{\Delta I}{I_{SS}}\right)^{\frac{1-\alpha}{\alpha}} + \left(1-\frac{\Delta I}{I_{SS}}\right)^{\frac{1-\alpha}{\alpha}}} \tag{3.8}$$

where

$$g_m = \frac{1}{2}\alpha I_{SS}\left(\frac{I_{SS}}{k}\right)^{-\frac{1}{\alpha}} \tag{3.9}$$

and

$$\frac{\Delta I}{I_{SS}} \approx \sqrt[\alpha]{1-\left[1-2^{1-\alpha}\times\left(\frac{g_m\Delta V_{IN}}{I_{SS}}\right)^{\alpha}\right]^{\alpha}} \tag{3.10}$$

Weak Inversion Operation: On the other hand, when the devices are pushed towards WI region, transconductance and differential output currents can be calculated by

$$G_m = \frac{g_m}{\cosh^2\left(\frac{\Delta V_{IN}}{2n_n U_T}\right)} \tag{3.11}$$

where $g_m = I_{SS}/(2n_n U_T)$ and

$$\frac{\Delta I}{I_{SS}} = \tanh\left(\frac{g_m\Delta V_{IN}}{I_{SS}}\right) = \tanh\left(\frac{\Delta V_{IN}}{2n_n U_T}\right) \tag{3.12}$$

Operating in subthreshold regime, the device transconductance strongly depends on temperature through U_T, while it does not depend on device sizes. Therefore, it is not possible to change the transfer curve by design parameters [12].

Voltage Swing: One of the main advantageous of SCL topology is the possibility of reducing the signal swing. Compared to the CMOS topology where the signal swing is equal to V_{DD}, in SCL topology voltage swing and hence the current needed for charging and discharging the parasitic capacitances is less.

Using as a logic circuit, the voltage swing at the input and output of the circuit should be high enough to make sure that the tail bias current will be completely switched to one of the two output branches. In other words, the voltage swing at the output node, i.e.:

$$V_{SW} = R_L \cdot I_{SS} \tag{3.13}$$

should be high enough to switch completely the input differential pair of the next stage:[3]

$$V_{SW} > V_{SW,min}. \tag{3.14}$$

[3] In subthreshold region it is not possible to completely steer the tail bias current to one branch; therefore, complete switching is not possible.

that is equivalent to say that the gain of each SCL circuit should be high enough to be used as a logic circuit with acceptable noise margin. The minimum acceptable voltage swing at the output of each SCL gate, i.e., $V_{\mathrm{SW,min}}$, depends on the region of operation of NMOS devices [17, 18]:

$$V_{\mathrm{SW,min}} = \begin{cases} \sqrt{2} \cdot n \cdot V_{\mathrm{DSsat}} & \text{in strong inversion,} \\ 4 \cdot n \cdot U_T & \text{in subthreshold} \end{cases} \tag{3.15}$$

where n is the subthreshold slope factor of NMOS devices. Biased in subthreshold regime, the minimum acceptable value for input swing can be reduced to $4 \cdot n \cdot U_T$, which is about 150 mV at room temperature (assuming $n = 1.5$).

Load Resistance: To implement the load resistances, passive resistors or PMOS devices biased in triode region can be used. Since PMOS transistors can add some extra parasitic capacitances to the output node, generally passive resistors are used for high frequency applications. If the parasitic effect associated with the PMOS load transistors could be tolerated, then PMOS loads are preferred mainly because of their smaller area and possibility of adjusting their resistivity.

It is required to control the resistivity of the load devices with respect to the tail bias current in order to keep the output voltage swing on a desired level. A simple approach to control the load resistivity is shown in Fig. 3.3. In this topology, the output voltage swing of a sample SCL gate is controlled by an amplifier inside a controlling loop. The control voltage generated for the load device M8, V_{BP}, is then applied to the other gates in a circuit. The controlling circuit in this approach is called *replica bias (RB)*. This system relies on matching between replica bias circuit and SCL gates used in the circuit.

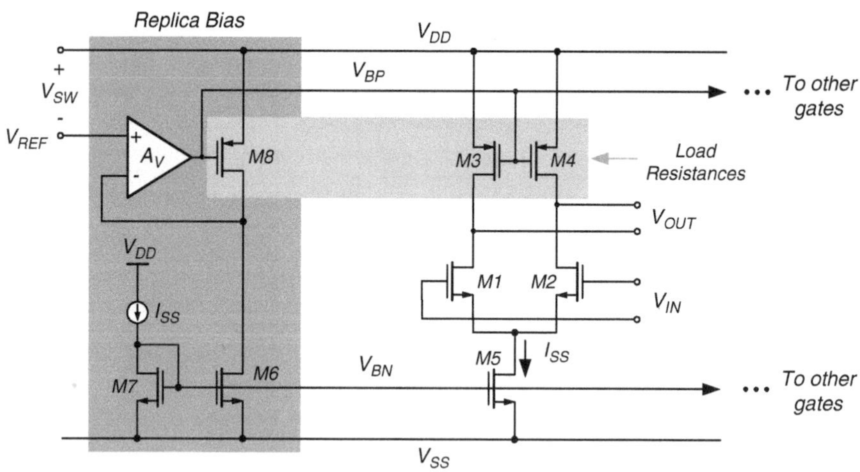

Fig. 3.3 Replica bias circuit used to control the resistivity of the load devices

Assuming that the load devices are biased in SI, then

$$I_{SD,M8} = I_{SS} = \mu C_{ox} \frac{W}{L_e} \cdot V_{SD} \cdot \left(V_{SG} - \mid V_{T,P} \mid - \frac{V_{SD}}{2} \right) \Bigg|_{V_{SD}=V_{SW}} \tag{3.16}$$

When the entire bias current flows through a PMOS load, the voltage drop across its source-drain is intended to be V_{SW}. Now, if there is any mismatch between the replica bias circuit and the SCL gate inside the circuit, the voltage swing at the output of this SCL gate will change as

$$\left(\frac{\Delta V_{SW}}{V_{SW}}\right)^2 = \left(\frac{1}{1+\frac{\beta \cdot V_{SW}^2}{2I_{SS}}}\right)^2 \times \left(\left(\frac{\Delta I_{SS}}{I_{SS}}\right)^2 + \left(\frac{\Delta\beta}{\beta}\right)^2 + \left(\frac{\beta V_{SW} \cdot \Delta V_{T,P}}{I_{SS}}\right)^2\right) \tag{3.17}$$

Regarding (3.17), to have an acceptable performance with required noise margin (NM), ΔV_{SW} should be kept as small as possible. This requires large enough NMOS tail bias transistors and PMOS load devices. Neglecting the mismatch due to β and adding amplifier offset, V_{OS}, the expression for ΔV_{SW} can be more simplified to

$$\left(\frac{\Delta V_{SW}}{V_{SW}}\right)^2 \approx \left(\frac{V_{OS}}{V_{SW}}\right)^2 + \left(\frac{1}{1+\frac{\beta V_{SW}^2}{2I_{SS}}}\right)^2 \times \left(\left(\frac{\Delta V_{T,P}}{V_{SG} - \mid V_{T,P} \mid - V_{SW}}\right)^2 + \left(\frac{\Delta I_{SS}}{I_{SS}}\right)^2\right) \tag{3.18}$$

In general, a high enough value for V_{SW} should be selected in order to compensate the effect of variation at the output voltage swing and keep the NM on acceptable level.

3.2.2 *Tradeoffs in Design of Strong-Inversion SCL Gates*

The main design parameters in SCL circuits are bias current and voltage swing which should be optimized for the required operating frequency. The design needs to be done for each gate in a circuit separately. Unlike subthreshold SCL circuits, the minimum required voltage swing in strong-inversion SCL depends on bias current. Hence voltage swing should be included in the design process.

Having minimum logic depth or maximum activity rate, as will be discussed later, will help to improve the power efficiency of system. For this reason, SCL topology is generally used for implementing very high speed and low complexity circuits [13]. Also, proper sizing of NMOS switching network is very important. For example, larger aspect ratio for NMOS devices results in lower gate overdrive voltage (V_{DSsat}), while the total input capacitance of the gate increases. In the following, a methodological approach for designing SCL gates in a chain is proposed [3].

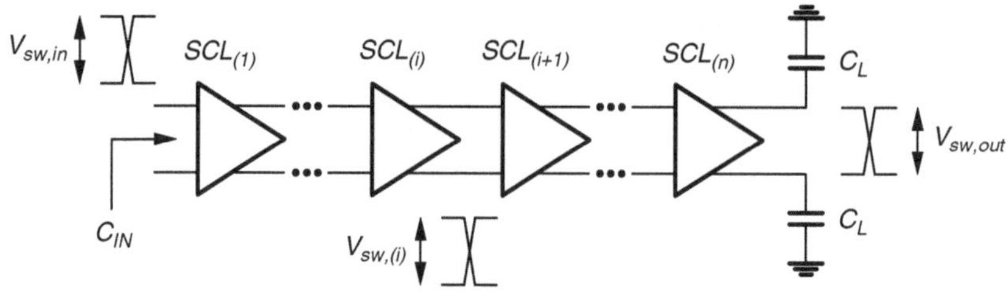

Fig. 3.4 SCL-based buffer chain to drive the load capacitance C_L at the desired data rate. The load resistance of the stage (i) is $R_{L,i}$ and C_i is the total capacitance seen by $R_{L,i}$

Consider that n consecutive SCL-based buffer stages have been utilized to drive a load capacitance C_L (Fig. 3.4). If the maximum acceptable input capacitance is $C_{\text{IN,Max}}$, then it is possible to determine the value of n for minimum possible power consumption. Assuming that the time constant at the output of ith stage is m_T times less than T_D which is the input data period, then:

$$R_{L,i} \cdot C_i \leq \frac{T_D}{m_T}, \qquad i \in \{1, \ldots, n\} \tag{3.19}$$

By applying this constraint to all the intermediate nodes, it can be shown that the input capacitance of each stage with respect to the input capacitance of the next stage can be presented by:

$$C_i = (P \cdot S \cdot D_i) \cdot C_{i+1} \tag{3.20}$$

in which P is a process-dependent constant defined as:

$$P = \frac{2L_{\text{min}}^2}{\mu_n} \tag{3.21}$$

The parameter S depends on the speed of operation as:

$$S = \frac{m_T}{T} \tag{3.22}$$

and D_i is:

$$D_i = \left(1 + (1 + \gamma_M) \cdot \frac{L_{\text{ov}}}{L_{\text{min}}}\right) \cdot \gamma_{\text{sat}}^2 \cdot \frac{V_{\text{sw},i}}{V_{\text{sw},i-1}^2} \tag{3.23}$$

Therefore, the total input capacitance can be found as:

$$C_{\text{IN}} = \left(P^n \cdot S^n \cdot \Pi_{i=1}^n D_i\right) \cdot C_L < C_{\text{IN,Max}}. \tag{3.24}$$

Regarding (3.23) and (3.24), it can be seen that larger voltage swing at the preceding stages leads to smaller input capacitance or in other words smaller number of stages needed to achieve the desired input capacitance. Meanwhile, (3.20) implies that to be able to reduce the total input capacitance by buffering, it is necessary that: $P \cdot S \cdot D_i < 1$. Assuming that all the stages have the same voltage swing ($V_{\text{sw},i} = V_{\text{sw}}$ for $i = 1$ to n), this criteria puts an upper limit on the maximum operation speed of the circuit by:

$$f_D < \frac{\mu_n}{2L_{\min} \cdot (L_{\min} + (1 + \gamma_M) \cdot L_{\text{ov}})} \cdot \frac{V_{\text{sw}}}{\gamma_{\text{sat}}^2} \cdot \frac{1}{m_T} \tag{3.25}$$

This equation means that the voltage swing at the intermediate stages should be maximized to achieve a higher speed of operation. The main reason is that by increasing the voltage swing at the input of each stage by a factor of k_V, it is possible to reduce the size of switching transistors of that stage by a factor of k_V^2 without affecting the switching process. This voltage scaling leads to k_V^2 times smaller input capacitance. Meanwhile, γ_{sat} should be selected as small as possible to increase the lower limit on f_D. The lower limit on γ_{sat} is $\sqrt{2}$ [19].

In addition, based on (3.25), m_T should be selected as small as possible. In a configuration with n identical stages, the total circuit bandwidth (BW_n) can be estimated by $\text{BW}_n = \text{BW} \cdot \sqrt{\sqrt[n]{2} - 1}$ (BW is the bandwidth of each stage) [20]. Then m_T should be high enough to satisfy the general requirement of $\text{BW}_n \geq 0.7 \times f_D$ [21].

To calculate the power consumption, one can show that:

$$I_i = k_{I,i} \cdot I_{i+1} = P \cdot S \cdot D_i \cdot \frac{V_{\text{sw},i}}{V_{\text{sw},i+1}} \cdot I_{i+1}. \tag{3.26}$$

This expression is derived by this assumption that the time constants of the all intermediate nodes are satisfying (3.19). Equation (3.26) also shows that the bias current in each stage depends on the voltage swing at the input ($V_{\text{sw},i-1}$) and output of that stage ($V_{\text{sw},i}$) as well as the voltage swing at the output of the next stage ($V_{\text{sw},i+1}$). Assuming a constant voltage swing for all the stages, the total current drawn from the supply voltage can be evaluated by:

$$I_{\text{tot}} = \frac{V_{\text{sw,out}} \cdot m_T \cdot C_L}{T_D} \cdot \frac{1 - k_I^n}{1 - k_I} \tag{3.27}$$

which would be dominated by the latest stages of the buffer chain and also increases by $V_{\text{sw,out}}$. Based on (3.25) and (3.27), choosing a low voltage swing for the last stage and at the same time higher voltage swing at the intermediate stages can help achieving a good speed-power consumption compromise.

Figure 3.5 shows the total current consumption calculated based on (3.27) for different number of stages and different voltage swing values. Based on Fig. 3.5, to get the desired input capacitance ($C_{\text{IN,Max}} = 50\,\text{fF}$) it is possible to increase the number of stages or increase the voltage swing at the intermediate stages. To have small n values, the only possibility is to increase the voltage swing to 0.5 V. Also, it can be seen that it is possible to reduce the total current consumption by increasing the voltage swing for high n values.

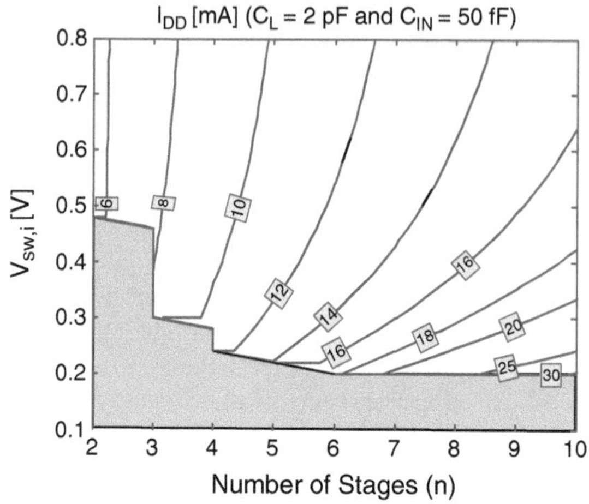

Fig. 3.5 Current consumption in an SCL buffer chain for different number of stages n and different voltage swing values at the intermediate nodes ($V_{sw,i}$) based on (3.27). In this simulation, $C_L = 2\,\text{pF}$, $V_{sw,in} = 0.4\,\text{V}$ and it is assumed that C_{IN} should be smaller than 50fF. Inside the gray area, it is not possible to achieve the desired C_{IN}

3.3 Ultra-Low-Power Source-Coupled Logic

In this work, some new techniques for implementing ULP SCL circuits are developed. The main goal is to study the possibility of using CMOS leakage current (which is unavoidable in CMOS topology) for successful logic operation in SCL topology. This requires to bias SCL circuits deeply in subthreshold regime and hence implement subthreshold SCL (STSCL) circuits.

3.3.1 High-Valued Load Device Concept

Regarding (3.15), the minimum acceptable voltage swing in subthreshold regime depends on technology only through the subthreshold slope factor n and is independent of the threshold voltage of the NMOS switching devices. This means that the switching operation of NMOS devices, and hence the speed of operation in subthreshold region has low dependence on process variations. Therefore, as long as the tail bias current I_{SS} is much higher than the junction leakage currents and also the output impedance of the devices is much higher than the load resistance, the proposed topology can operate properly as a logic circuit, even in aggressively scaled deep sub-micrometer technologies.

To maintain the desired output voltage swing at very low bias current levels, it is necessary to increase the load resistance value in inverse proportion to the reducing tail bias current as

$$R_L = \frac{V_{SW}}{I_{SS}}. \tag{3.28}$$

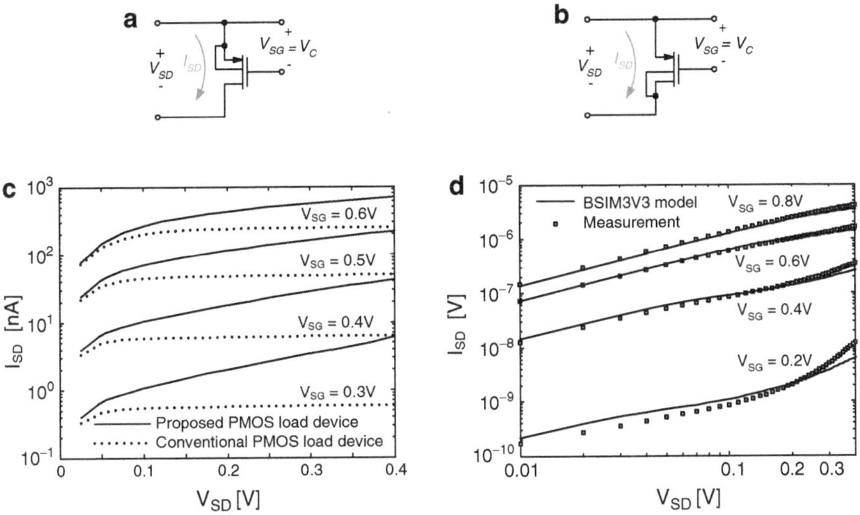

Fig. 3.6 (**a**) Conventional PMOS load device, (**b**) proposed load device, (**c**) I–V characteristics of the conventional PMOS load (*dotted*) in comparison to the proposed device (*solid line*), (**d**) measured I–V characteristics of the proposed load device in comparison to the BSIM model (all data obtained using 0.18 μm CMOS technology)

In subthreshold operation, the tail bias current would be in the range of few nA or even less. Therefore, to obtain a reasonable output voltage swing, the load resistance should be in the range of hundreds of $M\Omega$.

It is also essential to be able to control the load resistance value very precisely with respect to I_{SS} value. Hence, a well controlled high resistivity load device with a very small area occupation is required. For this range of resistivity, conventional PMOS devices biased in triode region, shown in Fig. 3.6a, can not be utilized since the required channel length of the transistor would be impractically large. Therefore, a new technique for implementing the load device is required.

Figure 3.6c (dotted line) shows the I–V characteristics of a PMOS device realized in 0.18 μm technology for different V_{SG} values, indicating that the configuration of Fig. 3.6a results in a current source with almost infinite impedance, even for deep sub-micron devices. Hence, the gain would not be limited, neither would be the amplitude.

Figure 3.6b shows a modified load device, where the drain of the PMOS device is connected to its bulk. As illustrated in Fig. 3.6c, the configuration shown in Fig. 3.6b produces a finite and controllable resistance, which, associated with the transconductance of the differential pair will provide a controlled, limited gain and amplitude at the output of SCL circuit. Hence, it is possible to implement a very high resistivity load device using a single minimum size PMOS device.

The measured DC I–V characteristics of this device are shown in Fig. 3.6d. For $V_{SD} > 0$ (bulk tied to the drain), the device operates as a very high resistivity element as expected. This plot also shows that the measurement results are very close the resistance values predicted by transistor level simulations.

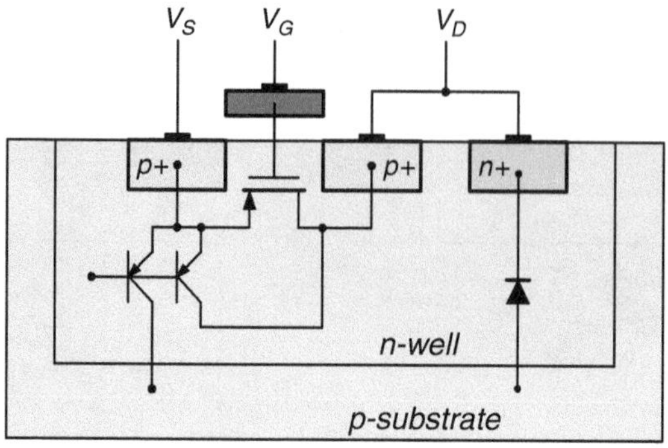

Fig. 3.7 Cross-section view of the proposed PMOS load device, showing the parasitic components that contribute to its operation in subthreshold regime

The cross-section view of the proposed PMOS load device can be seen in Fig. 3.7. Tieing the drain to the bulk of the PMOS load device connects the cathode of the n-well-to-substrate reverse-biased diode to the output node. This reverse-biased diode increases the capacitive loading at the output of the circuit and hence can reduce the circuit bandwidth (or logic cell speed). As the device is of minimum size, the parasitic capacitance associated with this diode is very small and can usually be neglected (in this design using 0.18 μm technology: $C_d < 1$ fF).

The other important parasitic element is the forward biased source-bulk diode. Unlike CMOS logic circuits where the subthreshold channel leakage current is the dominant leakage source, in the STSCL topology the main leakage current sources are the leakage currents due to the PN junctions of MOS devices. Illustrated in Fig. 3.7, this diode can limit the possible voltage swing at the drain of the device to 400–500 mV depending on the level of bias current. However, as the required voltage swing for subthreshold SCL gates is well below this value, the source-bulk diode does not influence the circuit operation.

As the bulk of each PMOS device is connected to its drain, a separate n-well for each device is required. Therefore, the area overhead due to each n-well region and also the minimum distance between n-wells should be studied. The fact that each individual PMOS load device must be confined in its own n-well also does not have a severe impact on area as will be demonstrated later.

3.3.1.1 DC Characteristics of the Load Devices

Using the EKV model [18], the I–V characteristics of the subthreshold PMOS device can be expressed by

$$I_{\mathrm{SD}} = I_0 \cdot \mathrm{e}^{\frac{V_{\mathrm{BG}}-V_{T0}}{n_p U_T}} \left(\mathrm{e}^{\frac{-V_{\mathrm{BS}}}{U_T}} - \mathrm{e}^{\frac{-V_{\mathrm{BD}}}{U_T}} \right) \tag{3.29}$$

in which $I_0 = 2n_p\mu C_{\mathrm{ox}} \cdot (W/L_e)U_T^2$. In the proposed configuration illustrated in Fig. 3.6b, $V_{\mathrm{BD}} = 0$, hence:

$$I_{\mathrm{SD}} = I_0 \cdot \mathrm{e}^{\frac{V_{\mathrm{DG}}-V_{T0}}{n_p U_T}} \left(\mathrm{e}^{\frac{V_{\mathrm{SD}}}{U_T}} - 1\right). \tag{3.30}$$

The output small signal resistance of the proposed load device can be calculated by

$$\begin{aligned} R_{\mathrm{SD}} &= \left(\frac{\partial I_{\mathrm{SD}}}{\partial V_{\mathrm{SD}}}\right)^{-1} \\ &= \frac{n_p U_T}{I_b} \cdot \left((n_p - 1) \cdot \mathrm{e}^{(n_p-1)v_{\mathrm{SD}}} + \mathrm{e}^{-v_{\mathrm{SD}}}\right)^{-1} \end{aligned} \tag{3.31}$$

$$R_{\mathrm{SD}} = \frac{n_p U_T}{I_{\mathrm{SD}}} \cdot \frac{\mathrm{e}^{V_{\mathrm{SD}}/U_T} - 1}{(n_p - 1)\mathrm{e}^{V_{\mathrm{SD}}/U_T} + 1} \tag{3.32}$$

in which $v_{\mathrm{SD}} = V_{\mathrm{SD}}/(n_p U_T)$ and $I_b = I_0 \cdot \mathrm{e}^{\frac{V_{\mathrm{SG}}-V_{T0}}{n_p U_T}}$. To complete the analysis, it is necessary to include the forward biased source-bulk diode into the calculations. Although the effect of this diode is negligible for low values of V_{SD}, in high values of V_{SD} or in very low I_{SD} values the current of this diode contributes considerably in the total current:

$$I_T = I_{\mathrm{SD}} + I_{F,D} \tag{3.33}$$

The diode forward bias current is

$$I_{F,D} = I_{\mathrm{sat}} \left(\mathrm{e}^{\frac{V_{\mathrm{SW}}}{\eta U_T}} - 1\right) \tag{3.34}$$

where, η is a process dependent parameter and I_{sat} is the saturation current of the drain-bulk PN junction which depends on the area and perimeter of this junction. It is specially important to include the diode current in very low bias current values.

As depicted by (3.32), R_{SD} can be controlled through the source-gate voltage (V_{SG}) of the device with respect to I_{SD}. Because of exponential dependence of the equivalent resistance of this device on V_{SG}, resistivity can be adjusted in a very wide range.

As explained before, to avoid process-related deviations, a replica bias generator can be employed. The wide tuning range of R_{SD} means that the proposed STSCL gate can be used in a very wide range of operating conditions without the need for modifying the size of devices. Meanwhile, as long as the matching requirements are respected, the frequency of operation would be linearly proportional to the bias current.

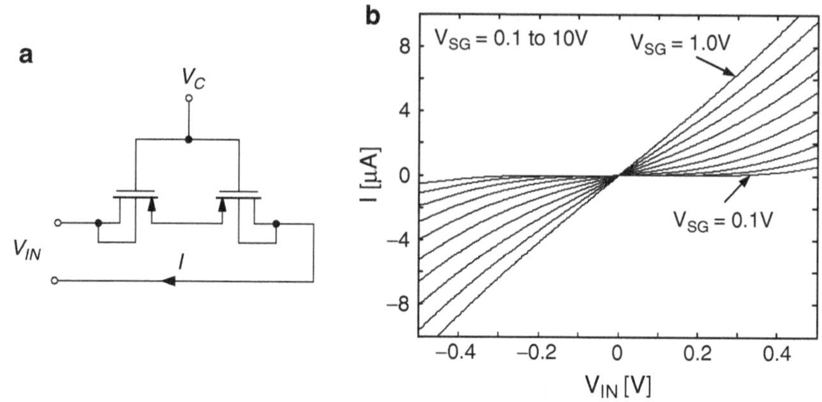

Fig. 3.8 A very high-valued floating resistor composed of two back to back PMOS devices: (**a**) circuit schematic and (**b**) measured I–V characteristics of the controlled floating resistor in CMOS 0.18 μm

3.3.1.2 Floating High-Valued Resistance

It is noticeable that when V_{SD} becomes negative, the current direction becomes reversed and the device switches to conventional configuration in which the bulk is connected to the source. In this case, the drain current will increase rapidly by increasing V_{DS}. This property can help to implement high valued *floating resistors* with a very wide adjusting range by connecting two PMOS transistors in series as shown in Fig. 3.8. The measured I–V characteristics of this floating resistor show moderate linearity in a relatively wide voltage range, which can be exploited in various analog circuit applications. In Chap. 7, the proposed floating high-valued resistance is used to implement continuous-time filters.

3.3.2 STSCL Gates

The proposed PMOS load device can be utilized to implement an SCL gate biased in subthreshold. Figure 3.9 shows the basic structure of the proposed STSCL gate. A simplified circuit diagram of the replica bias circuit used to control the output voltage swing is also shown. In this schematic, all the devices operate in subthreshold regime and the tail bias current can be reduced until it becomes comparable in magnitude to the leakage currents that exist in the circuit.

Since the input differential pair transistors are operating in subthreshold, it can be shown that the transconductance of the input differential pair is:

$$G_m = \frac{\partial I_{\text{OUT}}}{\partial V_{\text{IN}}} = \left(\frac{I_{\text{SS}}}{2n_n U_T}\right) \cdot \frac{1}{\cosh^2\left(V_{\text{IN}}/(2n_n U_T)\right)} \tag{3.35}$$

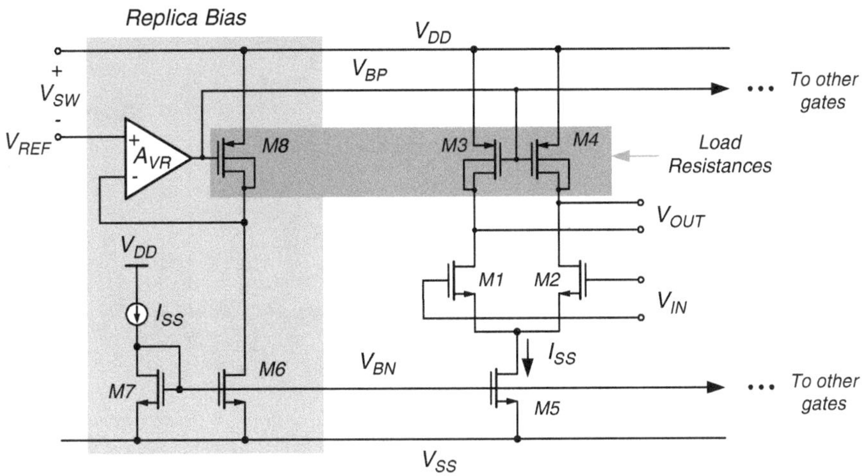

Fig. 3.9 A subthreshold SCL gate and its replica bias circuit used to control the output voltage swing

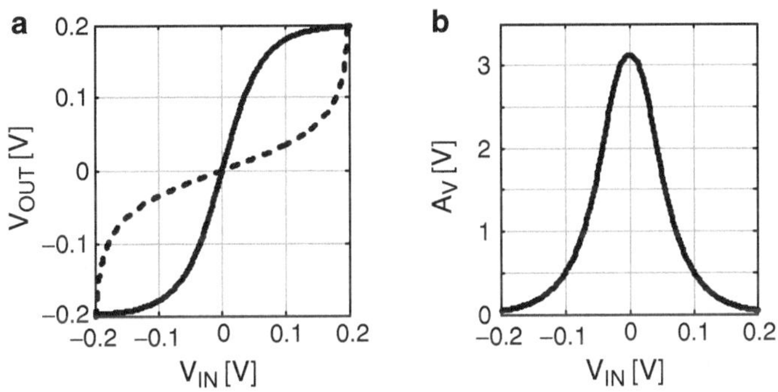

Fig. 3.10 DC transfer characteristics of a STSCL gate designed in 0.18-μm CMOS and biased with I_{SS} =100 pA, V_{SW} = 200 mV: (**a**) voltage transfer characteristic and (**b**) DC differential voltage gain

where V_{IN} indicates the input differential voltage and n_n is the subthreshold slope of NMOS devices. Based on (3.35), for $V_{IN} > 4n_nU_T$ the entire current will be switched to one of the branches. Therefore, a voltage swing of more than $4n_nU_T$ would be sufficient to make sure that the gain of STSCL circuit is high enough to be used as a logic gate. Combining (3.32) with (3.35) results in:

$$A_V = \frac{\partial V_{OUT}}{\partial V_{IN}} \leq A_V|_{V_{IN}=0} \simeq \frac{n_p}{n_n \cdot (n_p - 1)}. \tag{3.36}$$

Figure 3.10 illustrates the DC transfer characteristics of an STSCL gate as well as the stage gain. The simulated DC gain of 3.2 at the cross-over point is very close to the value estimated by (3.36) in 0.18-μm CMOS technology.

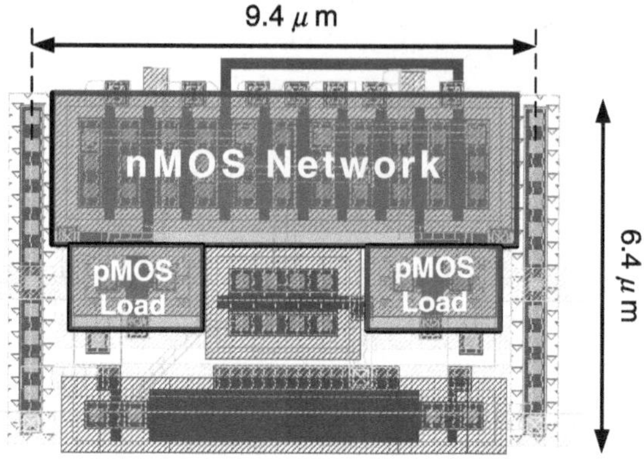

Fig. 3.11 Mask layout of a 3-input XOR gate showing the area occupied by the major components in CMOS 0.18 μm. Note that the PMOS load device with their isolated n-wells occupy a relatively small area compared to the NMOS logic network and biasing transistors

Meanwhile, based on (3.31) it can be shown that the equivalent output resistance of the PMOS load for $V_{SD} = 0\,V$ is finite and equal to:

$$R_{SD}|_{V_{SD}=0} = \frac{U_T}{I_0} \cdot e^{-\frac{V_{SG}-V_{T0}}{n_p U_T}} = \frac{U_T}{I_b} \tag{3.37}$$

which means the load devices are capable of pulling up the output node completely to V_{DD}.

Concerning the area overhead associated with the PMOS load devices, actual mask layout examples using 0.18-μm CMOS technology design rules provide an accurate assessment. The layout of a 3-input XOR gate is shown in Fig. 3.11 where the area required for the PMOS load devices is demonstrated to be small compared to the remaining parts of the circuit.

3.4 Design Issues and Performance Estimation

3.4.1 Power-Speed Tradeoffs in STSCL

The speed of operation in an SCL gate is mainly limited by the time constant at the output node which is

$$\tau_{SCL} = R_L \cdot C_L \approx \frac{V_{SW}}{I_{SS}} \cdot C_L. \tag{3.38}$$

and the power consumption of a single gate is:

$$P_{diss,STSCL,1} = V_{DD} I_{SS}. \tag{3.39}$$

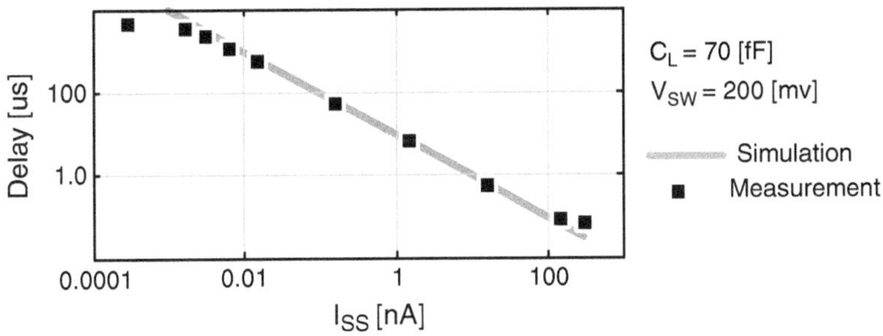

Fig. 3.12 Measured gate delay for different tail bias currents in 0.18-μm CMOS technology

Based on this, the propagation delay is inversely proportional to the tail bias current:

$$t_{d,\mathrm{SCL}} = \ln 2 \times \tau_{\mathrm{SCL}} = \ln 2 \times R_L \cdot C_L = \ln 2 \times \frac{V_{\mathrm{SW}}}{I_{\mathrm{SS}}} \cdot C_L. \tag{3.40}$$

Using (3.40), one can choose the proper I_{SS} value to operate at the desired frequency. Since the power consumption and delay of each gate depend only on I_{SS} which can be controlled very precisely, this circuit exhibits very low sensitivity to the process variations. Meanwhile, since the speed of operation in this case does not depend on threshold voltage (V_T) of the MOS devices, it is not necessary to use special process options to adjust the device threshold voltage as frequently is done for static CMOS. In Fig. 3.12, it can be seen that the gate delay is adjustable in a very wide range proportional to the tail bias current. This figure shows that the tail bias current can be reduced to about 10 pA where the forward bias current of the source-to-n-well diode of the PMOS load device becomes comparable to I_{SS}.

Considering (3.39), it can also be concluded that the power consumption is constant and independent of the operation frequency. Therefore, it is necessary to use the SCL circuits at their maximum activity rate to achieve maximum achievable efficiency. It is also noticeable that the gate delay does not depend on supply voltage while it varies with the tail bias current linearly. This property can be exploited for applications in which the supply can vary during circuit operation.

Based on (3.38) and (3.39), power-delay product (PDP) of each gate can be approximated by

$$PDP_{\mathrm{STSCL},1} \approx \ln 2 \times V_{\mathrm{DD}} V_{\mathrm{SW}} C_L \tag{3.41}$$

which is directly proportional to the supply voltage, the voltage swing at the output of the gate, and the total load capacitance while it is independent of I_{SS} [12, 13, 22]. Using $V_{\mathrm{DD}} = 0.5$ V and $V_{\mathrm{SW}} = 0.2$ V, for example, the PDP of an SCL gate can be as low as 70 aJ / fF / gate.

To have a better understanding of the power-speed tradeoff in SCL configuration, consider a simple SCL circuit constructed of N cascaded identical gates (indeed,

N is the logic depth) that is operating at frequency f_{op}. Using (3.38) and (3.39), it can be shown that the total power consumption of this chain will be:

$$P_{\text{diss,STSCL},N} \approx \ln 2 \times N^2 V_{DD} V_{SW} C_L f_{op} \tag{3.42}$$

which is increasing quadratically with the logic depth and linearly with the operation frequency.

However, compared to the conventional CMOS digital circuits, an SCL circuit with logic depth of $N > V_{DD}/V_{SW}$ exhibits higher *PDP* which is mainly due to the static current consumption of SCL gates (see [13]). In a digital SCL circuit with logic depth of N, the total delay is $t_{d,N} = N \cdot t_d$ and the total power consumption is $P = N V_{DD} I_{SS}$. Therefore, for an SCL digital circuit with a logic depth of N, the maximum operating frequency would be:

$$f_{op,N} \approx \frac{1}{t_{d,N}} = \frac{I_{SS}}{\ln 2 \cdot N V_{SW} C_L} \tag{3.43}$$

which is N times less than the maximum possible operating frequency of each SCL gate:[4]

$$f_{op,\text{Max}} \approx \frac{1}{t_d} = \frac{I_{SS}}{\ln 2 \cdot V_{SW} C_L}. \tag{3.44}$$

The main reason for this reduction is that the activity rate in a digital circuit with the logic depth of N is reduced by a factor of N, while the power consumption of each gate remains the same.

Defining the activity rate (or duty rate) as:

$$\alpha = \frac{f_{op}}{f_{op,\text{Max}}} \tag{3.45}$$

and regarding (3.42), one can show that the power-delay product with logic depth of N is:

$$\text{PDP}_{\text{SCL},N} = \ln 2 \cdot \frac{N}{\alpha} V_{DD} V_{SW} C_L. \tag{3.46}$$

Therefore, by increasing the activity rate it is possible to reduce the power-delay product of the proposed SCL circuit with a logic depth of N [23]. Comparing this result with the PDP of CMOS gates which is [13]:

$$\text{PDP}_{\text{CMOS},N} = \ln 2 \cdot N V_{DD}^2 C_L \tag{3.47}$$

it can be seen that increasing the activity rate of the STSCL topology can help to achieve a PDP performance which is at least as good as the PDP of conventional CMOS topology, with the additional benefit of keeping the output swing and the delay completely independent of the supply voltage.

---

[4] Here, we are neglecting the effect of incomplete settling when N is small.

Regarding (3.43), one can conclude that the delay (or the maximum operating frequency) in a STSCL gate depends on the tail bias current (I_{SS}), but not on V_{DD}. Therefore, the delay of a logic block can be controlled without influencing PDP, which is not possible in conventional CMOS topologies. More importantly, the speed and the operation (supply) voltage can be effectively decoupled in STSCL circuits as illustrated in Fig. 3.1.

To reduce the PDP of STSCL circuits as predicted in (3.46), α should be kept as large as possible. This observation does not contradict with the similar results for conventional CMOS, where

$$(P/f)_{\mathrm{CMOS}} = C_L V_{\mathrm{DD}}^2 \left(1 + \frac{2}{\alpha} \mathrm{e}^{-\frac{V_{\mathrm{DD}}}{nU_T}}\right) \tag{3.48}$$

as shown in [24]. Here, power-to-frequency is defined as:

$$(P/f) = \frac{P_{\mathrm{diss}}}{f_{\mathrm{op}}}. \tag{3.49}$$

However, the influence of V_{DD} on (P/f) is quite different in conventional CMOS, where an optimum V_{DD} value to minimize (P/f) can be found, especially for small α values, due to significant leakage in CMOS topology.

Therefore, assuming that the system clock frequency is dictated by the longest delay path between two consecutive register stages, and assuming that the activity rate depends inversely on the maximum logic depth between two registers, it is most beneficial to keep the logic depth as shallow as possible, and thus, increase α. This calls for very short (ideally one stage) pipelining in STSCL systems, which is demonstrated with an example in Chap. 5.

3.4.2 Noise Margin

Generally, robustness of a logic gate against external or internal perturbations is measured by noise margin (NM) [25, 26]. NM is measured in quasi-static operating conditions and represents the maximum perturbation amplitude in voltage units that does not influence the logic state of the circuit.

In a subthreshold SCL circuit with ideal load resistors, it can be shown that the NM is:

$$\frac{\mathrm{NM}}{V_{\mathrm{SW}}} = \sqrt{1 - \frac{1}{A_V}} - \frac{1}{A_V} \cdot \tanh^{-1}\left(\sqrt{1 - \frac{1}{A_V}}\right). \tag{3.50}$$

where A_V represents the DC voltage gain of the circuit. As DC voltage gain of STSCL circuit calculated in (3.36) is independent of the design parameters, the only parameter that can be used for improving NM is the voltage swing. In a real STSCL with bulk-drain shorted PMOS load devices, the DC gain is almost constant and equal to $A_V \approx n_p/(n_n \cdot (n_p - 1))$. For typical values of DC voltage gain of a STSCL circuit ($A_V \approx 3.24$), NM can be as high as approximately 40% of the whole output voltage swing.

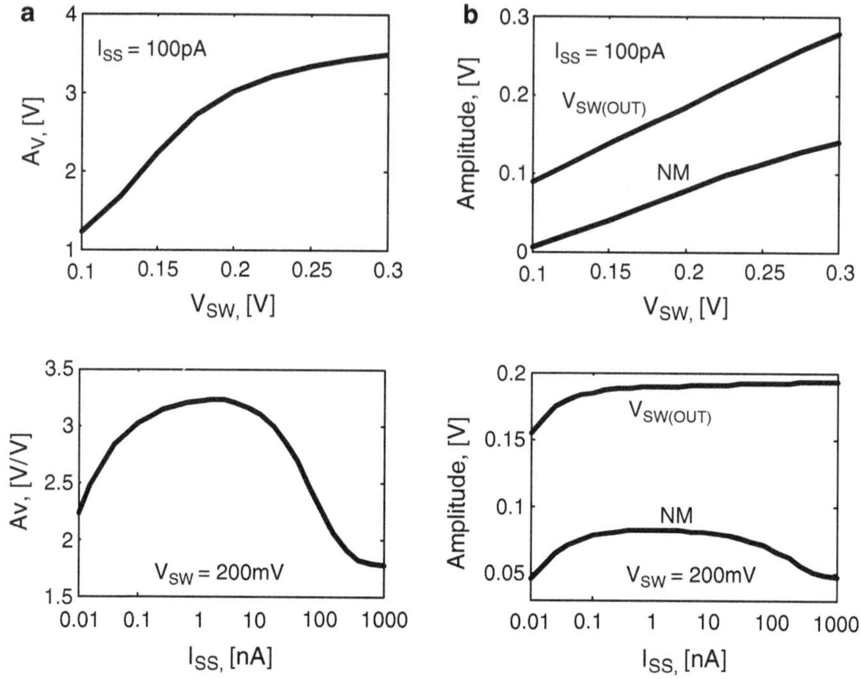

Fig. 3.13 DC transfer characteristics of an STSCL circuit designed in 0.18-μm CMOS technology. (**a**) Differential DC gain versus desired V_{SW} and tail bias current. (**b**) Noise margin and output voltage swing versus V_{SW} and tail bias current

The output voltage swing, peak gain value, and noise margin of an STSCL buffer versus V_{SW} and tail bias current are shown in Fig. 3.13. As illustrated in this figure, gain and NM are both improving by increasing V_{SW}. For voltage swing values higher than 200 mV, the gain improvement slows down. It should be mentioned that the output voltage swing should be always smaller than V_{SG} of NMOS differential pair devices. Otherwise, current switching in the differential pair circuit will not be completed. In high current densities, the devices enter into the medium and strong inversion. Hence, the gain of circuit degrades as well. In very low bias current values, the tail bias current becomes comparable to the leakage currents in the circuit. Therefore, the performance starts to degrade. It is noticeable that the noise margin is about 50 mV for tail bias current values of as low as 10 pA.

Increasing the length of differential pair transistors can help to improve the gain and noise margin by reducing the velocity saturation effect.

Mismatch Effect: Noise margin degrades due to device mismatch and process variations. Variation at the output voltage swing as well as voltage offset at the input of STSCL circuits are the two main causes of NM reduction in presence of device mismatch.

In practice and in presence of device mismatch, the noise margin can be estimated by:

$$\text{NM} \approx \text{NM}_0 - \left(\frac{\partial \text{NM}}{\partial V_{\text{SW}}}\right) \cdot \Delta V_{\text{SW}} - V_{\text{OS}}. \tag{3.51}$$

where V_{OS} is the equivalent input referred offset of the proposed STSCL gate and NM_0 is the NM without device mismatch. Variation of NM with respect to V_{SW} can be estimated using (3.50):

$$K_{\text{NM}} = \frac{\partial \text{NM}}{\partial V_{\text{SW}}} = \sqrt{1 - \frac{1}{A_V}}. \tag{3.52}$$

To calculate (3.52), it is assumed that the DC gain of an STSCL stage can be approximated by:

$$A_V \mid_{\Delta V_{\text{IN}}=0} \approx \frac{V_{\text{SW}}}{2n_n U_T}. \tag{3.53}$$

For random variations on offset voltage and voltage swing, NM degradation can be indicated by

$$\Delta \text{NM}^2 \approx (K_{\text{NM}} \Delta V_{\text{SW}})^2 + V_{\text{OS}}^2. \tag{3.54}$$

The input referred offset in a STSCL circuit can be estimated by

$$\sigma_{\text{OS}}^2 \approx \left(\frac{A_{\text{VT},N}^2}{W_N L_N}\right) + \left(\frac{A_{\text{VT},\,P}^2}{W_P L_P}\right) \times \left(\frac{n_n}{n_p}\right)^2 \tag{3.55}$$

where, A_{VT} represents the threshold voltage variation per unit mico-meter square area of gate, W and L are the width and length of the transistors, and n_n and n_p are subthreshold slope factor for differential NMOS and PMOS load devices, respectively.

Variation on V_{SW} can be caused by tail bias current mismatch and the mismatch between PMOS load devices of the STSCL circuits and the replica bias circuit:

$$\sigma_{\text{SW}}^2 \approx \left(\frac{n_p}{n_p - 1}\right)^2 \cdot \left(\frac{A_{\text{VT},N}^2}{n_n^2 W_B L_B} + \frac{A_{VT,P}^2}{n_p^2 W_P L_P}\right) \tag{3.56}$$

where W_B and L_B are the width and length of the tail bias transistors and W_P and L_P are the width and the length of the PMOS load devices.

Figure 3.14 shows the Monte Carlo simulation results in 65-nm CMOS technology for an STSCL gate. This figure shows the variation on output voltage swing, input referred offset of the STSCL circuit, and the voltage gain. In addition, the scattering plot in Fig. 3.15 depicts the relationship between variation on output voltage swing and noise margin and also offset voltage and noise margin.

There is a good agreement between the Monte Carlo simulation results and hand calculations in (3.54). Figure 3.15 shows the correlation between variation on NM and the input referred offset voltage and also between NM and variation at the output voltage swing which are both close to the estimated values in (3.54).

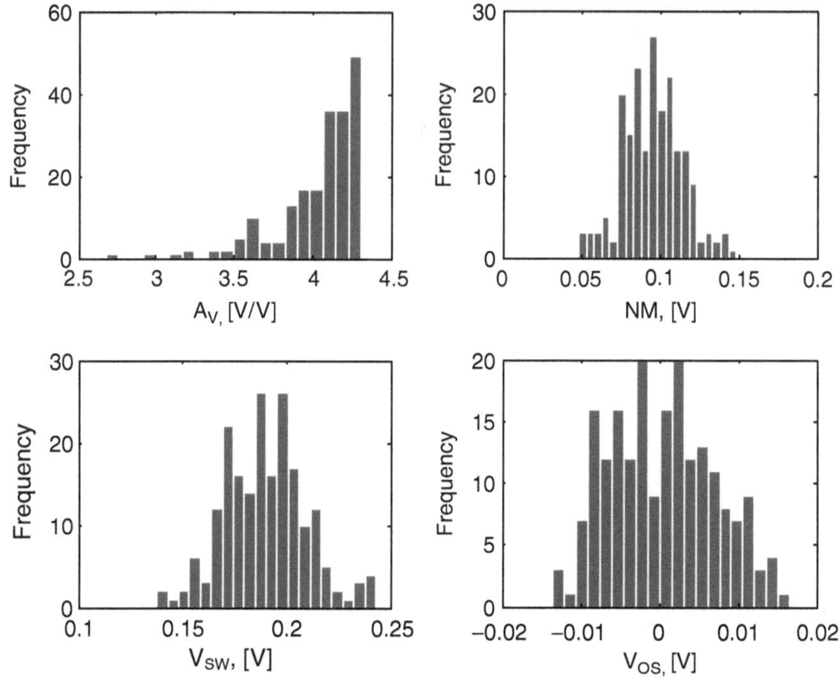

Fig. 3.14 Mismatch effect on STSCL gate performance. Variation on gain, NM, voltage swing, and input referred offset are shown. The value of NM depends highly on the output voltage swing. Here, $V_{SW} = 200\,\text{mV}$ and $I_{SS} = 100\,\text{pA}$ for 200 runs of Monte Carlo simulations

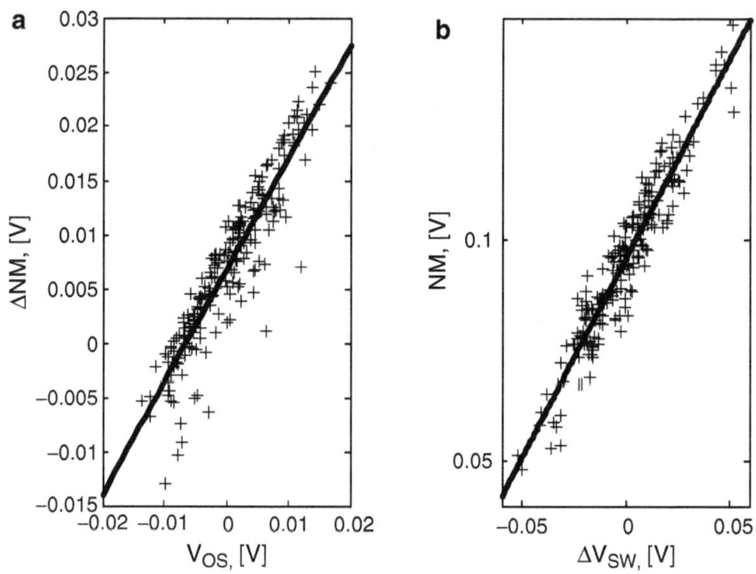

Fig. 3.15 Correlation between (**a**) variation on NM and offset voltage and (**b**) variation on NM and output voltage swing, based on Monte Carlo simulations in CMOS 65 nm

To have an approximate estimation of the variation on NM, assuming that $A_{\mathrm{VT},P} = A_{\mathrm{VT},N} = A_{\mathrm{VT}}$:

$$\left(\frac{\sigma_{\mathrm{NM}}}{A_{\mathrm{VT}}}\right)^2 \approx \frac{1}{S_N} + \frac{A_V(A_V - 1)}{S_B} + \frac{A_V(A_V - 1) + 1}{S_P} \tag{3.57}$$

where, $S = W \times L$ stands for gate area of transistor. From (3.57), it is clear that the size of biasing and PMOS load transistors are very important for having the desired NM. This expression clearly represents the relationship between cell area and NM in STSCL topology. Less variation on device threshold voltage results in smaller cell area.

As the circuit noise margin is defined to be the minimum value of NM_H and NM_L, and since NM_H and NM_L are statistically correlated, special techniques are required to calculate the precise value of NM [30]. In this case, as it is shown in [12], the mean value and variance of $\mathrm{NM} = \min\{\mathrm{NM}_H, \mathrm{NM}_L\}$ become:

$$\mu_{\mathrm{NM}} = \mu_{\mathrm{NM}_{H,L}} - \sigma_{\mathrm{NM}_{H,L}} \cdot \sqrt{\frac{1 - \rho_{\mathrm{NM}}}{\pi}} \tag{3.58}$$

$$\sigma_{\mathrm{NM}} = \sigma_{\mathrm{NM}_{H,L}} \cdot \sqrt{\frac{1 - \rho_{\mathrm{NM}}}{\pi}} \tag{3.59}$$

with ρ_{NM} defined as correlation factor between NM_H and NM_L:

$$\rho_{\mathrm{NM}} = \frac{\sigma_{\Delta\mathrm{NM}}^2}{2\sigma_{\mathrm{NM}}^2} \tag{3.60}$$

A Simple Remedy to Reduce NM Variation: One remedy to reduce the sensitivity of NM to variation is creating intentional mismatch between replica bias and the STSCL gates. If the bias current of each cell increases by about 20%, for example, then the voltage swing in STSCL gates will be by the same percentage more than V_{SW}. Therefore, the initial NM will be larger and hence more resistant against process variation. Of course this approach increases the circuit power dissipation by 20%; however, this effect can be compensated partially by using smaller devices. Analysis show that using slightly more current in STSCL gates compared to replica bias circuit reduces the variation on gain of cells respect to the process variation (i.e., $\partial A_V/\partial V_T$) considerably, which in turn makes the NM more resistant against process variation.

3.4.3 Replica Bias Circuit

A controlling circuit is necessary to keep the voltage swing at the output of SCL gates on a desired value. If V_{SW} decreases, NM will degrade and if V_{SW} increases, gate delay will increase proportional to that. Hence, V_{SW} should be selected close to its optimum value. Therefore, replica bias circuit needs to be precise enough.

A simple schematic for replica bias circuit has been shown in Fig. 3.9. The amplifier A_{VR} in Fig. 3.9 should provide enough gain with a very low offset to maintain the desired accuracy. In this work, a folded-cascode amplifier has been used to have a large output voltage swing and to be able to test the SCL gates in a very wide range of bias current values. Current-mirror based operational transconductance amplifier (OTA), is the other suitable topology for implementing this amplifier. This topology also provides a wide output voltage swing. Both topologies have a single dominant pole at the output node and hence higher load capacitance can make the feedback more stable.

The STSCL gate used inside replica bias circuit should be well matched to the SCL gates inside the circuit to have very low deviation at the desired operating point. Any mismatch between the bias current and the devices in STSCL gates and the corresponding devices in RB circuit will result in variation of the desired output voltage swing (ΔV_{SW}) and it can be shown that the sensitivity of this circuit to the mismatches is:

$$\left(\frac{\Delta V_{SW}}{U_T}\right)^2 \simeq \left(\frac{n_p}{n_p-1}\right)^2 \cdot \left(\left(\frac{\Delta I_{SD}}{I_{SD}}\right)^2 + \left(\frac{\Delta\beta}{\beta}\right)^2 + \left(\frac{\Delta V_{T0}}{n_p U_T}\right)^2\right) \tag{3.61}$$

in which $\beta = \mu C_{ox} W/L_e$. Amplifier offset should also be added to this estimation. Monte Carlo simulations show that for minimum size devices, ΔV_{SW} can be as high as 20–40 mV in a typical 0.18-µm process. To compensate the influence of device mismatch, V_{SW} should be selected a little larger than the minimum value.

Meanwhile, it can be shown that the voltage gain from gate to drain of transistor M8 in Fig. 3.9 is not very large:

$$|A_{V,\mathrm{MPR}}| = g_{m,M8} \cdot R_{SD} \simeq \frac{1}{n_p - 1}. \tag{3.62}$$

Therefore, in spite of the exponential relationship between $I_{SD,M8}$ and $V_{SG,M8}$, the gain of this stage is low and the RB circuit can be stabilized without difficulty. One single replica bias circuit can be used for a large number of STSCL gates. Therefore, its area overhead would be negligible in large scale applications.

3.4.4 *Minimum Operating Current*

The minimum operating current ($I_{SS,min}$) in STSCL topology is very important since it represents the minimum possible energy consumption of the circuit. There are different parameters determining the minimum bias current of an STSCL circuit.

To adjust the tail bias current at very low values, it is necessary to have a very precise current mirror. For bias currents in the range of pico-Ampere, tail bias transistor is deeply in subthreshold (weak inversion) region. Therefore, it is very difficult to control the operating condition of the tail bias transistor precisely. One possible remedy to construct a good current mirror is using high threshold voltage devices. Fortunately, the speed of operation in this configuration does not depend on the threshold voltage of the tail transistor. Thus, this technique in addition to using long channel devices can be helpful to implement precise enough current mirrors for pico-Ampere ranges.

The other important issue is the leakage current of the NMOS devices which are mainly due to the reverse-biased PN source-bulk or drain-bulk junctions. Also, we should include the bias current of the forward-biased PN junctions of the drain-bulk of the PMOS load devices. Indeed, this current is the main limiting factor for reducing the tail bias current and can be estimated by:

$$I_{F,D} = I_{\text{sat}}\left(\mathrm{e}^{\frac{V_{\text{SW}}}{\eta U_T}} - 1\right) \tag{3.63}$$

where η is a process dependent parameter and I_{sat} is the saturation current of the drain-bulk PN junction which depends on the area and perimeter of this junction. Therefore, it is expected that the leakage current due to this forward bias junction ($I_{F,D}$) reduces slightly by technology scaling.

Figure 3.16 shows the DC current of the load device for $V_{\text{SG}} = 0\,\text{V}$ versus temperature. In CMOS 90 nm, the leakage current is less than 10 pA in $100°C$ while at the same temperature, it is 60 pA in CMOS 130 nm. Therefore, by technology scaling, the portion of STSCL leakage current which is due to the forward-biased source-bulk PN junction is reducing. As this current is mainly due to the forward biased diodes, it does not change significantly with the process variation.

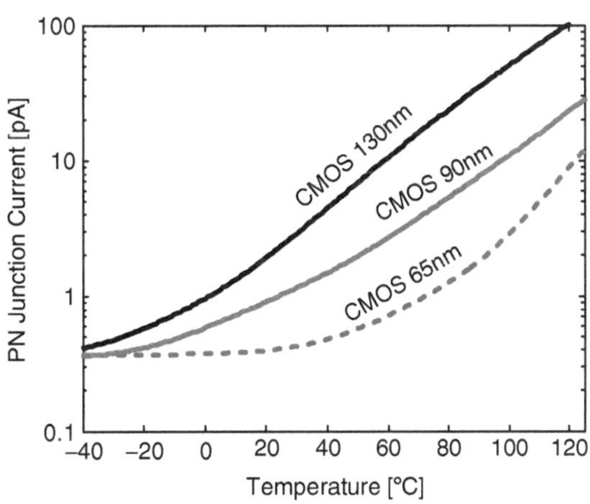

Fig. 3.16 Current of the load device when $V_{\text{SG}} = 0\,\text{V}$ versus temperature for CMOS 130, 90, and 65 nm technologies. This current is mainly due to the forward-biased source-bulk PN junction of the PMOS load device

As it can be seen in Fig. 3.16, the PN junction current increases by temperature. To calculate the temperature variation of $I_{F,D}$, the temperature dependence of I_{sat} needs to be included. As shown in [17]:

$$I_{\text{sat}} = \frac{qAn_i^2\overline{D_n}}{Q_B} \tag{3.64}$$

where n_i is the intrinsic minority-carrier concentration, Q_B is the total base doping per unit area, μ_n is the average electron mobility in the base, A is the area of emitter-base junction, $\overline{D_n}$ is the diffusion constant, and T is the temperature. Applying Einstein relationship[5] $\mu_n = q\overline{D_n}/(kT)$:

$$I_{\text{sat}} = Bn_i^2T\mu_n \tag{3.65}$$

where constant B does not depend on temperature [17]. Using $\mu_n = CT^{-n}$ and $n_i^2 = DT^3\mathrm{e}^{-\frac{V_{G0}}{U_T}}$:

$$I_{\text{sat}} = ET^{4-n}\mathrm{e}^{-\frac{V_{G0}}{U_T}} \tag{3.66}$$

where V_{G0} is the bandgap voltage of silicon extrapolated to 0°K and D and E are temperature independent parameters. Based on this, temperature dependence of $I_{F,D}$ can be represented by:

$$I_{F,D} = ET^{4-n}\mathrm{e}^{-\frac{V_{G0}}{U_T}} \times \left(\mathrm{e}^{\frac{V_{SW}}{nU_T}} - 1\right) \tag{3.67}$$

Adding the other sources of leakage, such as junction leakage in differential pair transistors, results in a minimum tail bias current slightly larger than the values shown in Fig. 3.16. Experimental results show that the tail bias current of each STSCL gate can be reduced to 10 pA in 0.18-μm CMOS technology. Based on simulations, this current can be reduced to about 5 pA in 90-nm and 65-nm technologies at room temperature.

3.4.5 Global Process and Temperature Variation

Considering (3.42), it can be concluded that the device parameters and especially threshold voltage does not influence the speed-power consumption tradeoff in SCL topology. As mentioned before, the replica bias circuit will compensate for the effect of temperature and process variations [27]. Therefore, this topology exhibits a very low sensitivity to PVT or global variations.

[5] Also known as Einstein–Smoluchowski relation revealed independently by Albert Einstein in 1905 [28] and by Marian Smoluchowski in 1906 [29].

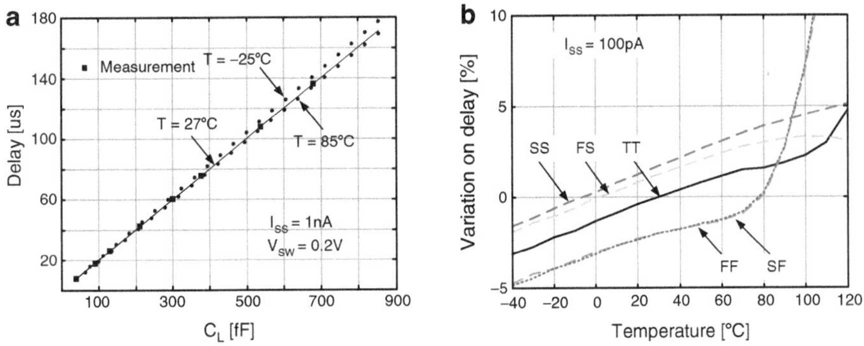

Fig. 3.17 (**a**) Variation on gate delay due to the temperature variations in 0.18 μm. (**b**) Delay variation over different corner cases for CMOS 65 nm

Figure 3.17a shows the simulated gate delay versus load capacitance in different temperatures. Simulations show that the variation on gate delay due to the temperature variations is less than 2%. Based on this figure, $t_d \approx 1.4 \times 10^8 C_L$ which is very close to the value predicted by (3.38), and also agrees very well with the measurement results. The delay variation due to process variation in CMOS 65 nm is shown in Fig. 3.17b. Here, the delay values are normalized to the typical gate delay in 27°C. Both of these two graphs depicts low sensitivity of the STSCL topology to the global process and temperature variations.

3.4.6 Effect of Mismatch on Delay

Gate delay can be varied from gate to gate due to the device mismatch effects. Mismatch on the tail bias current and the load resistance are the main sources of delay variation in STSCL topology. Assuming that the load resistance can be approximated by:

$$R_L \approx \frac{V_{SW}}{I_{SS}} \tag{3.68}$$

then, the variation on STSCL gate delay can be expressed by:

$$\frac{\Delta t_d}{t_d} \approx \frac{\Delta V_{SW}}{V_{SW}} - \frac{\Delta I_{SS}}{I_{SS}}. \tag{3.69}$$

where variation on load capacitance has been ignored. Using (3.30), one can show that:

$$\Delta V_{SW} \approx \frac{\Delta V_{T,P}}{n_p - 1} + \frac{n_p U_T}{n_p - 1} \times \frac{\Delta I_{SS}}{I_{SS}} \tag{3.70}$$

Therefore,

$$\left(\frac{\Delta t_d}{t_d}\right)^2 \approx \left(\frac{\Delta I_{SS}}{I_{SS}}\right)^2 \times \left(\frac{n_p}{n_p-1}\cdot\frac{U_T}{V_{SW}}-1\right)^2 + \left(\frac{\Delta V_{T,P}}{V_{SW}}\cdot\frac{1}{n_p-1}\right)^2 \quad (3.71)$$

where the variation on tail bias current is:

$$\frac{\Delta I_{SS}}{I_{SS}} \approx \frac{\Delta V_{T,N}}{n_n U_T} \quad (3.72)$$

Any mismatch on tail bias current is affecting the voltage swing at the output. By reduction (increase) of tail bias current, the output voltage swing will also reduce (increase) which in turn reduces (increases) the gate delay. However, at the same time available current for discharging the output parasitic capacitance will be reduced (increased) which results in delay increase (reduction). Therefore, variation on tail bias current has two opposite effects on delay which partially cancel out each other. Although the variation can be quite large, however, still it is much less than the gate delay variation in CMOS topology.

To have a very approximate estimation of the delay variation in STSCL topology due to the device mismatch, let us assume that $A_{VT,N} = A_{VT,P}$ and area of PMOS load device ($S_P = W_P L_P$) and tail bias device ($S_B = W_B L_B$) are equal ($S = S_B = S_P$). Then:

$$\frac{\Delta t_d}{t_d} \approx \frac{A_{VT}}{n_n U_T}\cdot\frac{1}{\sqrt{2S}}. \quad (3.73)$$

Figure 3.18 shows the approximate variation of gate delay for different gate area values. As can be seen, the delay variance for minimum size devices can be as high as 100% for minimum size devices.

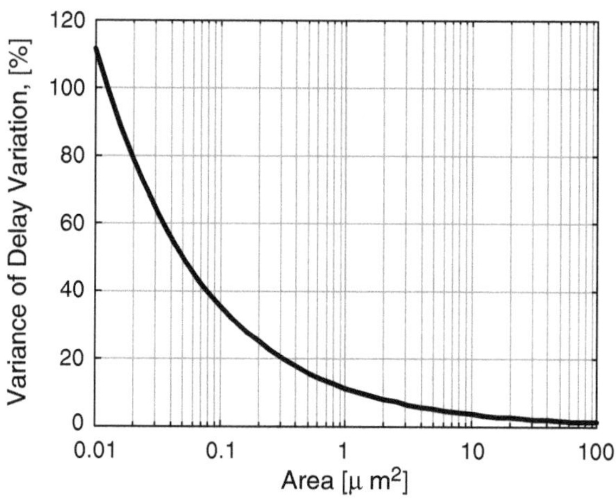

Fig. 3.18 Delay variation due to the device mismatch based on (3.73). Here, it is assumed that $A_{VT} = 5$[mV·μm] and gate area of PMOS load and tail bias NMOS devices are both equal to S

3.4.7 Minimum Supply Voltage

Since all the devices are biased in weak inversion, it is possible to use high-threshold-voltage (HVT) devices is STSCL circuits without affecting the speed of operation. The minimum supply voltage of a STSCL gate is (Fig. 3.9):

$$V_{DD,min} = V_{CS} + V_{GS1} \tag{3.74}$$

where V_{CS} is the required headroom for the current source. Since all the devices are in subthreshold, therefore, $V_{CS} \approx 4U_T$. Meanwhile, $V_{GS,1} = V_{T0} + n_n U_T \ln I_{SS}/I_0$ (V_{T0} stands for the threshold voltage of M1-M2 and $I_0 = 2n_n(W/L_{eff})U_T^2$) [18]. Notice that for a complete switching $V_{GS,1}$ should always be larger than V_{SW} to make sure that $V_{DS} \geq 0$:

$$V_{GS,1} > V_{SW}. \tag{3.75}$$

Therefore, assuming $V_{SW} \approx 6U_T$, the absolute supply voltage will be:

$$V_{DD,min} \approx 10U_T. \tag{3.76}$$

Measurements show that it is possible to reduce the supply voltage of an (8×8) multiplier implemented based on STSCL topology down to 300 mV [27].

The other limiting factor for reducing the supply voltage is the required headroom for biasing PMOS load devices. When the tail bias current increases, the required V_{SG} to keep the resistivity of the PMOS load devices will also increase. Therefore, supply voltage needs to be increased with increasing the tail bias current. In this case, the minimum supply voltage which should be larger than V_{SG} which increases proportional to the logarithm of the bias current:

$$V_{DD,min} > V_{SG} + V_{DSsat,amp} \tag{3.77}$$

where $V_{DSsat,amp}$ is the required headroom for the amplifier used in the replica bias circuit and is shown in Fig. 3.9.

3.5 Experimental Results

In this chapter, STSCL topology has been introduced and its main characteristics and specifications have been studied. In the following, some experimental results will be presented to justify the performance of this type of circuits.

3.5.1 Basic Building Blocks

In order to measure the I–V characteristics of the proposed PMOS load device and also test the characteristics of simple STSCL gates, a test circuit has been

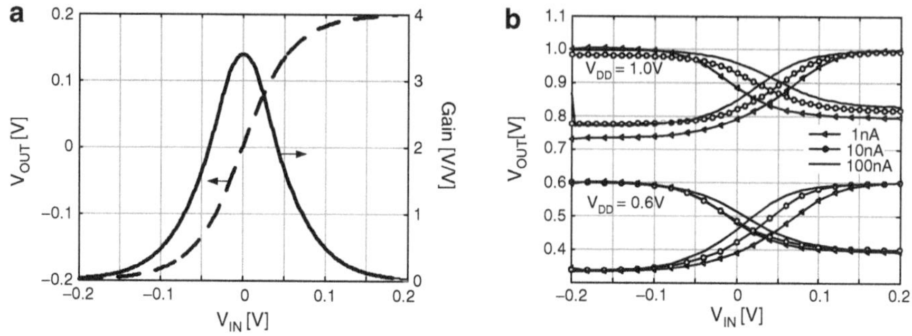

Fig. 3.19 (**a**) Simulated DC transfer characteristics and DC gain of an STSCL gate biased at $I_{SS} = 1$ nA. (**b**) Measured transfer characteristics of an STSCL adder stage for two different supply voltages ($V_{DD} = 0.6$ V and 1.0 V) and different bias currents ($I_{SS} = 1, 10,$ and 100 nA). The test circuit has been implemented in 0.18-μm CMOS

fabricated in 0.18-μm CMOS technology. The first test chip included an STSCL buffer (inverter) circuit and a single bit full adder gate. To have a full control on the test circuit, all the input and output nodes of the proposed gates have been connected directly to the test pads.

Using probe station, extensive DC measurements to characterize the load device as well as the gates have been performed. Figure 3.6d shows the measured I–V characteristics of the load device which exhibits a very good agreement with the BSIM model. Meanwhile, measurement results for the high-valued floating resistance constructed based on the concept shown in Fig. 3.8a is depicted in Fig. 3.8b.

Simulated DC characteristics of an STSCL gate is shown in Fig. 3.19a. Based on this graph, the gain of an STSCL circuit can be as high as 3.2. The input–output DC characteristics of an STSCL adder gate are shown in Fig. 3.19b based on measurements in three different tail bias currents and two different supply voltages. In these measurements, as the probes are directly connected to the circuit through a very simple ESD[6] protection circuit, it has been very difficult to reduce the tail bias current below 1 nA. The leakage current of the ESD protection circuit constructed by the reverse biased PN junctions caused some displacement in the output DC characteristics.

The basic DC measurements approve that the performance of the proposed high-valued load device concept is very close to the expected performance (Fig. 3.19b), and can be successfully used for implementing STSCL circuits.

3.5.2 *Ring Oscillator and Frequency Divider*

To study the delay versus power consumption for the proposed STSCL topology, a second test chip has been designed and fabricated in conventional 0.18 μm CMOS

[6] Electro-static discharge.

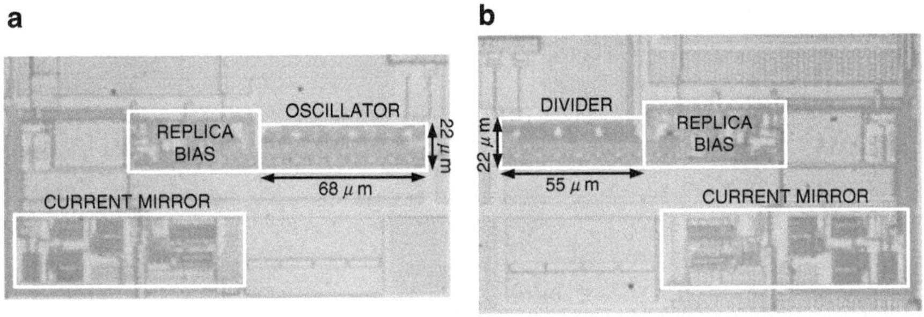

Fig. 3.20 Microphotograph of the test circuits: (**a**) ring oscillator and (**b**) frequency divider

technology. The test structures consist of an 8-stage ring oscillator and a frequency divider (divide-by-8) circuits, both of which are implemented based on a 2-input multiplexer (MUX) STSCL gate. The microphotographs of the test circuits are shown in Fig. 3.20.

To control the operation of the test circuits, the tail bias current of the SCL gates can be adjusted externally. Internal current mirrors with current gain of $I_{SS}/I_{EXT} = 0.01$ have been used to simplify the process of tail bias current control during the measurements. The supply voltage of the test blocks are directly accessible to measure the total power consumption of each block separately using HP4156A Semiconductor Parameter Analyzer.

An internal replica bias circuit has been employed to control the voltage swing at the output of the STSCL gates. As described in Sect. 3.4.3, the output voltage swing should be larger than 150 mV in room temperature. The die-to-die variation of the gate bias voltage (ΔV_{BP}) required to ensure a fixed voltage swing of 150 mV at a given tail current was found to be less than $\pm 8\%$, in conventional 0.18-μm CMOS technology.

3.5.2.1 Ring Oscillator Test Circuit

Figure 3.21 illustrates the measured oscillation frequency of an 8-stage ring oscillator with differential STSCL NAND gates (which are constructed based on 2-input MUX gates) in comparison to the simulation results. The conventional CMOS oscillator used for comparison is built with 2-input standard NAND gates in the same 0.18-μm CMOS technology with driving strength of $\times 1$.

As depicted in this figure, the measurement results of the STSCL oscillator are very close to the simulation results, and consistent over a range of several orders of magnitude. Meanwhile, PDP is very well predictable by (3.46).

Depicted in Fig. 3.21, the oscillation frequency of the STSCL oscillator can be adjusted over a very wide range (below 1 kHz to more than 1 MHz). Corresponding to that, the tail bias current can be adjusted from about 10 pA to close to 1 μA with a linear power versus oscillation frequency relationship. The oscillation frequency

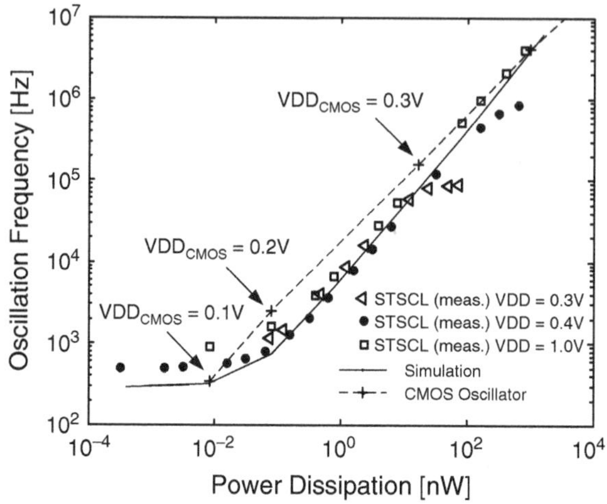

Fig. 3.21 Measured oscillation frequency versus power dissipation of the 8-stage ring oscillator based on the proposed STSCL topology for $V_{DD} = 0.3$, 0.4, and 1.0 V. Corresponding power-speed curves for a CMOS ring oscillator is shown as well

has a very small dependence on supply voltage. Based on this figure, as the supply voltage is reducing, the upper oscillation frequency is also decreasing and oscillator is saturating for lower controlling current values. This saturation behavior is because by increasing the tail bias current, required V_{SG} for the load PMOS devices needs to be increased to control the load resistance on desired value. Therefore, more voltage headroom is required. It is interesting to notice that the supply voltage of STSCL circuit could be reduced to 300 mV with tuning range of almost two decades for operating frequency.

This figure also shows the results for the CMOS ring oscillator, operating in subthreshold regime with different supply voltage values between 0.1 and 0.4 V. Comparing the results, CMOS ring oscillator exhibits less PDP which is mainly because of low activity rate of this circuit. It is expected that in more advanced CMOS technologies where leakage current of CMOS circuit grows, the power efficiency of CMOS topology degrades considerably.

The proposed ring oscillator has been also used to measure the gate delay versus tail bias current (Fig. 3.17) and also gate delay versus load capacitance (Fig. 3.12) to justify (3.40).

3.5.2.2 Divider Test Circuit

The divide-by-8 circuit has been realized using the source-coupled latch structure shown in Fig. 3.22. The measured maximum operating (input) frequency of the divider is plotted against power dissipation in Fig. 3.23a at $V_{DD} = 0.4$ V and $V_{DD} = 1.0$ V, comparing the results with the performance of an optimized CMOS

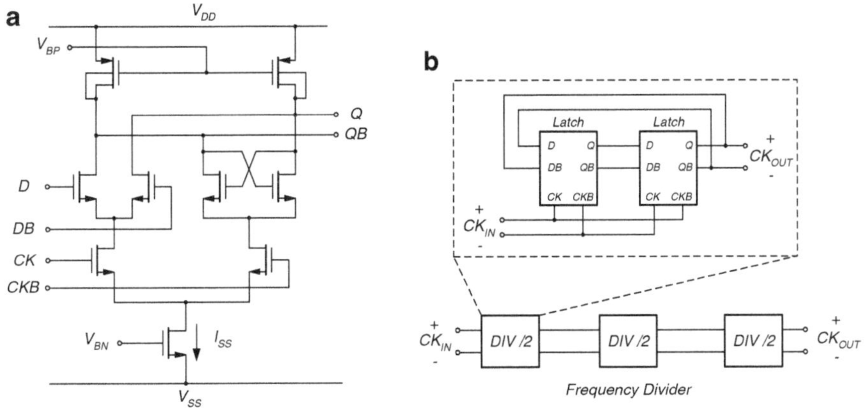

Fig. 3.22 (**a**) STSCL latch circuit schematic and (**b**) the topology of the divide-by-8 circuit used for measurement

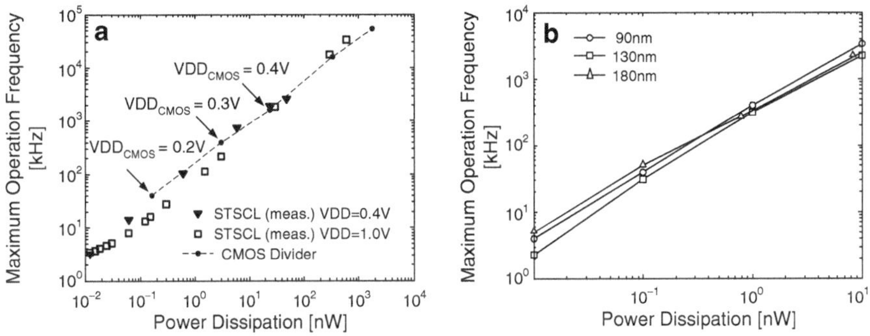

Fig. 3.23 (**a**) Measured maximum frequency of operation versus power dissipation of the divide-by-8 frequency divider shown in Fig. 3.22 for $V_{DD} = 0.4$ V and 1.0 V. (**b**) Simulated maximum operating frequency of STSCL divider in different technologies (CMOS 90, 130, and 180 nm)

frequency divider operating in subthreshold regime. While the CMOS divider cannot sustain correct operation below 200 mV supply voltage, the SCL divider with the bulk-drain connected PMOS load continues its operation down to 10 pA/gate of tail current, and 3 kHz of input frequency. Therefore, it has been possible to scale down the power consumption by one order of magnitude more for STSCL topology. The resulting measured PDP corresponds to less than 1 fJ/gate.

To compare the performance of the STSCL gates at scaled technology nodes, the maximum operating frequency of a divide-by-8 circuit has been simulated using technology parameters for 90-nm, 130-nm and 0.18-μm CMOS processes (Fig. 3.23b). Here, it is assumed that the DFF gates are loaded with the same amount of interconnect capacitance, and all leakage components are taken into account. It can be seen that the STSCL frequency divider exhibits very similar performance in different technology nodes. It is possible to reduce the tail bias current of the circuit down to 10 pA both in 130-nm and 90-nm technologies, whereas the subthreshold leakage current would be very different to control in conventional CMOS logic circuits.

Considering the results presented in Figs. 3.21 and 3.23, it can be observed that the STSCL solution can successfully extend the range of operation by two orders of magnitude along the power axis, while allowing completely separate control of voltage swing and power dissipation.

3.5.3 Multiplier Circuit

To illustrate the use of the proposed circuit topology for more complex functions, another test chip containing an (8×8) bit parallel carry–save multiplier has been designed and fabricated using 0.18-μm CMOS technology [31, 32]. Shown in Fig. 3.24, the proposed test chip includes also a similar CMOS (8×8) parallel carry–save multiplier which is used as the reference circuits, and a controlling unit. The controlling unit compares the outputs of the STSCL multiplier with the outputs of the CMOS multiplier to detect the errors. For further analysis, the outputs of both multipliers are accessible from outside the chip. SCL-to-CMOS and CMOS-to-SCL level converters are used to convert the signal levels at the input and output of STSCL multiplier. The size of STSCL multiplier is 2.4 times larger than the corresponding CMOS multiplier area.

Figure 3.25a shows the measured input-to-output delay of the STSCL-based multiplier, operating at $V_{DD} = 0.3$, 0.4, and 1.0 V, in comparison to the simulation results. It can be seen that the performance of the STSCL multiplier is accurately predicted by the simulations.

The supply voltage can be reduced down to 0.3 V while the circuit remains operational over a very wide range of tail bias current values. The saturation behavior of

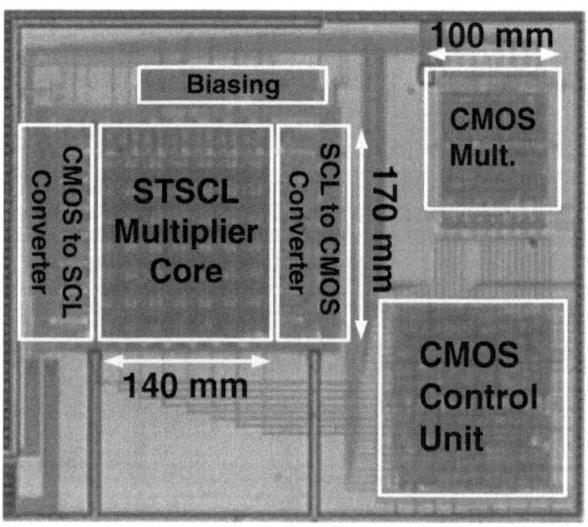

Fig. 3.24 Photomicrograph of the measured STSCL-based (8×8) bit Carry–Save multiplier

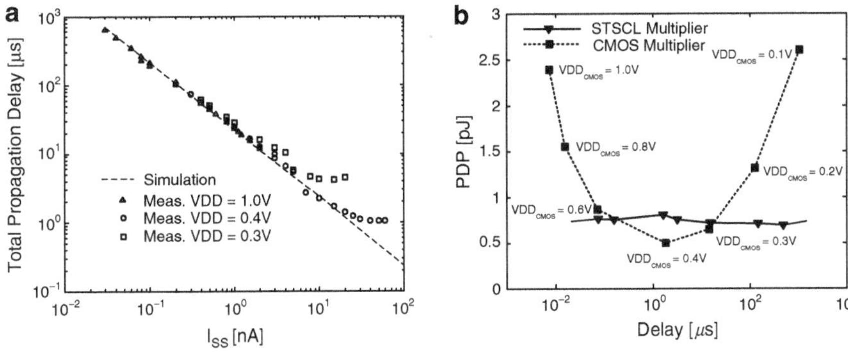

Fig. 3.25 (**a**) Measured total propagation delay of the proposed STSCL multiplier versus tail bias current (I_{SS}) for different supply voltages in comparison to the simulation results. (**b**) Comparing the power-delay product versus delay for two (8 × 8) bit Carry–Save multiplier circuits built with conventional CMOS and STSCL components

the delay at higher bias currents is mainly due to the limited swing of the replica bias circuit that is used to produce the proper gate voltage for the PMOS load devices.

To illustrate the independent control of the delay and the voltage swing, the power delay product (PDP) versus the delay of the STSCL multiplier circuit is plotted in Fig. 3.25b for different bias current levels, and compared with the variation of PDP of an equivalent CMOS multiplier circuit, also operating in sub-threshold regime. In this example, the power supply voltage and the output voltage swing of the STSCL circuit is kept at 0.35 V and 0.15 V, respectively, resulting in nearly constant PDP over the entire operating range. The PDP of the CMOS circuit, on the other hand, varies significantly with V_{DD}, due to the quadratic dependence of PDP on V_{DD}, and increasing dominance of leakage at low V_{DD} values. As the leakage current of CMOS circuits in CMOS 0.18 μm is very small, it is expected that in more advanced technologies, the benefit of using STSCL topology for lowering the energy consumption becomes more visible.

3.6 Conclusion

In this chapter, after a short overview on conventional SCL topology, subthreshold SCL (STSCL) circuits for ultra-low-power applications have been introduced. The proposed topology is based on a novel load device concept which makes it possible to use close to minimum size PMOS devices to construct very high-valued resistances.

The power-speed tradeoffs in conventional and subthreshold SCL circuits have been analyzed. Meanwhile, the performance of SCL and CMOS topologies has been very briefly compared to show the capabilities and benefits of using each topology. More extensive analysis and comparison is provided in Chap. 6

Confirmed by the measurement results, the proposed circuit topology can be used for bias current levels as low as tens of pico-Amperes. This is especially interesting when subthreshold leakage current in conventional CMOS topology precludes reducing the power consumption below a certain level.

In the next two chapters, implementing standard STSCL cell libraries and also some techniques for improving the performance of STSCL circuits will be described.

References

1. F. M. Wanlass and C. T. San, "Nanowatt logic using field-effect metal-oxide semiconductor triodes," in *IEEE Solid-State Circuit Conference (ISSCC)*, pp. 32–33, Feb. 1963
2. K. Roy, S. Mukhopadhyay, and H. Mahmoodi-Meimandi, "Leakage current mechanisems and leakage reduction techniques in deep-submicrometer CMOS circuits," in *Proceeding of the IEEE*, vol. 91, no. 2, pp. 305–327, Feb. 2003
3. A. Tajalli and Y. Leblebici, "A slew controlled LVDS output driver circuit in 0.18 μm CMOS technology," *IEEE J. Solid-State Circuits*, vol. 44, no. 2, pp. 538–548, Feb. 2009
4. D. W. Murphy, "High speed non-saturating switching circuits using a novel coupling technique," *ISSCC Dig. Tech. Papers*, pp. 48–49, Feb. 1962
5. J. A. Narud, W. C. Seelbach, and N. Miller, "Relative merits of current mode logic microminiaturization," in *IEEE Solid-State Circuit Conference (ISSCC)*, pp. 104–105, Feb. 1963
6. M. I. Elmasry and P. M. Thompson, "Analysis of load structure for current-mode logic," *IEEE J. Solid-State Circuits*, pp. 72–75, Feb. 1975
7. L. G. Heller, W. R. Griffin, J. W. Davis, and N. G. Thoma, "Cascode voltage swing switch logic: a differential CMOS logic family," in *IEEE Solid-State Circuit Conference (ISSCC)*, pp. 16–17, Feb. 1984
8. M. Cooperman, "High speed current mode logic for LSI," in *IEEE Transactions on Circuits and Systems*, vol. 27, no. 7, pp. 626–635, Jul. 1980
9. M. I. Elmasry, "Nanosecond NMOS VLSI current mode logic," *IEEE J. Solid-State Circuits*, vol. 12, no. 2, pp. 411–414, Apr. 1982
10. A. Tajalli, P. Muller, and Y. Leblebici, "A power-efficient clock and data recovery circuit in 0.18-μm CMOS technology for multi-channel short-haul optical data communication," *IEEE J. Solid-State Circuits*, vol. 42, no. 10, pp. 2235–2244, Oct. 2007
11. A. Tanabe, M. Umetani, I. Fujiwara, T. Ogura, K. Kataoka, M. Okihara, H. Sakuraba, T. Endoh, and F. Masuoka, "0.18-μm CMOS 10-Gb/s multiplexer/ demultiplexer ICs using current mode logic with tolerance to threshold voltage fluctuation," *IEEE J. Solid-State Circuits*, vol. 36, no. 6, pp. 988–996, Jun. 2001
12. S. Badel "MOS current-mode logic standard cells for high-speed low-noise applications," *PhD Dissertation*, Ecole Polytechnique Fédérale de Lausanne (EPFL), Switzerland, 2008
13. J. M. Musicer and J. Rabaey, "MOS current mode logic for low power, low noise CORDIC computation in mixed-signal environment," in *Proceedings of International Symposium on Low Power Electronics and Design (ISLPED)*, pp. 102–107, 2000
14. Y. Taur and T. H. Ning, *Fundamentals of Modern VLSI Devices*, Cambridge University Press, 1998
15. T. Sakurai and A. R. Newton, "Alpha-power law MOSFET model and its applications to CMOS inverter delay and other formulas," *IEEE J. Solid-State Circuits*, vol. 25, pp. 584594, Apr. 1990
16. T. Sakurai and A. R. Newton, "A simple MOSFET model for circuit analysis," in *IEEE Transactions on Electron Devices*, vol. 38, pp. 887894, Apr. 1991
17. P. R. Gray, P. J. Hurst, S. H. Lewis, and R. G. Meyer, *Analysis and Design of Analog Integrated Circuits*, Wiely, Fourth Ed., 2000

18. C. C. Enz and E. A. Vittoz, *Charge-based MOS Transistor Modeling*, Wiley, 2006
19. C. H. Doan, "Design and implementation of a highly-integrated low-power CMOS frequency synthesizer for an indoor wireless wideband-CDMA direct-conversion receiver," *Master Dissertation, Electrical Engineering and Computer Science Department*, University of California at Berkeley, 2000
20. B. Razavi, *Design of Integrated Circuits for Optical Communications*, Mc-Graw Hills, 2004
21. T. Gabara, and et al., "LVDS I/O buffers with a controlled reference circuit," in *Proceedings of IEEE ASIC Conference*, pp. 311–315, Sep. 1997
22. M. Alioto and G. Palumbo, "Power-aware design techniques for nanometer MOS current-mode logic gates: a design framework," in *IEEE Circuits and Systems Magazine*, vol. 6, no. 4, pp. 40–59, 2006
23. A. Tajalli, E. J. Brauer, and Y. Leblebici, "Ultra low power 32-bit pipelined adder using subthreshold source-coupled logic with 5fJ/stage PDP," *Elsevier Microelectron. J.*, vol. 40, no. 6, pp. 973–978, Jun. 2009
24. E. Vittoz, "Weak Inversion for Ultimate Low-Power Logic", in *Low-Power Electronics Design*, Editor C. Piguet, CRC, 2005
25. J. R. Hauser, "Noise margin criteria for digital logic circuits," in *IEEE Transactions on Education*, vol. 36, Nov. 1993
26. J. Lohstroh, E. Seevinck, and J. De Groot, "Worst-case static noise margin criteria for logic circuits and their mathematical equivalence," *IEEE J. Solid-State Circuits*, vol. 18, Dec. 1983
27. A. Tajalli, E. J. Brauer, Y. Leblebici, and E. Vittoz, "Sub-threshold source-coupled logic circuit design for ultra low power applications," *IEEE J. Solid-State Circuits*, vol. 43, no. 7, pp. 1699–1710, Jul. 2008
28. A. Einstein, "ber die von der molekularkinetischen Theorie der Wrme geforderte Bewegung von in ruhenden Flssigkeiten suspendierten Teilchen," *Annalen der Physik*, no 17, pp. 549560, 1905
29. M. Smoluchowski "Zur kinetischen Theorie der Brownschen Molekularbewegung und der Suspensionen," Annalen der Physik, no. 21, pp. 756780, 1906
30. S. Nadarajah and S. Kotz, "Exact distribution of the max/min of two gaussian random variables," in *IEEE Transactions on Very Large Scale Integration (VLSI) Systems*, vol. 16, no. 2, pp. 210–212, Feb. 2008
31. M. Mercaldi "Ultra-low power computational logic systems," *Master Thesis*, Ecole Polytechnique Fédérale de Lausanne (EPFL), Switzerland, 2007
32. B. Ray "Power efficient computational logic systems," *Master Thesis*, Ecole Polytechnique Fédérale de Lausanne (EPFL), Switzerland, 2007

Chapter 4
STSCL Standard Cell Library Development

4.1 Introduction

In Chap. 3, subthreshold source-coupled logic (STSCL) circuits have been introduced and their performance have been analyzed. In this chapter, standard cell based approach for implementing complex STSCL digital circuits will be described. The main goal is to automatize the design, synthesis, and place and route (PAR) steps for application-specific integrated circuit (ASIC) designs. In the proposed semi-custom approach, a set of custom primitive cells will be developed that can be used for constructing digital systems with the aid of specific automation tools.

In a typical semi-custom design flow, a standard-cell library including at least basic logic and storage gates with few driving strengths are required. To be able to estimate the system performance, the transient behavior of the cells should be provided. This can be done by characterizing the cells in different conditions and process corners. Then, using a hardware description language (HDL),[1] the proposed system can be constructed and then using the primitive components in the library it can be synthesized. The final design can be optimized using the cell specifications by a proper CAD tool.[2] For this purpose, different design constraints such as speed, power, and area can be applied to the design. Finally, the physical implementation will be produced using a PAR tool.

The main issue with the STSCL circuits is that all the logic signals are differential. As the existing tools cannot handle the differential signal routing, some novel techniques have been developed in [1] to overcome this problem. In this work, the same approach is adopted for STSCL circuits [2]. To handle the differential signal routing based on the approach proposed in [1], two sets of standard cells need to be developed and characterized. The first group consists of differential STSCL gates with different driving strengths. The second group is exactly similar to the first group while the gates are assumed to be single ended. Indeed, the synthesis

[1] HDL languages such as Verilog or VHDL.

[2] Computer aided design tool.

A. Tajalli and Y. Leblebici, *Extreme Low-Power Mixed Signal IC Design: Subthreshold Source-Coupled Circuits*, DOI 10.1007/978-1-4419-6478-6_4,

and the initial placement and routing will be done using single ended library cells. In the last step, using special techniques, the single ended routing will be converted to differential routing [1–3]. Two sample FIR filters have been implemented using STSCL standard cells as demonstration circuits.

4.2 Standard Cell Library

4.2.1 Background

A library of digital primitive blocks includes a minimum set of logic cells called standard cells. Each standard cell consists of a set of transistors and their connections implementing a specific boolean logic or a storage cell. Although it is possible to generate any boolean function using only a NAND (or a NOR) gate, most of the libraries include many different types of logic gates to make the final design more area and power efficient. A rich library of different types of cells with different driving capabilities helps to implement more efficiently complicated digital systems. The primitive gates such as buffer, inverter, NAND, NOR, XOR, and memory cells are often found in any standard library while more powerful libraries contain additional gates and sub-blocks with higher complexity such as adders.

The initial design of a standard cell begins with implementing the functionality of the cell at the transistor level. The schematic view of a cell is used for this purpose. In addition, schematic views are widely used for simulating and debugging the circuits. The schematic of a cell can be represented by symbol view which consists of the input and output ports of the cell as well as some text information.

Standard cell libraries contain another view which is called layout view. Designing the layout view of a cell is compulsory since the netlist is useful for simulation purposes but not for fabrication purpose. The layout of a cell represents what will be physically placed on a chip. Each layout consists of several base layers which form the structures of the transistors and interconnect lines. Designing area efficient layouts, which could meet the required power and timing constraints, is still a challenging task.

The designed cell layouts must be checked very carefully to insure that no design rules have been violated (DRC).[3] Then it is necessary to compare the schematic with respect to the layout using Layout Versus Schematic (LVS) tool in order to verify compatibility of the layout with corresponding schematic. After LVS, post layout simulations can be performed by extracting the parasitic components.[4] The next step is to prepare the set of cells and feed them to the design tool. In the following these steps will be explained in more detail.

[3] Design Rule Check or DRC.

[4] RCX extraction.

4.2.2 Cell Types

In this work, two STSCL cell libraries have been developed, one in CMOS 0.18 μm and the other one in CMOS 90 nm [2–5]. In both libraries, different logic and storage gates with different driving strengths have been implemented. The designed cell libraries contain buffer (inverter), OR, AND, XOR, half adder (HA), full adder (FA), MUX, and DFF (with and without reset signal). Two types of AND, OR, and XOR logic gates with two and three inputs have been developed (OR2/OR3, AND2/AND3, XOR2/XOR3).

The cells in the 0.18-μm library, except for the flip-flops, come with five different driving strengths: ×1, ×2, ×4, ×8, and ×16. The HA, FA, and flip-flop cells have only one driving strength (×2). In Sect. 4.3, the two different strategies used to implement area efficient cells will be described.

4.2.3 Cell Layout

Common Signals: Each STSCL gate has four biasing pins: V_{DD}, V_{SS}, V_{BN} for biasing NMOS tail bias transistors, and V_{BP} for biasing the PMOS load devices. The nodes that can be shared among all the cells could be placed at the same position in the layout of all the cells. In addition, they could be placed somehow to be connected automatically when the cells are placed next to each other. In this way, the routing process can be simplified considerably. A sample layout for such a cell is shown in Fig. 4.1. In this cell, the area inside the dashed line can be shared between the two consecutive blocks which helps to reduce the area. Some signals such as supply lines can be shared among all the cells. When the cells are arranged in rows, these signals automatically will be connected. Therefore, for this type of signals the pins do not need to be on the grids. The other pins that need to be routed by the tool, as will be explained, need to be placed on grids.

Routing Grid: Routing grids are where the router routes the pins over the cells. The grid spacing for different routing layers should be selected very carefully to simplify the routing process and to avoid errors or incomplete routings. The grid spacing should be larger than minimum metal pitch number which is allowed in a technology. Meanwhile, both vertical and horizontal routing grids can be shifted by one-half of a grid with respect to the origin of the cell layout, as illustrated in Fig. 4.2. This half a pitch shift helps to increase the number of grids inside the cell and hence increase the number of nodes that are available for placement of pins.

Layout Cautions: The connections near the borders of each cell need to be placed very carefully. These type of connections must have sufficient spacing from the boundary to prevent DRC errors when the cells are placed concatenated. For example, the distance of n-well to border should be at least half of the allowed n-well-to-n-well distance (with different potential values). In this case, when two cells are placed next to each other, the n-well-to-n-well distance will not be smaller than the minimum acceptable value.

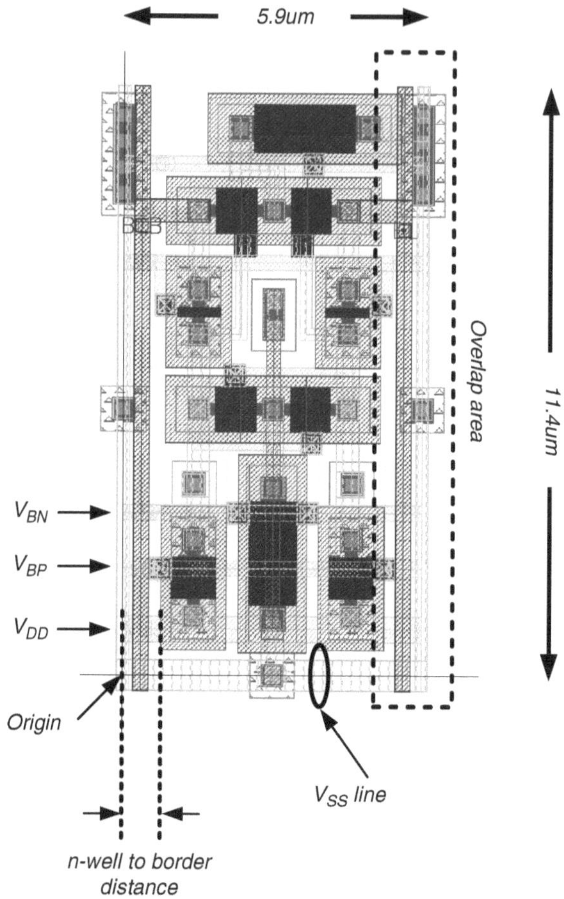

Fig. 4.1 Sample layout of an STSCL gate

Since the cell layout will be used by the automatic PAR tool, it is necessary to put the input/output (IO) pins on the intersection of minor grids, as depicted in Fig. 4.2. Using only few lowest levels of metal layers inside the cells (e.g, only poly, PO, and metal one, M1, layers if possible) helps the tool to do the top level routing more easily. Since the top level routing deals with intra cell interconnections, the possibility of vertical and horizontal access to each pin inside the cell should also be guaranteed.

Differential Routing: The current design automation tools are not able to handle the routing of circuit with differential input output ports. Therefore, some modifications need to be done in the conventional PAR flow.

From logic point of view, one of the two differential signals is sufficient to represent each signal. Hence, it is possible to do the synthesis and initial steps of placement and routing using single ended blocks. For this purpose, at the first steps differential IO ports need to be treated as single ended signals [1].

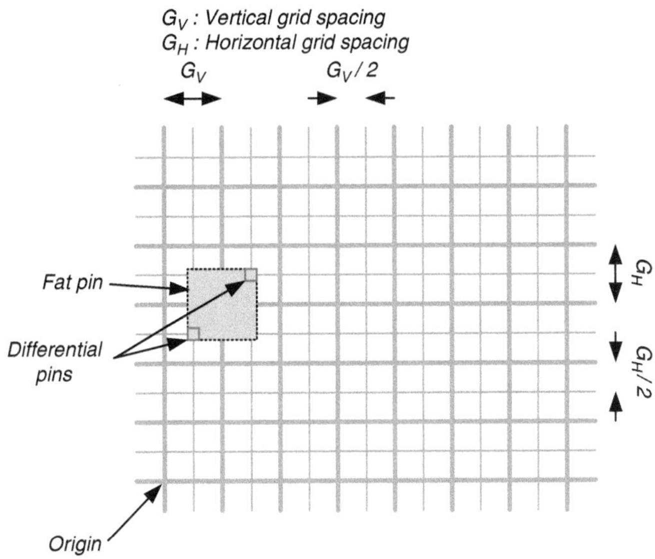

Fig. 4.2 The template for placing the cell and fat pins [1,2]

After placing the cells, the fat pins and fat lines which are replaced for differential pins and differential signals, will be routed. The fat pins are created on each pair of differential pins using a specific layer, which can be called, for example, *fat ME1*. This layer covers the entire differential pin. One sample fat pin and its placement inside a cell is shown in Fig. 4.2. After successfully routing the fat pins, each fat pin and fat interconnect needs to be split to the corresponding differential pins and interconnects [1].

4.2.4 Characterization

The transient characteristics of all the cells in a library need to be evaluated in different operating conditions. This information will be used later on for estimating the system performance and also optimizing the system specifications. For this purpose, an extensive characterization step is required. For example, the gate delay (t_d), rise time (t_r), and fall time (t_f) for combinational gates and in addition settling time (t_{ss}), hold time (t_h), and delay for sequential gates need to be extracted [1]. The specifications of each gate needs to be evaluated at different corner cases, temperatures, and supply voltages if necessary. Meanwhile, each parameter needs to be evaluated in a wide range of different load capacitance values (starting from few femto-Farad up to few hundreds of femto-Farad). The entire set of information will be collected in a database to be used by the CAD tool for synthesis and simulation purposes.

4.2.5 LEF File

To perform the placement and routing using *SoC Encounter* tool, it is necessary to generate an appropriate description of the cells. In this tool, this description is represented by LEF[5] files. This file includes all necessary information needed for PAR and is generated from *abstract* view of each cell and the technology files. The LEF of a cell does not contain all the layout of the cell but the layers and vias that are important from a routing point of view.

There are two types of LEF files: the first type is technology LEF and the second type is generated by the abstract generator. The abstract generator uses the technology LEF file to generate the other one. A LEF file contains the technology, site, and macros. Macro cell definition includes description, dimensions, blockages, layout of all the pins, and capacitances of a cell.

The technology LEF file is provided by the foundry and contains all the technology specifications including the layers, vias, and design rules. Layers are defined in process order from bottom to top and each layer consists of several attributes such as type, width, direction, resistance and capacitance per unit square, spacing rules, and antenna factor.

An abstract view is also generated by the abstract generator which will be used by *Silicon Ensemble* for placement and routing. The abstract view of a cell contains information such as routing obstructions, and the name, orientation, and PR boundary of a cell as well as the name, direction, type, and metal layers of the pins. In case of STSCL circuits, the LEF files must be generated for both differential and fat libraries [1].

4.2.6 Template Generation

The logic function in STSCL circuits is realized by an N-level NMOS switching network. This network can be modeled by a Binary Decision Diagrams (BDD). All possible N-level BDDs topologies are called footprints. The footprints of 1–2 levels network are shown in Fig. 4.3 [1]. A 1-level network can only be mapped to the Buffer and Inverter gates while for networks of 1–3 levels, 19 unique footprints exist and can be mapped to a large number of cells like XOR3, AND3, etc. Generation of the footprints is discussed in detail in [1].

Each of these footprints corresponds to a different physical network. The number of the nodes in an N-level footprint is between N and $2N + 1$. The footprints correspond to the function with the maximum number of inputs that can be realized with this network. Obviously, the functions with fewer inputs can also be realized by assigning the inputs to more than one node in the network. All boolean functions that can be realized by a specific footprint are simply obtained by trying all possible variable assignments [1].

[5] Library exchange format.

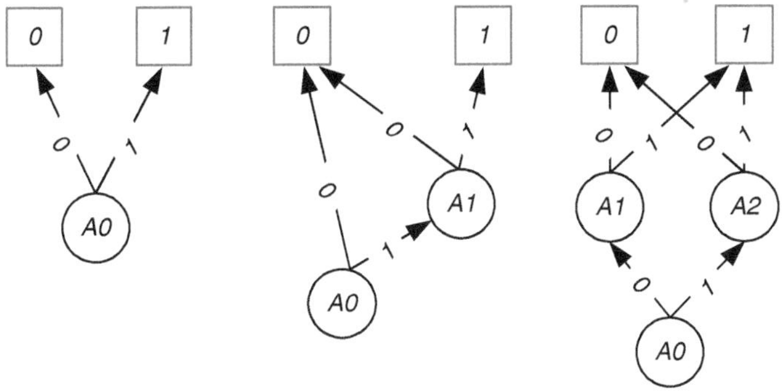

Fig. 4.3 Footprints of the 1-level and the 2-level networks [1]

The templates are generated out of the footprints by trying different input assignments. This way, a rich cell library is created with only a limited number of physical cells. A unique function may be realized using different templates, and therefore the function can be physically implemented with different networks. The different implementations of a same function are called variations which might have different electrical properties.

One important aspect of STSCL circuits is that all inputs and outputs are differential and therefore, inverted signals are always available. A new set of cells can be created by inverting the inputs and outputs of the cells in all possible combinations ($2N + M$ possible combinations for a cell with N inputs and M outputs) [1]. The new set of cells enables the synthesizer to select a gate with any combination of inputs and outputs. In this way, the synthesizer does not need to explicitly invert a signal when a signal has to be inverted. As a result, a significant number of inverters will be reduced in a large design which improves the delay as well as reducing the area. The drawback of this approach is that the number of cells in the library will be increased dramatically [1].

4.3 Design Strategies

One of the main issues in design of standard cell libraries is the area of the cells. Larger cell area not only results in larger chip size, but can also cause speed reduction. As the size of cells increases, the parasitic capacitance of the interconnects will also increase, and hence the logic cells need to drive more parasitic capacitances which results in lower speed. Therefore, it is necessary to reduce the size of each cell as much as possible.

One important issue with the STSCL logic cells is that driving strength can be scaled only by scaling the tail bias current. Therefore, for a cell with driving strength of N the size of tail bias NMOS transistor needs to be N times larger than a cell with unit driving strength. The scaling of the current driving by scaling the tail bias

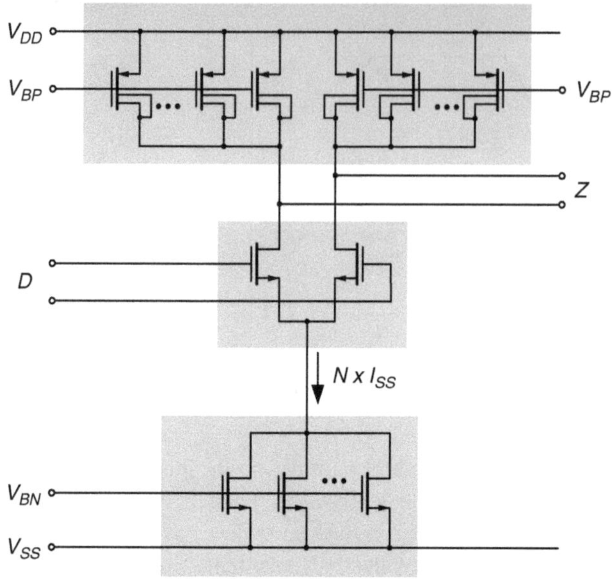

Fig. 4.4 Improving the cell driving strength by multiplying the tail bias current

current is shown in Fig. 4.4. To keep the output swing constant while the tail bias current is scaling, N parallel PMOS load devices should be used to reduce the load resistance by the same factor. Therefore, PMOS load devices will occupy N times larger area compared to the PMOS loads in a cell with unit driving capability. As the NMOS switching devices are in subthreshold region, there is no need to scale these devices with scaling the driving current. Therefore, the area of each cell will be scaled approximately proportional to the driving strength, and hence the cell area for large driving capabilities such as ×16 and ×32 could be very large.

4.3.1 Series–Parallel Tail Bias Transistors

To mitigate this problem, two different approaches have been proposed in this work. Based on the first approach, a combination of parallel and series NMOS transistors have been used to scale the tail bias current. As depicted in Fig. 4.5, as an example, the cell with driving strength of ×4 uses a single transistor to generate the proper tail bias current. To increase the bias current and hence the driving strength, parallel transistors can be used. On the other hand, to scale down the bias current, NMOS transistors could be put in series. In this way, the cell with driving strength of ×4 has the minimum tail transistor area occupation while the area of this part of circuit increases for higher and lower driving strengths.

As can be seen, the ratio between maximum and minimum areas for tail bias transistors in this approach is four instead of 16 in conventional approach shown in

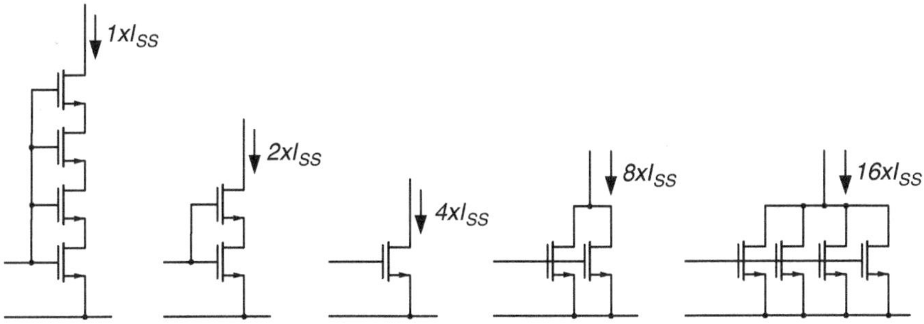

Fig. 4.5 Scaling the tail bias current using parallel and series configurations

Fig. 4.4. It is clear that this approach is less efficient in a design that cells with low driving strengths are mostly used. Here, the reference cell with one single tail bias transistor has been selected to be the gate with driving strength of ×4, however; the reference driving capability can be changed and selected properly with respect to the design issues.

As explained before, the size of NMOS switching network transistors can be kept constant for different driving strengths. Also, the PMOS transistors need to be scaled similar to the conventional approach shown in Fig. 4.4.

4.3.2 Constant Area Scaling

Figure 4.6 describes the second approach for scaling the driving strength. In the proposed approach, the size of all the devices in a cell and hence the area of a specific cell with different driving strengths are kept constant. Therefore, there is no area penalty by scaling the driving strength. To scale the driving strength in this approach, the bias voltage of NMOS and PMOS devices will be changed. Regarding the required driving strength, the bias voltages, V_{BN} and V_{BP}, need to be connected to the proper nodes as illustrated in Fig. 4.6. For example, in Fig. 4.6, driving strengths of ×16, ×2, and ×4 are implemented only by connecting the corresponding V_{BN} and V_{BP} nodes to the appropriate voltages.

This approach is very area efficient since the area of the cells remains unchanged with scaling of driving strength. The main penalty that should be paid is the need for extra routing of the different V_{BN} and V_{BP} voltages, which also require some extra effort and some more area.

In the next section, some test circuits implemented based on these two approaches will be demonstrated. The test libraries are implemented in 0.18-μm and 90-nm CMOS technologies. In each case, the performance of STSCL test circuit is compared to the corresponding CMOS implementation.

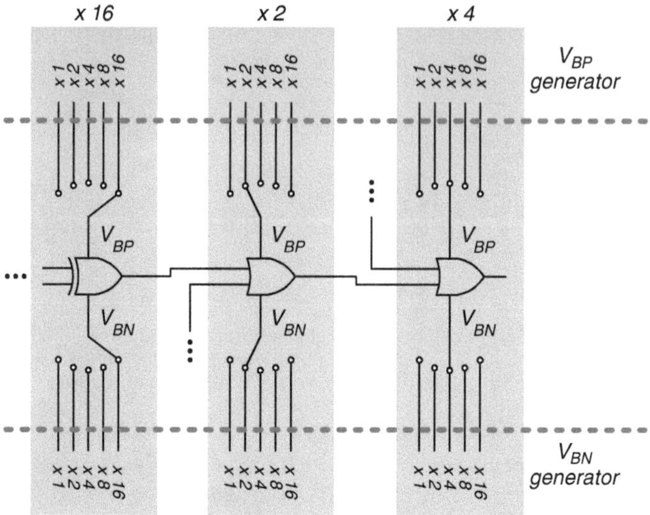

Fig. 4.6 Scaling driving strength by changing the bias voltages

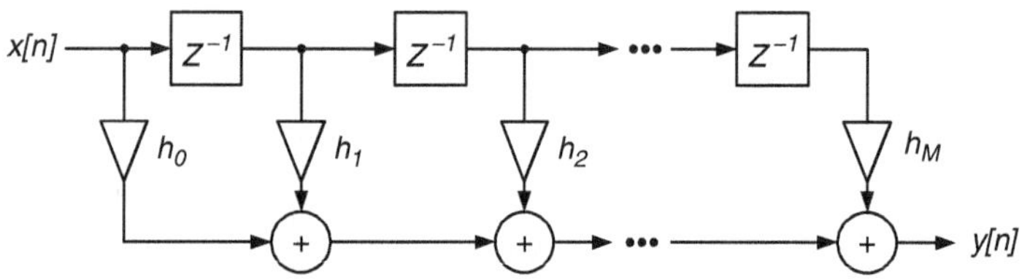

Fig. 4.7 Signal flow graph of an FIR filter with $N = M + 1$ taps

4.4 Demonstration Circuits

4.4.1 FIR Filter Topology

Finite impulse response (FIR) topology is one of the popular types of filters used in digital signal processing systems. Each FIR filter consists of one or multiple delay elements, multipliers, and adders. The output is the sum of delayed inputs multiplied with their respective filter coefficients. The following equation describes the operation of an FIR filter with $N = M + 1$ taps [6]:

$$y[n] = \sum_{i=0}^{M} x[n-i] \cdot h_i \tag{4.1}$$

where, $y[n]$ is the output at moment n, h represents the filter coefficients, and x is the sequence of the input samples. The corresponding signal flow graph of this filter is shown in Fig. 4.7.

4.4.2 Sample FIR Filter Demonstrator Circuit

An 8-bit, 9-tap low-pass FIR filter is synthesized to verify functionality of the STSCL cell libraries. The specifications of the filter are given in Table 4.1. The sampling frequency of the filter is chosen to be low since the cells in the library are characterized for a very low bias current (here: $I_{SS} = 100\,\text{pA}$). By scaling the bias current, the sampling frequency can be scaled up. This filter is designed to have more than 30 dB attenuation in the stop-band.

4.4.2.1 FIR Filter in CMOS 0.18 μm

The STSCL standard cell library that has been developed in 0.18-μm CMOS technology is based on the technique introduced in Fig. 4.5. In this approach, the tail bias transistors are configured in parallel and series structured in order to balance the cell area in different driving strengths. The layout of the inverter cells with different driving strengths developed based on this technique are shown in Fig. 4.8. As depicted in this figure, the cell area remains fairly constant for driving strengths of up to ×4. This is mainly because of reducing the number of NMOS tail bias transistors and at the same time increasing the number of PMOS load devices moving from ×1 towards ×4. The area slightly increases for driving strength of ×8 and almost doubles for ×16. The area ratio between the maximum and minimum driving strengths is about 2.

Since the NMOS switching devices are biased in subthreshold regime, their size is kept constant for different driving strengths. Long devices have been used for tail bias transistors to ensure having acceptable current matching among the cells. Also, larger than minimum size PMOS devices have been used to reduce the mismatch on the output voltage swing.

Figure 4.9 shows the layout of the implemented FIR filters based on STSCL and CMOS topologies. The area of STSCL circuit is about 3.5 times larger than the CMOS one. The larger area of STSCL circuit is mainly because of inherently larger cells in STSCL library compared to the CMOS one. Meanwhile, CMOS library benefits a large variety of different components while the proposed STSCL library has very limited number of elements. Also, a good portion of the total area belongs

Table 4.1 Specifications of the FIR filter

Specification	Value
Type	Low pass
order	8
Number of taps	9
Cut off frequency	10 Hz
Sampling frequency	100 Hz
Signal resolution	16 bits
Coefficient quantization	8 bits
Stop-band attenuation	30 dB

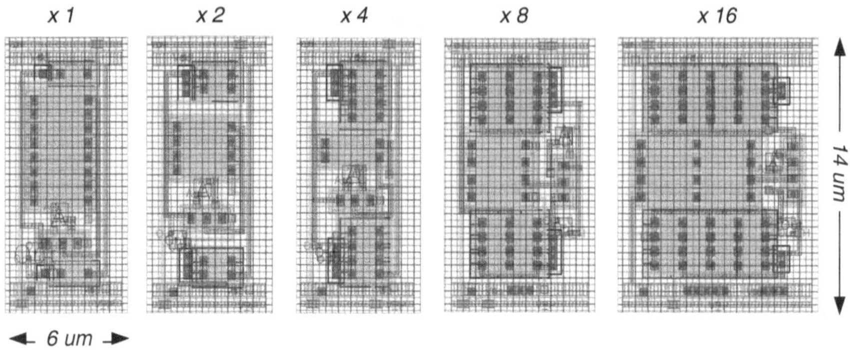

Fig. 4.8 The layout of STSCL buffer/inverter gates with different driving strengths in CMOS 0.18 μm [2–5]. To scale the driving strength of a cell, number of parallel PMOS loads needs to be increased proportional to the driving strength. Also, the number of series NMOS tail bias transistors needs to be reduced up to driving strength of ×4, and then for higher current driving, the number of parallel NMOS devices needs to be increased

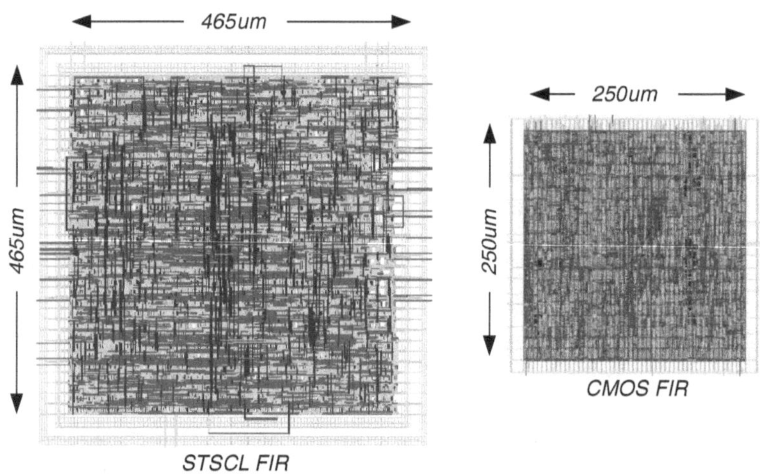

Fig. 4.9 The layout of the proposed FIR filter implemented in CMOS 0.18 μm technology based on STSCL and CMOS topologies

to the DFFs (about 60% of this circuit). Thus, making more area efficient flip-flops or using memory cells instead of flip-flops for storing the intermediate results can help to reduce the area of this circuit considerably.

In the STSCL layout, in addition to the supply rails (V_{DD} and V_{SS}), two extra rails for bias voltages, V_{BN} and V_{BP}, have been created that can be seen in Fig. 4.9.

Figure 4.10 shows the post-layout simulation results for the two FIR filters in CMOS 0.18 μm. As shown in Fig. 4.10a, the power consumption of both CMOS and STSCL circuits are very well matched with the estimated values shown with the

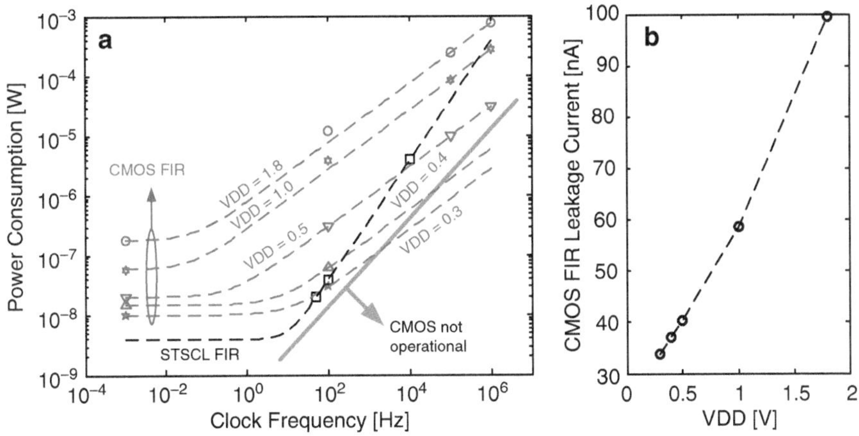

Fig. 4.10 (**a**) Simulated power consumption versus operation frequency of the STSCL and the CMOS FIR filters in 0.18 μm CMOS. Dashed lines are representing the estimated power consumption based on the methodology introduced in Chaps. 2 and 5. Here, the supply voltage of STSCL circuit is set to be 0.5 V. (**b**) Simulated leakage current of the CMOS FIR filter in different supply voltage values

dashed lines.[6] Based on these simulation results, STSCL FIR filter consumes less power for clock frequencies less than 10 kHz. It is expected that in more advanced CMOS technologies where leakage current is more pronounced, the comparison becomes in favor of STSCL topology even in higher clock frequencies. While the minimum total bias current for STSCL circuit is about 8 nA, in CMOS FIR filter the leakage current is between 35 nA (at $V_{DD} = 0.3$ V) and 100 nA (at $V_{DD} = 1.0$ V), as illustrated in Fig. 4.10b.

4.4.2.2 FIR Filter in CMOS 90 nm

The standard-cell library that has been developed for CMOS 90 nm is based on the constant area scaling technique illustrated in Fig. 4.6. Here, a single cell for different driving capabilities has been used. To scale the driving strength, bias voltage of NMOS tail bias transistors and correspondingly bias voltage of PMOS load devices need to be connected to appropriate voltage levels.

Figure 4.11 shows some of the cells that have been developed for this library. The height of all the devices is set to be 5 μm. In the design of the cells, relatively large size devices have been used in order to keep the noise margin of the cells on an acceptable level even in the presence of device mismatch.

[6] The methodology used to estimate the power consumption versus operating frequency for CMOS and STSCL topologies are explained in Chaps. 2, and 5.

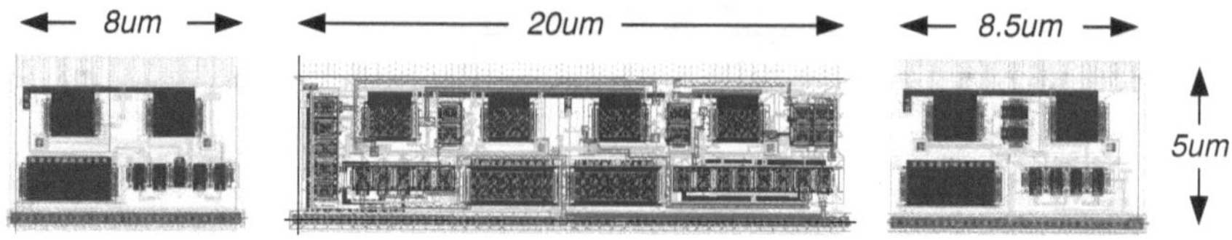

Fig. 4.11 Layout of AND2, full adder (FA), and XOR2 (from left to right) implemented in CMOS 90 nm. The same cell is used for different driving capabilities

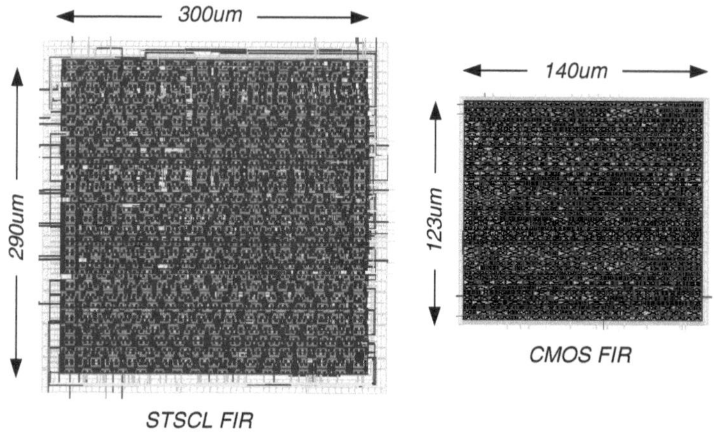

Fig. 4.12 Layout of the proposed FIR filter implemented in CMOS 90 nm using STSCL (*left*), and CMOS (*right*) topologies

The FIR filter that has been implemented based on this library is shown in Fig. 4.12 in comparison to the same circuit implemented based on CMOS topology. The area of the STSCL circuit is 5 times larger than the CMOS one. Scaling from 0.18-μm to 90-nm technology helps to reduce the size of CMOS FIR circuit by a factor of three, while this ratio is two for a STSCL circuit. Of course, since two different techniques have been used for implementing STSCL FIR filters in these two technologies, the area scaling of STSCL circuit cannot be fairly compared.

4.5 Conclusion

In this chapter, two different approaches for implementing STSCL cell libraries have been proposed. The goal in these two approaches has been to implement very small area cells with reduced area overhead due to the scaling of driving strength of the cells.

The standard cell libraries have been implemented in 0.18-μm and 90-nm CMOS technologies. The library in 0.18 μm is based on the first approach in which series–parallel configurations have been used for the tail bias transistors to have a balanced cell area in different driving strengths. A different approach has been developed for implementing standard cell library in 90nm technology. Based on this approach, the same cell is used for all different driving strengths. In this case, different bias voltages need to be applied to different driving strengths. Therefore, the cell area does not change with the driving strength.

An 8-bit, 9-tap low-pass FIR filter has been implemented using both STSCL libraries, and their performance and area are compared to their CMOS counterparts.

References

1. S. Badel, "MOS current-mode logic standard cells for high-speed low-noise applications," *PhD Dissertation*, Ecole Polytechnique Fédérale de Lausanne (EPFL), Switzerland, 2008
2. M. Beikahmadi, "Developing a standard cell library for subthreshold source-coupled logic," *Master Thesis*, Ecole Polytechnique Fédérale de Lausanne (EPFL), Switzerland, 2009
3. P. Vietti, "Design of MCML standard-cell library and differential routing methodology," *Master Thesis*, Ecole Polytechnique Fédérale de Lausanne (EPFL), Switzerland, 2007
4. C. Cakir, "STSCL standard library cell design," *Internship Report*, Ecole Polytechnique Fédérale de Lausanne (EPFL), Switzerland, 2008
5. B. Erbagci, "Performance comparison study between STSCL and CMOS logic styles," *Internship Report*, Ecole Polytechnique Fédérale de Lausanne (EPFL), Switzerland, 2008
6. A. V. Oppenheim and R. W. Schafer, *Discrete-Time Signal Processing*, Prentice-Hall, Second Ed., 1999
7. E. Brunvand, *Digital VLSI Chip Design with Cadence and Synopsys CAD Tools*, Addison-Wesley, 2009

Chapter 5
Subthreshold Source-Coupled Logic Performance Analysis

5.1 Introduction

Unlike conventional digital CMOS circuits where there is no static power consumption (neglecting the leakage current), in source-coupled logic (SCL) topology each cell consumes a specific amount of constant bias current. During each transition, this current is charging or discharging the load capacitance. Hence, more static bias current directly translates into faster transitions at the output nodes of an SCL circuit. When there is no transition at the input of an SCL gate, on the other hand, the static bias current of the gate is only used to preserve the output voltage levels on the desired values. Therefore, there is specific amount of static power consumption even during static operating conditions which is not used for processing purpose. Regarding that, as the circuit activity rate or duty rate reduces, the power efficiency of SCL topology degrades quickly. Under these conditions where the activity rate is low, CMOS circuits can offer a better power compromise.

This argument is correct while the static power consumption of CMOS topology is negligible. In advanced ultra-deep sub-micron (UDSM) technologies, however, the static power consumption of CMOS logic circuits constructs a considerable part of the total power consumption, and hence it cannot be ignored anymore. Therefore, under certain conditions subthreshold SCL (STSCL) topology with very low bias current levels can be preferred for having a better power efficiency.

In the rest of this chapter, an extensive analytical comparison between CMOS and STSCL topologies will be provided. Based on this analytical approach, the conditions that the STSCL topology offers a better power efficiency compared to the CMOS topology have been exploited. In addition, several techniques have been introduced to improve even more the power–delay performance of STSCL circuits. In each case, experimental results have been provided to show the benefits of these techniques in practice.

A. Tajalli and Y. Leblebici, *Extreme Low-Power Mixed Signal IC Design: Subthreshold Source-Coupled Circuits*, DOI 10.1007/978-1-4419-6478-6_5,

5.2 Comparison with the CMOS Topology

Comparing the performance of static CMOS and STSCL topologies in a general form is very difficult. Here, a simple test structure is used for comparing these two topologies while the argument can be generalized to more complicated systems. In the following section, the proposed approach is explained step by step. Since the main goal of this work is implementing ULP systems, first a brief review on the main challenges for controlling the dynamic and the static power consumption of CMOS digital circuits is provided.

5.2.1 Ultra-Low-Power Requirements

To optimize the power consumption of digital systems implemented based on static CMOS topology, different approaches have been proposed in literature [1]. These techniques, e.g., multiple threshold voltage devices or various power management techniques, can be used to optimize the system power dissipation with respect to the work load [1,2].

In ultra-low power applications, where the power dissipation is a crucial parameter, supply voltage (V_{DD}) is generally reduced below the threshold voltage (V_T) of MOS devices [3,4]. Reducing the supply voltage or choosing high threshold voltage (HVT) devices results in a smaller effective gate voltage, $V_{eff} = V_{DD} - V_T$, and hence less dynamic power consumption [5]. At the same time, lower supply voltage helps to reduce the subthreshold and gate oxide leakage currents. However, reducing V_{eff}, reduces the ratio of the on current of a logic gate (I_{ON}) to its leakage current (I_{OFF}) as shown in Fig. 5.1 for different process corners. Reduction in

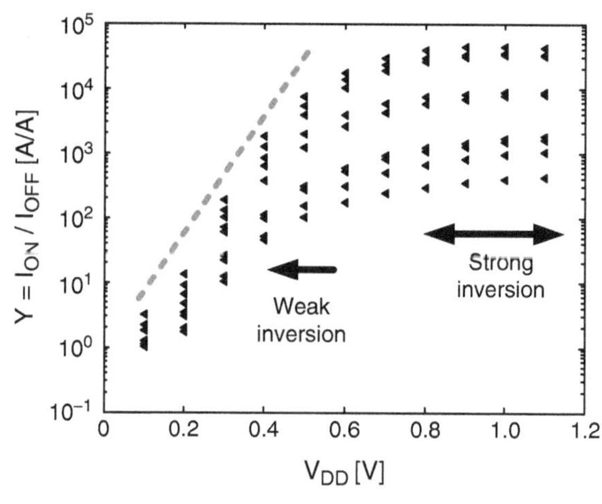

Fig. 5.1 Simulated turn-on to turn-off current ratio ($\gamma = I_{ON}/I_{OFF}$) of a static CMOS inverter gate implemented in 65-nm CMOS technology in different corner cases

$\gamma = I_{\mathrm{ON}}/I_{\mathrm{OFF}}$ results in degradation of reliability and power efficiency of the circuit, and hence special design techniques are required to implement reliable logic circuits [4].

Wide variation of circuit characteristics, such as speed of operation, noise margin, and power dissipation, due to the process, supply voltage, and temperature (PVT) variation is the other important issue in design of ultra-low power digital circuits in modern nanometer scale technologies [6]. The effects of this type of variations becomes more evident when the devices are biased in subthreshold regime for having less power consumption. Operating in this regime, I–V characteristics of devices are exponential and hence any small variation on threshold voltage can change the current levels considerably. For this reason, most of the time operating in subthreshold regime is avoided. Figure 5.1 depicts the variation of γ for different process corner parameters using CMOS 65 nm technology.

In addition, in subthreshold regime the operation frequency and power consumption both depend exponentially on the supply voltage. Therefore, a very accurate control on V_{DD} is required [7]. The design of such high-precision supply voltage control systems becomes more challenging in for example battery operated systems where the power budget is very restricted, and also battery voltage reduces by time.

Subthreshold source-coupled logic (STSCL) topology is an alternative approach for implementing ultra-low power circuits [8, 9]. The accurate control on the power consumption of each gate makes this topology very suitable for operating at very low bias current levels where the conventional static CMOS circuits are limited by their subthreshold leakage current. Meanwhile, the gate delay in this configuration does not depend on supply voltage, and hence there is a very low sensitivity to the supply voltage variations. The sensitivity to the PVT variations is also much less in this type of circuits compared to the static CMOS topology, as will be shown later.

5.2.2 Power-Speed Tradeoff in STSCL

As mentioned before, each STSCL gate draws a constant bias current of I_{SS} from supply source (Fig. 3.9). Therefore, the power consumption of each STSCL gate can be calculated by

$$P_{\mathrm{diss,STSCL,1}} = V_{\mathrm{DD}} I_{\mathrm{SS}}. \tag{5.1}$$

Meanwhile, the time constant at the output node of each STSCL gate, i.e.,

$$\tau = R_L \cdot C_L \approx \frac{V_{\mathrm{SW}}}{I_{\mathrm{SS}}} \cdot C_L \tag{5.2}$$

is the main speed limiting factor in this topology (C_L is the total output loading capacitance). Based on (5.2), one can choose the proper I_{SS} value to operate at the desired frequency. Since the power consumption and delay of each gate only depend on I_{SS} which can be controlled very precisely, this circuit exhibits very low sensitivity to the process variations. Meanwhile, since the speed of operation in this case does not depend on device threshold voltage, it is not necessary to use special process options to have low threshold voltage devices as frequently used for static CMOS circuits. Shown in Fig. 3.12, it can be seen that the gate delay is adjustable in a very wide range proportional to the tail bias current. It is also noticeable that the tail bias current can be reduced to about 10 pA where the forward bias current of the source-bulk diode of the PMOS load devices becomes comparable to I_{SS}.

Considering (5.1), it can also be concluded that the power consumption is constant and independent of the operation frequency or delay. Therefore, it is necessary to use the STSCL circuits at their maximum activity rate to achieve the maximum achievable efficiency. It is also important to note that the gate delay does not depend on the supply voltage while it varies with the tail bias current linearly. This property can be exploited for applications in which the supply can vary during the operation.

Based on (5.1) and (5.2), power–delay product (PDP) of each gate can be approximately calculated by

$$PDP_{\mathrm{STSCL},1} \approx \ln 2 \times V_{\mathrm{DD}} V_{\mathrm{SW}} C_L \tag{5.3}$$

which is directly proportional to the supply voltage, the voltage swing at the output of the gate, V_{SW}, and the total load capacitance. To have a better understanding of the power-speed tradeoff in STSCL configuration, consider a simple STSCL circuit constructed of N cascaded identical gates (indeed, N is the logic depth) that is operating at frequency of f_{op}. Using (5.1) and (5.2), it can be shown that the total power consumption of this chain is:

$$P_{\mathrm{diss,STSCL},N} \approx \ln 2 \times N^2 V_{\mathrm{DD,STSCL}} V_{\mathrm{SW}} C_L f_{\mathrm{op}} \tag{5.4}$$

which is increasing quadratically with the logic depth and linearly with the operation frequency.

5.2.3 *Performance Analysis of CMOS Logic Circuits*

Static CMOS topology has been widely used for implementing digital systems for different applications and different specifications [10]. The main concentration in this section is to study the performance of this topology and developing a proper analytical description to compare this topology with STSCL topology.

The required power consumption of a chain of N STSCL gates operating at a frequency of f_{op} was calculated in 5.4. Similar to that case, consider a chain of identical CMOS gates. As shown in Chap. 1, the total RMS power consumption of a chain of CMOS gates can be calculated by:

$$P_{\mathrm{diss,CMOS},N} = V_{\mathrm{DD}} \cdot \sqrt{\frac{1}{T}\int_0^T i_{\mathrm{DD}}^2(t)\mathrm{d}t}. \tag{5.5}$$

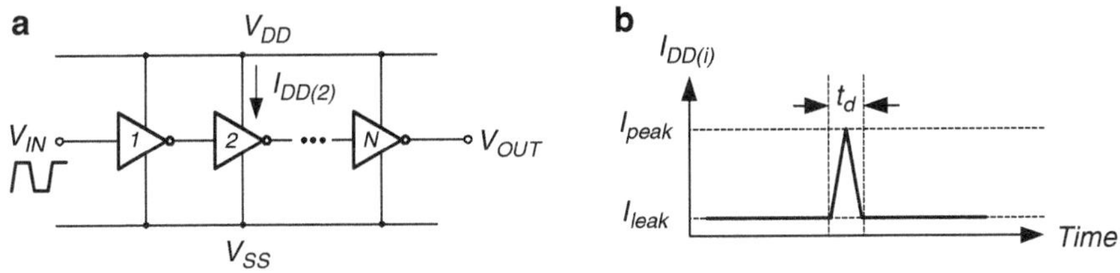

Fig. 5.2 (**a**) A chain of CMOS gates with logic depth of N. (**b**) Current drawn from supply source by one of the gates

Fig. 5.3 Power consumption of a chain of CMOS gates versus activity rate (α)

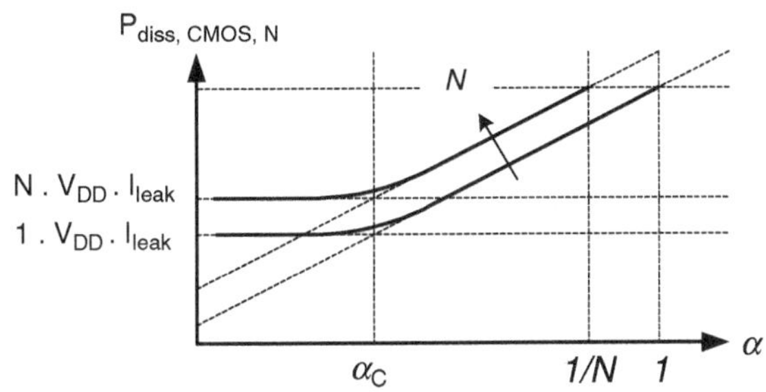

Regarding Figs. 5.2a and b, the total rms power consumption of the circuit is:

$$P_{\text{diss,CMOS},N} \approx N I_{\text{leak}} V_{\text{DD}} \sqrt{1 + \frac{\alpha \cdot \eta}{6} \left(\frac{\gamma^2}{N^2} + \frac{\gamma}{N} - 2 \right)} \tag{5.6}$$

where, $\alpha = f_{\text{op}}/f_{\text{Max}}$ represents the activity rate of the proposed circuit, $f_{\text{Max}} = 1/(2t_d)$ is the maximum operation frequency of a single gate, $\gamma = I_{\text{peak}}/I_{\text{leak}}$, $f_{\text{op}} = 1/T$, and $\eta = [(N+1)/2]$. As expected, the minimum power consumption of the circuit is determined by the leakage current when activity rate is very low ($\alpha \approx 0$). At higher operating frequencies where the dynamic power consumption becomes dominant, the power dissipation is proportional to the square root of the operating frequency or activity (duty) rate.

Figure 5.3 illustrates the power consumption versus speed of operation (or activity rate) as predicted by (5.6). By increasing the logic depth, the total power consumption scales up proportionally while the maximum speed of operation reduces by the same factor. Based on (5.6), it can be found that for activity rates smaller than "critical activity rate" which is defined by

$$\alpha_C \approx \frac{6N}{\eta \cdot \gamma^2} \approx \frac{12}{\gamma^2} \tag{5.7}$$

the subthreshold leakage power consumption is dominant, while for higher activity rates, the dynamic power consumption constructs the main part of power consumption. Since α_C is proportional to: $1/\gamma^2 = (I_{\text{leak}}/I_{\text{peak}})^2$, it increases

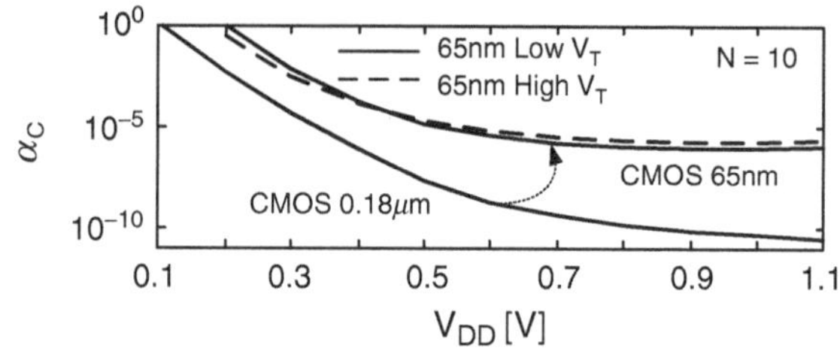

Fig. 5.4 Variation of the critical activity rate (α_C) as a function of V_{DD} for different technology nodes

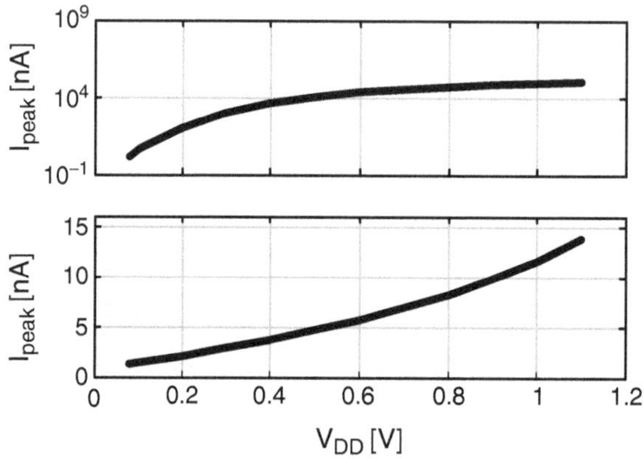

V_{DD}	γ
0.2	4.4
0.4	369
0.6	2607
0.8	5686
1.0	7582

Fig. 5.5 Peak current and leakage current of a CMOS inverter gate as a function of V_{DD} in 65-nm technology

quadratically with reducing γ. This means that in more advanced CMOS technologies, the contribution of leakage current will be more dominant and hence α_C will be higher. As illustrated in Fig. 5.4, α_C increases considerably by moving towards technologies with smaller feature sizes. While in 0.18-μm CMOS technology $\alpha_C \approx 10^{-4}$ for $V_{\text{DD}} = 0.2\,\text{V}$, it increases by almost four orders of magnitude at 65-nm CMOS technology with the same supply voltage. As depicted in this figure, even using high V_T devices does not help very much to reduce α_C.

Based on Fig. 5.2b, the maximum operating frequency of a CMOS gate (f_{Max}) can be estimated by:

$$f_{\text{Max}} \approx \frac{I_P}{2V_{\text{DD}}C_L}. \tag{5.8}$$

Although this is a simplified relationship, it can predict f_{Max} with a good accuracy. To complete the calculations, it is necessary to estimate the peak and leakage currents. The EKV model can provide a general expression for drain current of MOS devices operating in different regions and different supply voltages [11]. Using the EKV model, it is possible to calculate the peak and leakage currents in $|\ V_{\text{GS}}\ | = V_{\text{DD}}$ and $|\ V_{\text{GS}}\ | = 0\,\text{V}$, respectively.

Figure 5.5 depicts the peak and leakage currents for a CMOS inverter gate designed in 65-nm technology. It is noticeable that the leakage current does not reduce considerably by reducing the supply voltage when the devices operate in subthreshold. This implies that reducing the supply voltage does not help very much to reduce the leakage current.

The other important parameter is $\gamma = I_{\text{peak}}/I_{\text{leak}}$ which is an indicator of the power efficiency in CMOS topology. While $\gamma \approx 10^4$ for $V_{\text{DD}} > 0.6\,\text{V}$, it reduces rapidly when the supply voltage is reduced and ultimately it gets close to unity for very low supply voltages.

In addition to (5.6), the EKV model provides the necessary information in order to estimate the power consumption versus speed of operation in CMOS topology.

5.2.4 Performance Comparison

Using (5.4) and (5.6), it is possible to compare the power consumption of two chains of identical gates with logic depth of N that are constructed based on CMOS and STSCL topologies. Based on this comparison, the maximum logic depth for which the STSCL topology exhibits lower power consumption compared to the CMOS topology, is:

$$N_{\max} \approx \begin{cases} \dfrac{I_{\text{leak}} V_{\text{DD}}}{\ln 2 V_{\text{SW}} C_L f_{\text{op}} V_{\text{DD,STSCL}}} & \text{if } \alpha << \alpha_C, \\ \sqrt[3]{\dfrac{\alpha}{6} \cdot \left(\dfrac{I_{\text{peak}} V_{\text{DD}}}{\ln 2 V_{\text{SW}} C_L f_{\text{op}} V_{\text{DD,STSCL}}} \right)^2} & \text{if } \alpha >> \alpha_C. \end{cases} \tag{5.9}$$

where V_{DD} is the supply voltage of CMOS circuit.

Figure 5.6a compares the power consumption of CMOS and STSCL XOR gates for logic depth of 20 as a function of operation frequency based on simulation results in CMOS 65 nm. It can clearly be seen that the power consumption of CMOS gates cannot be reduced beyond a certain level due to leakage (both for LVT and HVT case), whereas the STSCL topology offers smaller power consumption below the cross-over frequency points.

The maximum logic depth for which an STSCL circuit with operating frequency f_{op} consumes less power compared to its CMOS counterpart, is shown in Fig. 5.6b, for 65-nm CMOS technology. The comparison has been made for XOR gates for both HVT and LVT devices. As expected, increasing the logic depth reduces the efficiency of the STSCL topology. However, at low CMOS supply voltages which is intended for low operation frequencies and where the leakage current is more evident, STSCL starts to exhibit better performance. This can be also concluded from (5.9).

At high frequencies, $N_{\max}$ grows with activity rate. This means that STSCL (or SCL) topology needs to be employed in high activity rates. On the other hand, Fig. 5.6 and (5.9) imply that as the operation frequency reduces, $N_{\max}$ increases and hence power efficiency of STSCL increases in comparison to CMOS. In other words, in nanometer scale technologies where subthreshold leakage current

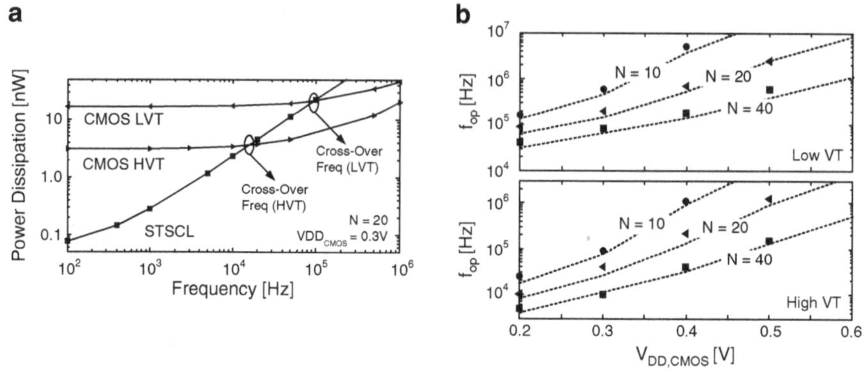

Fig. 5.6 (**a**) Simulated power consumption versus operation frequency for CMOS and STSCL XOR gates with logic depth of $N = 20$. Note that CMOS power consumption cannot be reduced beyond a certain level due to leakage. (**b**) Maximum logic depth for which STSCL topology exhibits less power consumption compared to the CMOS topology based on (5.9) (*dashed lines*) in comparison to the simulation results. The results are shown for both low V_T (*top*) and high V_T devices (*bottom*) in 65-nm CMOS technology. XOR logic gates are used for this comparison. Here, $V_{\mathrm{DD,STSCL}} = 400\,\mathrm{mV}$ and $V_{\mathrm{SW}} = 200\,\mathrm{mV}$

in CMOS topology is more evident, STSCL topology can offer a more power efficient solution, even at low activity rates (or equivalently, for higher logic depths). This is in addition to the superior power–delay performance of SCL topology at very high activity rates or very high frequencies [9].

Figure 5.6b also shows that with HVT devices the power efficiency of CMOS topology improves. However, the main issue with HVT devices is that they can not be used in very low supply voltages mainly because of the reliability issues.

Figure 5.7 shows the measurement results for two (8×8) array multipliers designed based on CMOS and STSCL topologies (see Chap. 3). The test circuits are implemented in 0.18-μm CMOS technology where the leakage current of CMOS circuitry is much less than in CMOS 65 nm. As depicted in Fig. 5.7, for frequencies below 80 kHz, the STSCL topology consumes less power and exhibits less variations due to the process and temperature differences. As predicted in Fig. 5.6, it is expected that in more advanced technologies, the cross-over frequency increases.

5.3 Performance Improvement Techniques

In the last section, the general behavior of STSCL and CMOS topologies have been compared. The comparison has been made using simple STSCL topology. In the following sections, some techniques are proposed to improve the performance of this type of circuit and reduce the circuit power–delay product which is predicted

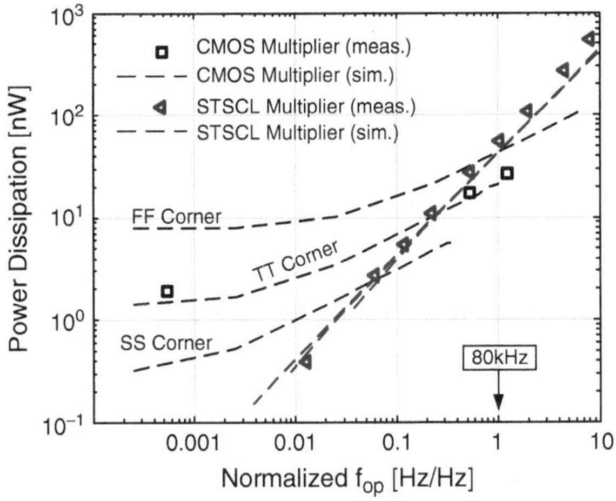

Fig. 5.7 Measured power consumption versus operating frequency for two (8×8) STSCL and CMOS array multipliers. The simulations for both topologies are plotted for different process corners and temperatures

by (5.4). Current re-use (or compound logic style), pipelining, and using output buffer are the main approaches which can be used to improve the performance of STSCL circuits.

5.3.1 *Compound Logic Style*

Compound SCL gates with merging two or more logic operations in a single gate can provide the possibility of reducing the circuit power consumption and improving the speed of operation simultaneously. Figure 5.8a shows an example in which an AND gate and an XOR gate are merged together to construct the proposed compound logic operation. Using this technique, it is possible to have only one pair of output load devices and also only one single tail bias transistor, and hence reduce the area in addition to halving the total current consumption. Assuming that the time constant at the output nodes of each SCL gate is equal to

$$\tau_L = R_L C_L = \frac{V_{SW} C_L}{I_{SS}} \tag{5.10}$$

then the total equivalent time constant of a simple N stage SCL gate will be:

$$\tau_{tot,A} \approx N \times \frac{V_{SW} C_L}{I_{SS}} \tag{5.11}$$

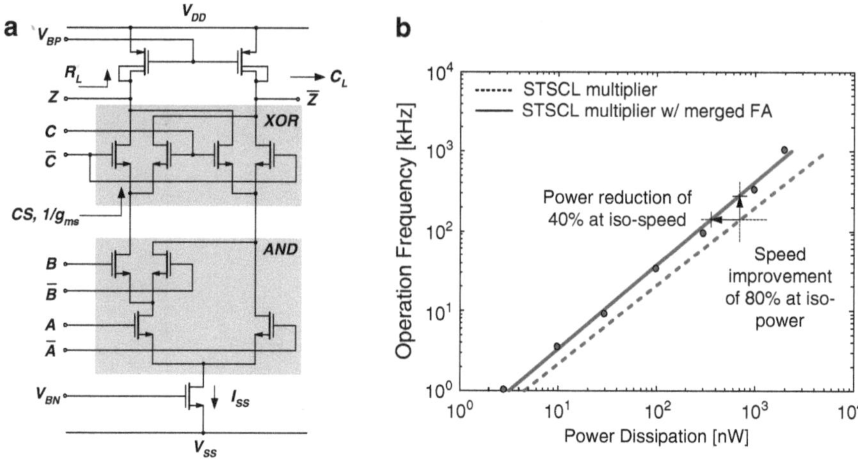

Fig. 5.8 (**a**) Compound STSCL gate (AND operation followed by XOR gate). (**b**) Performance improvement in an (8×8) multiplier circuit using compound STSCL gates

On the other side, in a compound STSCL gate with M stacked levels of NMOS differential pairs (e.g., in Fig. 5.8a: $M = 3$), the total time constant of the circuit is

$$\tau_{\text{tot},A} \approx \left(\frac{V_{\text{SW}}C_L}{I_{\text{SS}}}\right) + M\left(\frac{C_S}{g_{\text{ms}}}\right) \tag{5.12}$$

where $R_L \approx V_{\text{SW}}/I_{\text{SS}}$, $g_{\text{ms}} = I_{\text{SS}}/U_T$, and C_S is the parasitic capacitance seen from the source of each NMOS differential pair. Here, it is assumed that the time constant at the intermediate nodes of a compound SCL gate is $\tau_i = C_S/g_{\text{ms}}$ (see Fig. 5.8) and the total time constant can be calculated by $\tau_{\text{tot}} = \tau_L + \sum_{i=1}^{M} N\tau_i$ [12]. Comparing (5.11) and (5.12) it can be concluded that as far as $MU_TC_S << (N-1) \times V_{\text{SW}}C_L$, or

$$M << (N-1)\cdot\frac{V_{\text{SW}}C_L}{U_TC_S} \tag{5.13}$$

stacking differential pair stages will not degrade the speed of operation considerably. Simulations show that the proposed technique can reduce the power dissipation of an (8×8) multiplier by about 40% and at the same time improve the speed of operation. Figure 5.8b depicts this improvement for different operating frequencies.

In this approach, it is necessary to make sure that the stacking of M differential pair stages will not affect the correct current switching behavior of the circuit. In other words, with M stacked transistors, the differential pair transistors should be able to switch the current completely to one of the output branches with the specified input voltage swing, V_{SW}. As in this case, there are M series transistors, it is possible to show that the inversion coefficient, IC, of ON transistors based on the EKV model would be:

$$IC = \frac{I_{\text{SS}}}{2n_n\mu_nC_{\text{ox}}\frac{W_N}{M\times L_N}U_T^2} = \frac{M\times I_{\text{SS}}}{2n_n\mu_nC_{\text{ox}}\frac{W_N}{L_N}U_T^2}. \tag{5.14}$$

This equation implies that W_N/L_N (aspect ratio of differential NMOS transistors) should be large enough to keep their inversion coefficients small and make sure that V_{SW} is sufficient to switch the tail bias current to one of the output branches:

$$M \leq IC \cdot \frac{2n_n\mu_n C_{ox}\frac{W_N}{L_N}U_T^2}{I_{SS}} \quad (5.15)$$

which puts an upper limit on M and should be taken into account in design of stacked topologies.

5.3.2 *Using Source-Follower Buffer*

As explained in Chap. 4, for implementing a complex digital system using STSCL topology, it is necessary to build a library of different logic functions (or cells) with different driving strengths, which can then be used in a top-down synthesis flow followed by automated placement and routing [13]. To design different types of logic functions, it is possible to use binary decision diagram (BDD) configuration in the differential NMOS switching network (see Fig. 4.3). Meanwhile, to construct logic cells with different driving strengths, the tail bias current of each cell as well as the size of PMOS load devices must be scaled. This scaling needs to be proportional to the required driving strength which will scale the power consumption and also the cell area proportional to the driving strength. Therefore, to achieve larger driving strength, each cell will have to occupy more area which in turn reduce the power efficiency of the gate because of increased parasitic capacitances.

In this section, a technique for improving the performance of STSCL circuits and that can also help simplify the design of standard-cell library will be proposed.

5.3.2.1 Proposed Topology

To avoid scaling the area of each cell proportional to the driving strength, we are proposing the configuration shown in Fig. 5.9a. Here, each STSCL gate uses a pair of simple source-follower buffers (SBFs), one each at both of its complementary outputs. The added output buffer will isolate the load capacitance from the core SCL gate. Since the output impedance of the source-follower buffer $(1/g_{m6,s})$ is very small compared to the output impedance of SCL gate (R_L), an improvement in total gate speed is expected. On the other hand, since the load capacitance in this topology is driven by the output buffers not the core STSCL circuit, in order to have different driving strengths, it is sufficient to change only the bias current of the SFB part not the core STSCL part. This means that the core STSCL gate does not need to be scaled, and the area and power consumption of this part remains unchanged for different driving strengths.

To scale the driving capability of the output buffer, it is necessary to scale the tail bias current of the output stage (i.e., scaling I_B in Fig. 5.9a). Since the

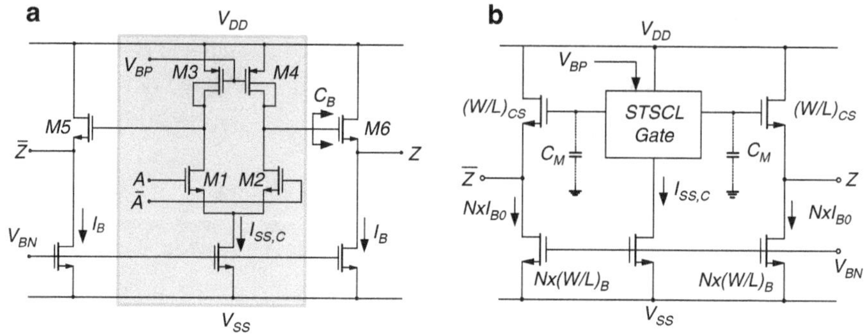

Fig. 5.9 (**a**) Generic STSCL gate uses source follower buffer at the output (SCLSFB) to improve the power–delay product of the gate. (**b**) Design of standard library cells with different driving strengths based on SCLSFB topology. C_M stands for the total parasitic capacitance seen by each output node of the STSCL core

common-drain transistors (M5 and M6) are biased in subthreshold, their size can be kept unchanged for different driving strengths. Therefore, the topology shown in Fig. 5.9a offers a more power and area efficient implementation of the STSCL cells for creating digital library cells.

5.3.2.2 Performance Analysis

The output load capacitance seen by any gate in a complex design is generally due to the interconnections and can be as high as hundreds of fF. In this case, using a simple buffer stage can relax the power–delay tradeoff in the SCL circuits considerably. As illustrated in Fig. 5.9a, in this case the SCL core drives the parasitic capacitances due to M1–M3 and M2–M4, as well as the input capacitance of the buffer stage.

Note that this capacitance is composed of the gate-drain overlap capacitance and the gate-source contribution of M5–M6, and hence it can be very small. Operating at very low bias current levels, the size of devices used in SFB can be kept small so the output stage would have a very small loading effect on the STSCL core. Therefore, the dominant time constant at the circuit topology shown in Fig. 5.9a is expected to be at the output node:

$$\tau_{\mathrm{SFB}} \approx \frac{C_L}{g_{m6,s}} \tag{5.16}$$

which is valid for small signal variations. In a real case when the output swing is in the order of several hundreds of mV, however, this equation will no longer be valid. Indeed, at each rising edge more current will flow into the proposed common-source device. In this case, the time constant of the node would even be smaller than the value predicted in (5.16). On the other hand, for falling transitions, the common-source transistor will be turned off and the only path for discharging the output node will be I_B (Fig. 5.10a). Therefore, the output will slew down with a slope of I_B/C_L. This means that the improvement predicted by (5.16) can be expected only at the

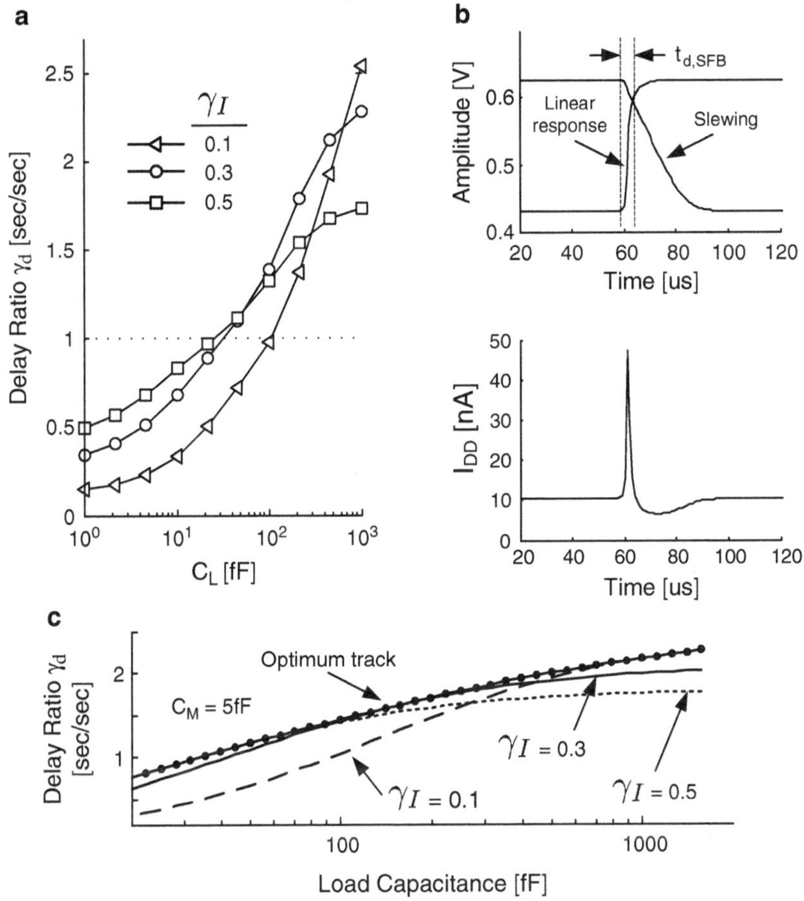

Fig. 5.10 (**a**) Total delay improvement using source-follower buffer at the output of STSCL circuit in equal total power consumption based on transistor level simulations. Data points with a delay ratio of larger than unity represent delay improvement (reduction). (**b**) Transient simulation results: output waveforms (*top*) and supply current (*bottom*) for an SCLSFB topology ($I_{SS} = 10\,\text{nA}$). (**c**) Delay reduction (γ_d) for different γ_I values compared to the $\gamma_{d,\text{Max}}$ calculated based on (5.20)

rising edges. Neglecting the delay of STSCL core and assuming typical conditions ($V_{SW} = 200\,\text{mV}$ and in room temperature), it can be shown that the slew mode will increase the total delay to

$$t_{d,\text{SFB}} \approx 1.6 \times \tau_{\text{SFB}}. \tag{5.17}$$

Here, it is assumed that M5 and M6 will turn off very quickly at the falling edges. Therefore, one output slews toward lower voltage level with a slew rate of I_B/C_L, and the other output increases toward higher voltage level by the time constant of τ_{SFB}. This assumption can be acceptable when the time constant at the output of STSCL gate is much less than the time constant at the output of SFB stage.

Including the delay of STSCL core to the total delay, and assuming $t_{d,\text{STSCL-SFB}} \approx t_{d,\text{STSCL}} + t_{d,\text{SFB}}$, it can be shown that the delay improvement (reduction) ratio is

$$\gamma_d = \frac{t_{d,\text{STSCL}}}{t_{d,\text{STSCL-SFB}}} \approx \frac{\ln 2 \cdot C_L R_{L1}}{\ln 2 \cdot C_M R_{L2} + 1.6 C_L / g_{m6,s}} \tag{5.18}$$

where, C_M is the total parasitic capacitance at the output of SCL stage as shown in Fig. 5.9b, $R_{L1} = V_{\text{SW}}/(I_{\text{SS},C} + 2I_B)$ is the load resistance of a simple STSCL gate and $R_{L2} = V_{\text{SW}}/(I_{\text{SS},C})$ is the load resistance of SCL core in Fig. 5.9a. Replacing $g_{m6,s} = I_B/(n_n U_T)$, then

$$\gamma_d = \frac{\gamma_I}{1+\gamma_I} \cdot \frac{1}{\gamma_C + \frac{3.2 n_n U_T}{\ln 2 \cdot V_{\text{SW}}} \gamma_I} \tag{5.19}$$

in which $\gamma_C = C_M/C_L$ and $\gamma_I = I_{\text{SS},C}/(2I_B)$ (see Fig. 5.9a). Here, it is assumed that the total bias current in both topologies are equal, i.e,: $I_{\text{SS}} = I_{\text{SS},C} + 2I_B$. This equation also implies that by properly choosing the value of γ_I with respect to γ_C, it is possible to achieve a balanced design for different load capacitance values. This property is especially useful for the design of digital library cell elements as will be explained later. It is also interesting to notice that for very large load capacitance values: $\gamma_d \approx 2.25/(1+\gamma_I) \approx 2.25$. Therefore, using SFBs, it is possible to reduce the delay (or PDP) of STSCL circuits by a factor of approximately 2.25 for the same amount of power consumption.

Figure 5.10a shows the total delay improvement using SFB stage at the output of STSCL gates compared to a simple STSCL gate, under the assumption that both circuit solutions are dissipating the same amount of power. The comparison is shown for different load capacitances and for different ratios of the bias currents between the core and buffers.

For low load capacitances (less than 20 fF), the simple STSCL gate without the SFB stage shows smaller total delay. However, as the load capacitance increases, the topology shown in Fig. 5.9a exhibits less delay compared to a simple STSCL gate. In complex digital systems where the output load is dominated by interconnect capacitance, an improvement in the PDP by a factor of more than two can be observed.

Figure 5.10b depicts the transient response of the circuit. While the proposed STSCL–SFB gate exhibits a considerable improvement in rising edges, the falling edge does not improve very much. This is mainly due to the fact that the source-follower stage turns off very quickly after a high-to-low input transition. Consequently, the charge on the output capacitance will be discharging by the constant bias current of I_B. The estimated value of $t_{d,\text{SFB}}$ in (5.17) which is slightly higher than τ_{SFB} is based on this behavior. This figure also shows that unlike a simple STSCL circuit, the supply current, I_{DD}, in SCLSFB topology is no more constant.

To keep the noise margin of SCLSFB gates as high as that of STSCL gates (which is about NM = 100 mV for $V_{\text{SW}} = 200$ mV) it is necessary to increase the voltage swing (V_{SW}) of the core SCL gate in SCLSFB topology. This is mainly for compensating the gain of source-follower stage which is less than unity. Since the gain of source-follower stage is very close to unity, an increase of about 10–15% on voltage swing is sufficient to compensate this effect.

The other main issue is the mismatch between the gates and replica bias circuit and also the mismatch between the source-follower buffers inside a cell. As discussed in Chap. 3, it is possible to control the mismatch effect among the gates and the replica bias circuit by proper sizing of devices and also selecting V_{SW} high enough. The size of source-follower transistors needs to be selected large enough to make sure that the offset between them does not affect the proper logic operation of the gate. This can put a lower limit on the size of devices and hence C_B in Fig. 5.9a. Minimizing C_B helps to maximize the PDP improvement as will be discussed later.

5.3.2.3 Optimized Design

In a complex digital system, the parasitic capacitance due to the interconnections will be the dominant part of the C_L, resulting in relatively high values such as $C_L >$ 30 fF. Therefore, SFB stages can improve the PDP of the complex STSCL digital circuits by a factor of two or even higher.

The choice of the output buffer topology also reflects a careful balance between circuit complexity and performance. Using a more complex output stage, can result in more improvement in delay. For example, a push–pull output stage would reduce the sensitivity to the load capacitance even further. However, in this case the circuit complexity would increase rapidly and controlling the power consumption and voltage swing would be very difficult. Using a push–pull output stage can also increase the sensitivity to the supply voltage variations.

The simple SFB stage output buffer technique can simplify the design of library cells. Based on this approach, it is sufficient to design a single logic cell and provide the required driving strength by using different SFB stages as shown in Fig. 5.9b. Illustrated as an example in Fig. 5.9b, a single STSCL boolean gate together with different SFB stages with different bias or driving capabilities can provide the required specifications. Here, $I_{SS,C}$ can be kept constant for all STSCL gates while N can be changed to achieve different driving capabilities.

Since all the devices are biased in subthreshold regime, it is sufficient to change the bias current in the SFB stage without changing the size of source follower devices (i.e., W_{CS} and L_{CS} can be kept constant) to implement different driving strengths. Therefore, the only required modification is changing the size of tail bias transistors at the output buffer stage.

It is possible to use (5.19) to determine the proper bias current for the SFB stage with respect to the load capacitance (C_L). By solving $\partial \gamma_d / \partial \gamma_I = 0$, it can be shown that the optimum value for γ_I at a given γ_C is:

$$\gamma_{I,rmopt} = \sqrt{\ln 2 \cdot V_{SW} \cdot \frac{\gamma_C}{3.2 U_T}} \qquad (5.20)$$

which indicates that for larger load capacitances (i.e., a smaller γ_C), a smaller portion of the total current budget should be dissipated in the STSCL core (i.e., smaller γ_I should be selected). Regarding (5.20), it can be also concluded that for increasing the driving capability of each gate by a factor of S, it is sufficient to keep the bias current of the core constant and to increase the bias current of the SFB stage by a factor of $\sqrt{S}$ which is always smaller than S for $S > 1$.

Defining $\eta = 3.2U_T/(\ln 2V_{\text{SW}})$ and using (5.20), the maximum improvement that can be achieved is

$$\gamma_{d,\text{Max}} = \frac{1}{\left(\sqrt{\gamma_C} + \sqrt{\eta}\right)^2} \tag{5.21}$$

Therefore, to have $\gamma_{d,\text{Max}} > 1$ (or better performance for SCL–SFB configuration compared to STSCL), then

$$C_L > \frac{C_M}{\left(1 - \sqrt{\eta}\right)^2} \tag{5.22}$$

Using the optimum value for γ_I and using nominal values in the proposed design, it can be shown that STSCL gates that are using source follower buffer show a better performance for $C_L > 11C_M$. Using minimum size devices and a compact layout, it is possible to reduce the size of C_B to only a few fF. Therefore, using a careful design strategy it is possible to have superior performance for load capacitances as low as 30 fF using SCLSFB topology. For $C_L < 11C_M \approx 30$ fF, simple STSCL topology will exhibit a comparable or better performance. However, it is not possible to have a design mixed of simple STSCL gates and SCLSFB gates in a design mainly because of voltage drop on source follower stage. Since this limit (i.e., $C_L < 11C_M \approx 30$ fF), is very low, it is expected that even in not very complex designs the proposed topology provides considerable advantages from the power-speed points of view.

Figure 5.10c shows the delay reduction factor for different load capacitance values and also for three different γ_I values. To maximize the improvement it is necessary to use different γ_I values with respect to the load capacitance as depicted by (5.20). This figure also illustrates the maximum achievable improvement in different load capacitance values and corresponding $\gamma_{I,\text{opt}}$.

5.3.3 *Pipelining Technique*

One possible approach for increasing the activity rate is to use a simple two-phase *pipelining* technique [14]. Figure 5.11 shows one possible approach to implement two-phase latch-based pipelining where the output of each gate is latched during one clock phase, and passed on to the next stage during the other clock phase, effectively reducing the maximum logic depth to two consecutive gates.

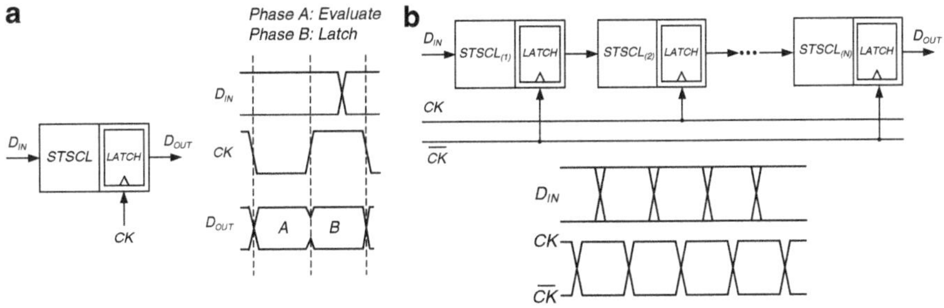

Fig. 5.11 Pipelining technique for improving the activity rate in STSCL topology. (**a**) Single stage pipelined gate and timing diagram. (**b**) Multi-stage pipelined logic

The topology of a single stage pipelined gate is shown in Fig. 5.11a. When clock is low, the latch is disabled and the gate is evaluating the output value based on the input data. In this period, as the gate is evaluating the output, the input data should remain constant.

When clock is high, on the other hand, the output is latched and the following stages can start their evaluation step. Since in this period the output of this stage is kept constant by the latch, input data can gets its new value. Therefore, the input data rate can be increased theoretically to $f_D = 1/(2t_d)$. The input data rate does not reduce if the logic depth increases (Fig. 5.11b) since during the evaluation phase of each gate, its inputs are kept constant by the latch of the previous stages, and hence does not change. Without pipelining the entire system needs to wait until all the gates in the chain complete their evaluation; hence, the maximum data rate is limited to $f_D = 1/(Nt_d)$. As a conclusion, pipelining can theoretically helps to improve the speed of operation by a factor of $N/2$.

Instead of using explicit latch stages, such two-phase pipelining can be achieved by increasing and reducing the tail bias current of alternating stages, using the gate terminal of the tail current bias transistor of each stage as the *clock* input. This can be done by applying clock signal to V_{BN} in Fig. 5.12a.

In the proposed approach, as illustrated in Fig. 5.12a for example of an STSCL full adder (FA) gate, the current bias of odd stages is reduced to a low (yet non-zero) level to retain (hold) their output while the current bias of even stages is raised to the nominal operating value to enable evaluation.

Very simple cross-coupled "keeper" stages connected to each gate output ensure that the output levels do not degrade significantly during the "hold" phase. Since the keeper stage is used to maintain the latest state of the output of each gate, it does not need to be very fast. Therefore, the bias current of keeper stage ($I_{SS,L}$) can be chosen as low as 1% of the nominal bias current of the main gate (I_{SS}). This means that the power overhead of the keeper stages is virtually negligible.

Meanwhile, since the bias current of half of the gates is almost zero in each clock phase, the overall power consumption of the system will be reduced by a factor of two. Figure 5.12b shows the transient simulation results for the output of an adder stage in a chain of adders. In this figure, it is possible to see the hold and evaluation phases for $I_{SS,L} = 0.01 I_{SS}$ at $V_{SW} = 0.2$ V.

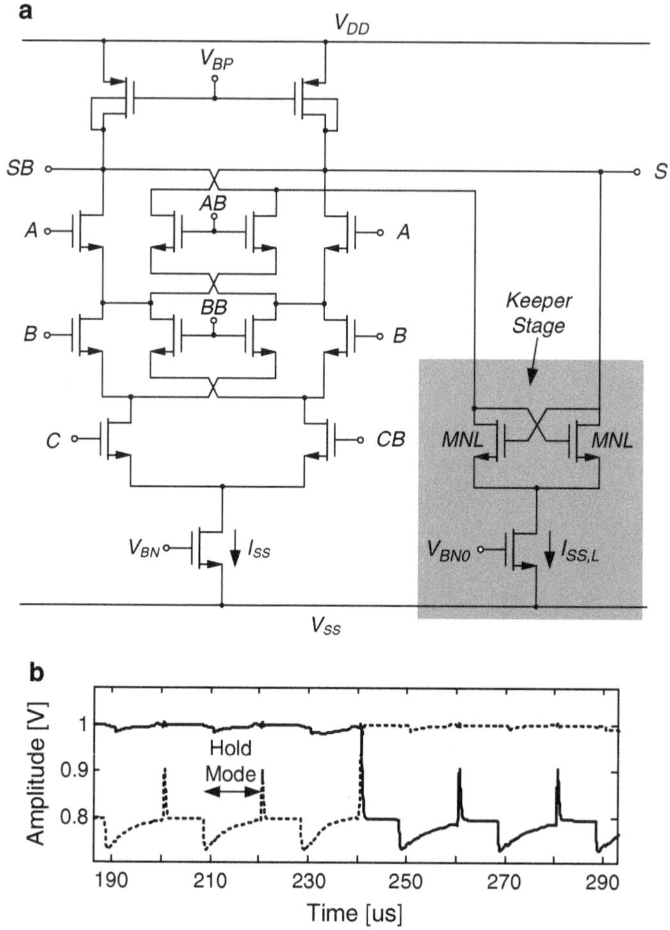

Fig. 5.12 (**a**) STSCL full adder and keeper stage. Here, the tail current bias V_{BN} is switched according to CK (or $\overline{CK}$) while V_{BNO} is kept as a constant bias. (**b**) Simulated output of the pipelined FA chain showing the holding and tracking modes of operation

Assuming that the delay of each gate is t_d, theoretically it is possible to increase the input data rate in Fig. 5.11 to approximately $1/(2t_d)$. Therefore, the power–delay product of a pipelined STSCL system can be calculated as

$$\mathrm{PDP}_{\mathrm{SCL},N,\mathrm{Pipe}} = 2\ln 2 \cdot N V_{\mathrm{DD}} V_{\mathrm{SW}} C_L. \tag{5.23}$$

Regarding (5.4) and (5.23), it can be seen that pipelining helps to reduce the system power–delay product by a factor of approximately $N/2$ which is a considerable improvement. In practice, the improvement in power–delay product is less than this value because of increased loading at the output nodes as well as power consumption of the keeper stage.

In a real case, it is necessary to switch V_{BP} according to V_{BN} value in each clock phase. In this way, when V_{BN} is low (high) and tail bias current is low (high), V_{BP} needs to be high (low) to increase (reduce) the resistance of the load devices. This can increase the complexity of the circuit with some power overhead.

5.4 Experimental Results

This section provides some experimental results to show the efficiency of the proposed techniques described in this chapter.

5.4.1 STSCL with Source-Follower Buffer

A test chip has been fabricated in a conventional digital 0.18-μm CMOS technology to verify the performance of STSCL topology with and without source-follower buffers in each stage. For this purpose, two ring oscillators have been implemented; one using simple STSCL MUX (multiplexer) gates configured as buffer stages and the other one using the same configuration where each MUX gate is followed by a source-follower buffer. Each ring oscillator has a capacitor bank to be able to change the loading capacitance in all intermediate nodes of the oscillator. In this way, it is possible to study the delay of cells for different capacitance load values. The chip photomicrograph is shown in Fig. 5.13a.

The measured PDP for the ring oscillators depends on the load capacitance and the results agree with the simulation results within ±20% accuracy. For simple STSCL based topology, the measured PDP per unit capacitance is approximately 0.125 J·F^{-1} or PDP = 0.7 fJ for $C_L = 6$ fF. The measured oscillation frequency is depicted in Fig. 5.13b. This figure also shows the simulated oscillation frequency for different temperatures. Thanks to the internal replica bias circuit, variations on oscillation frequency due to the temperature variations can be kept very low.

Figure 5.13c shows the measured delay ratio (γ_d) for the two ring oscillators in two different total bias currents of 1 nA and 10 nA per stage (i.e., the total current consumption of the ring oscillators is 8 nA and 80 nA, respectively). Both oscillators are connected to the same supply voltage and are consuming the same amount of power. In these measurements, $V_{\text{DD}} = 0.7$ V, $V_{\text{SW}} = 0.2$ V, and the total power consumption (excluding the replica bias circuit) is 5.6 nW and 56 nW for $I_{\text{SS}} = 1$ nA and 10 nA, respectively. This figure shows the results for three different γ_I values ($\gamma_I = 0.1, 0.3, 0.5$). It can be seen that the measured improvement in delay, agrees well with the analysis result derived in Sect. 5.3.2.2. The higher cross-over point (where $\gamma_d = 1$) in Fig. 5.13c compared to the analysis means that the C_M (see Fig. 5.9a) value in practice is higher than the expected value. For supply voltages lower than 0.7 V, the gain of amplifier used in the replica bias circuit starts to reduce and hence there is less precise control on the output voltage swing, in this case.

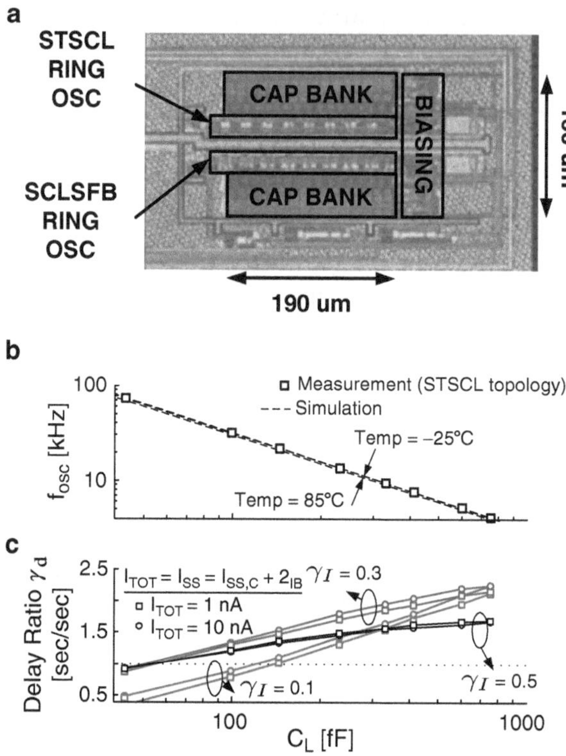

Fig. 5.13 (**a**) Photomicrograph of the test chip implemented in 0.18-μm technology. (**b**) Measured oscillation frequency of STSCL ring oscillator in comparison to the simulation results at different temperatures. (**c**) Total delay improvement for total bias current per stage of 1 nA and 10 nA. Each ring oscillator is constructed of 8 delay cells. Data points with a delay ratio of larger than unity represent delay improvement (reduction)

5.4.2 *Pipelined Adder Chain*

A test chip fabricated in digital 0.18-μm CMOS technology consisting of a 32-bit pipelined adder chain, and a conventional (non-pipelined) 32-bit ripple-carry adder as the comparison block, both designed with STSCL topology, have been used for this measurements. Figure 5.14a shows the test chip photomicrograph. Internal current mirrors are used to control the bias current of the gates and the keeper stage separately. Each adder chain is followed by an SCL-to-CMOS level converter circuit and an output driver. Two phase V_{BN} and V_{BP} signals have been generated externally. Therefore, the power and area overhead due to this part has not been included in the estimations.

Figure 5.14a, b shows the measured output of the pipelined FA chain in comparison to the input data and clock. The latency is equal to $NT_{CK}/2$ which in this figure is 320 μs. It is possible to measure the total delay in the simple non-pipelined 32-bit

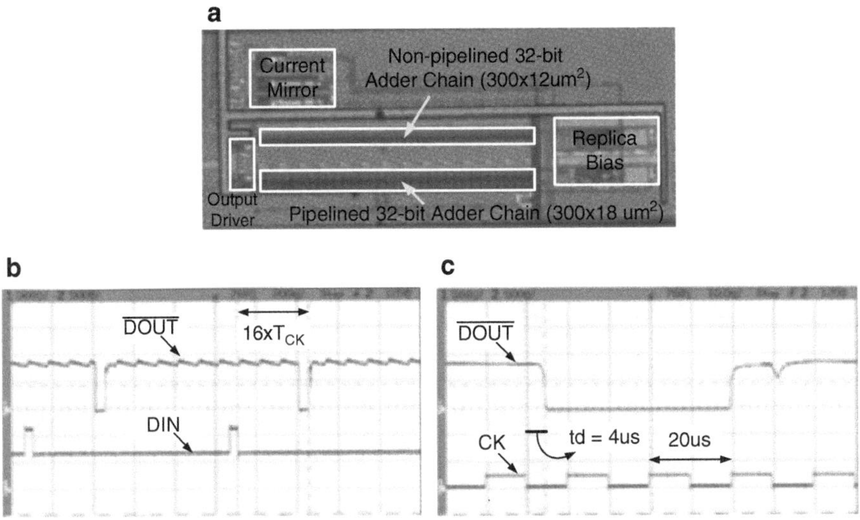

Fig. 5.14 (**a**) Test chip photomicrograph. Measured output of the pipelined full adder chain in comparison to the (**b**) input data and (**c**) reference clock. Here, $V_{DD} = 1$ V, $V_{SW} = 0.2$ V, $I_{SS} = 1$ nA

adder and also the delay of a single gate for the pipelined 32-bit adder. The measurement results are shown in Fig. 5.15a as delay versus tail bias current. The delay of both circuits can be adjusted linearly by changing their tail bias current in a very wide range which is about three orders of magnitude in these measurements. Note that the time delay between two consecutive inputs can be reduced by a factor of 14 with pipelining (maximum theoretical improvement would have been by a factor of $N/2 = 16$, as explained above).

The measured power–delay product for the two topologies are shown in Fig. 5.15b. Both topologies show a relatively constant PDP over their tuning range. The average PDP for simple and pipelined FA chains are 2.6 pJ and 0.18 pJ, respectively, which corresponds to an improvement factor of about 14.

Measurements for pipelined adder chain have been performed for two different bias current of $I_{SS,L}$: $I_{SS,L} = I_{SS}/10$ and $I_{SS,L} = I_{SS}/100$. As can be seen in Fig. 5.15b, the results for two different bias currents of the keeper stage are very close. Therefore, it is possible to reduce the bias current of the keeper stage to $I_{SS}/100$ and hence minimize the power overhead of this stage. This result is very close to the estimation made in (5.23).

5.4.3 *Pipelined Multiplier*

As already discussed, the power-to-frequency ratio of STSCL circuits (i.e., the power efficiency to operate at a given frequency) can be significantly improved by

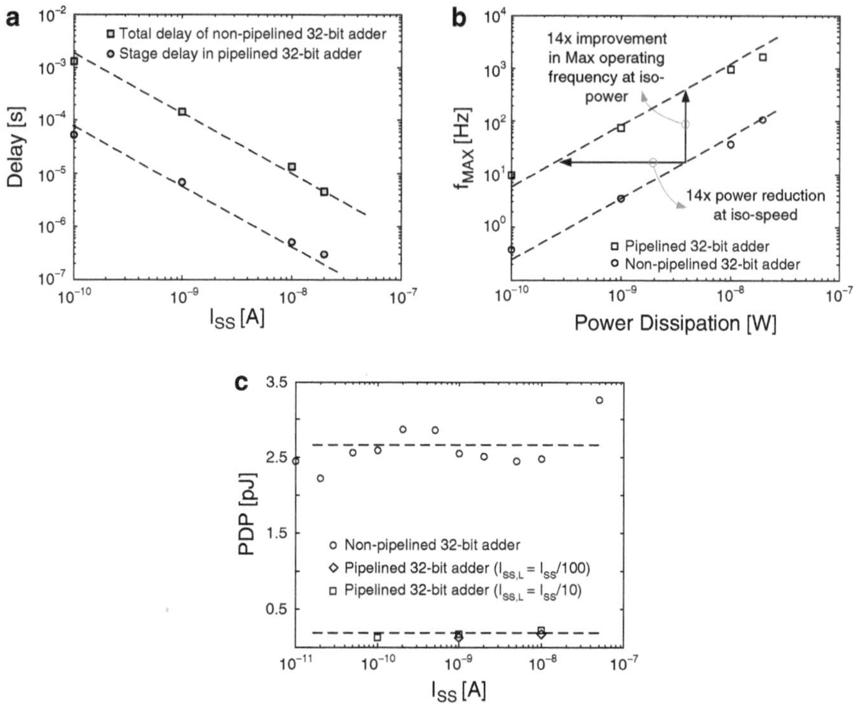

Fig. 5.15 (**a**) Measured delay versus tail bias current: total delay of simple adder chain and stage delay in pipelined adder chain. In both cases, the delay figure corresponds to the time period between two consecutive inputs. The effective operating frequency improves by a factor of 14 with pipelining. (**b**) Measured power–delay product for the two adder topologies. The pipelined adder topology achieves a very significant reduction of PDP, over a wide range of operating frequencies. (**c**) Power–frequency improvement achieved by pipelining technique

increasing the activity rate using shallow pipelining and by reducing logic depth, as much as possible. One possibility is to implement two-phase latch-based pipelining where the output of each gate is latched during one clock phase, and passed on to the next stage during the other clock phase–effectively reducing the maximum logic depth to two consecutive gates. Instead of using explicit latch stages, such two-phase pipelining can be achieved by increasing (and reducing) the source (tail) current bias of alternating stages, using the gate terminal of the tail current bias transistor of each stage as the "clock" input.

In this approach, illustrated in Fig. 5.16 for the example of the carry-save multiplier architecture, the current bias of odd stages is reduced to a low (yet non-zero) level to retain (hold) their output while the current bias of even stages is raised to the nominal operating value to enable evaluation. Very simple cross-coupled "keeper" stages connected to each gate output ensure that the output levels do not degrade significantly during the "hold" phase. Figure 5.16a shows the circuit topology of an adder (sum generator) stage and the output keeper stage, where the pulsed tail

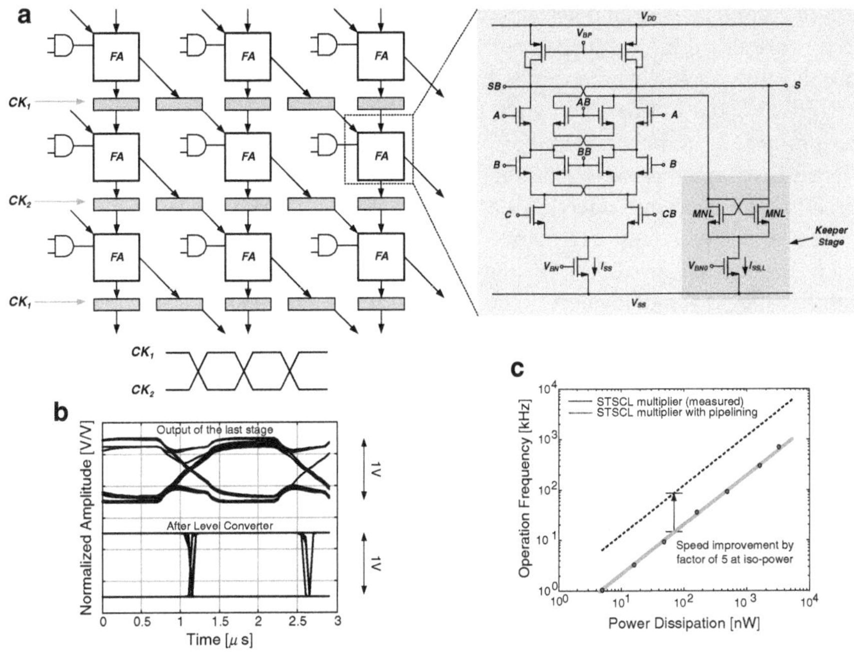

Fig. 5.16 (**a**) Section of the parallel multiplier where the signal flow is regulated using two-phase micro-pipelining technique for improving the performance of SCL gates. Note that every FA stage output is followed by a keeper/latch stage. (**b**) Eye diagram of the output of the multiplier circuit. This plot shows the output after SCL-to-CMOS level converter circuit. Input is a $2^7 - 1$ pseudo random bit stream (PRBS). Here, the period of input data is $T_p = 1.5\,\mu\text{s}$, $I_{SS} = 10\,\text{nA}$, and $I_{SS,L} = 100\,\text{pA}$; i.e., the keeper stages dissipate only 1% of the power dissipated by the FA stages. (**c**) Power–frequency improvement that can be achieved in the (8×8) carry-save multiplier circuit, by using shallow pipelining with keeper-latch stages

bias achieves a very robust dynamic latching effect, augmented by the output keeper with a tail bias current of 100 pA. In an (8×8) bit carry-save multiplier circuit, taking into account the additional power overhead of pipelining (which is 1% only), shallow pipelining using keeper-latch stages will result in an overall improvement of the (P/f) by a factor of 5 (Fig. 5.16c).

The pipelining technique described above can certainly be applied in combination with the gate-merging approach to improve the power–frequency performance of subthreshold SCL circuits considerably.

5.5 Conclusions

Source-coupled logic (SCL) circuits are traditionally used for high activity rate and high frequency applications [13, 15]. Comparing to the conventional CMOS topology, because of static power consumption of SCL circuits, their power efficiency is less in complicated digital systems where activity rate is generally low. Analytical results presented in this chapter show that in the presence of subthreshold leakage current, this argument is no more precise. It has been shown that under specific conditions, even in low activity rates, SCL circuits can exhibit better power–delay performance in comparison to the conventional CMOS topology.

In this chapter, some techniques for improving the power efficiency of SCL circuits have been introduced. It has been shown that using stacked SCL gates or current re-use technique can help to reduce the power consumption and area without degrading the speed of operation [18]. In addition, using output buffers helps to improve the power–delay performance of SCL circuits and at the same time help to simplify the design of standard cell library [16, 17].

Pipelining is another technique that can improve the performance of SCL circuits considerably [18]. Here, a very efficient technique with little area and power overhead has been introduced that can guarantee reliable performance of pipelined SCL circuits operating in subthreshold regime.

Finally, measurement results have been provided to illustrate the performance of proposed techniques in practice. In the next chapter, performance of STSCL circuits for low activity rate systems and memory circuits will be explored.

References

1. M. Pedram and J. Rabaey, *Power Aware Design Methodologies*, Kluwer, 2002
2. H. Soeleman, K. Roy, and B. C. Paul, "Robust subthreshold logic for ultra-low power operation," in *IEEE Transactions on Very Large Scale Integration (VLSI) Systems*, vol. 9, no. 1, pp. 90–99, Feb. 2001
3. B. Nikoliȅ, "Design in the power-limited scaling regime," in *IEEE Transactions on Electron Devices*, vol. 55, no. 1, pp. 71–83, Jan. 2008
4. B. H. Calhoun, and A. Chandrakasan, "Ultra-dynamic voltage scaling (UDVS) using subthreshold operation and local voltage dithering," *IEEE J. Solid-State Circuits*, vol. 41, pp. 238–245, Jan. 2006
5. M. Anis and M. Elmasry, *Multi-Threshold CMOS Digital Circuits, Managing Leakage Power*, Kluwer, 2003
6. N. Verma, J. Kwong, and A. Chandrakasan, "Nanometer MOSFET variation in minimum energy subthreshold circuits," in *IEEE Transactions on Electron Devices*, vol. 55, no. 1, pp. 163–174, Jan. 2008
7. E. Alon and M. Horowitz, "Integrated regulation for energy-efficient digital circuits," *IEEE J. Solid-State Circuits*, vol. 43, no. 8, pp. 1795–1807, Aug. 2008
8. A. Tajalli, E. Vittoz, Y. Leblebici, and E.J. Brauer, "Ultra low power subthreshold current mode logic utilizing a novel PMOS load device," in *IEE Electronics Letters*, vol. 43, no. 17, pp. 911–913, Aug. 2007
9. A. Tajalli, E. Vittoz, Y. Leblebici, and E. J. Brauer, "Ultra-low power subthreshold current-mode logic ulitising PMOS load device concept," *IET Electronics Letters*, vol. 43, no. 17, pp. 911–913, Aug. 2007
10. S.-M. Kang and Y. Leblebici, *CMOS Digital Integrated Circuits*, McGraw-Hill, 2003
11. C. C. Enz and E. A. Vittoz, *Charge-based MOS Transistor Modeling*, Wiley, 2006
12. P. R. Gray, P. J. Hurst, S. H. Lewis, and R. G. Meyer, *Analysis and Design of Analog Integrated Circuits*, Wiely, Fourth Ed., 2000

13. S. Badel, "MOS current-mode logic standard cells for high-speed low-noise applications," *PhD Dissertation*, Ecole Polytechnique Fédérale de Lausanne (EPFL), Switzerland, 2008
14. M. Mizuno, and et al., "A GHz MOS adaptive pipeline technique using MOS current-mode logic," *IEEE J. Solid-State Circuits*, pp. 784–791, vol. 31, no. 6, Jun. 1996
15. J. M. Musicer and J. Rabaey, "MOS current mode logic for low power, low noise CORDIC computation in mixed-signal environment," in *Proceedings of International Symposium on Low Power Electronics and Design (ISLPED)*, pp. 102–107, 2000
16. A. Tajalli, F. K. Gurkaynak, Y. Leblebici, M. Alioto, and E. J. Brauer, "Improving the power–delay product in SCL circuits using source follower output stage," in *Proceedings of International Symposium on Circuits and Systems (ISCAS)*, pp. 145–148, Seattle, USA, May 2008
17. A. Tajalli, M. Alioto, and Y. Leblebici, "Power–delay performance improvement of subthreshold SCL circuits," in *IEEE Transactions on Circuits and Systems-II: Express Briefs*, vol. 56, no. 2, pp. 127–131, Feb. 2009
18. A. Tajalli, E. J. Brauer, and Y. Leblebici, "Ultra low power 32-bit pipelined adder using subthreshold source-coupled logic with 5fJ/stage PDP," *Elsevier Microelectron. J.*, vol. 40, no. 6, pp. 973–978, Jun. 2009

Chapter 6
Low-Activity-Rate and Memory Circuits in STSCL

6.1 Introduction

As already discussed in Chap. 3, reduced voltage swing, fast current domain switching speed, and fully differential topology of SCL circuits make them very suitable for high frequency applications. In addition, SCL circuits exhibit very low sensitivity to common-mode noise sources with very low noise injection to substrate or supply lines [1, 2].

Traditionally, SCL topology has been used in very high speed systems (e.g., in the range of Gbit/s) where it is impractical or less efficient to employ conventional CMOS topologies [2, 3]. Since SCL circuits are continuously consuming a constant power from supply voltage, it is necessary to use this type of circuits at their maximum possible activity rate[1]. Otherwise, the power efficiency of this type of circuits degrades rapidly. This explains why SCL circuits have been only used in high speed applications with high activity rates or equivalently in systems with low average logic depth.

It is shown that CMOS circuits exhibit a superior power–delay performance compared to SCL circuits as the activity rate reduces (or logic depth increases) [4]. This argument is based on negligible static power consumption of CMOS circuits. By scaling the technology, however, static (leakage) power consumption of CMOS circuits becomes more and more evident. Therefore, the static power consumption of this type of circuits is no more negligible and the power dissipation will be dominated by the subthreshold channel residual (leakage) current [5].

The main concentration of this chapter is on low-activity-rate circuits. Based on this, the performance of CMOS and SCL families will be studied, and the conditions in which STSCL exhibits a better performance will be explored. In low-activity-rate conditions, the power consumption of the CMOS circuits is mostly dominated by the leakage current and the aim is to explore how SCL topology can help to reach to lower energy consumption levels.

---

[1] Activity rate is defined as the ratio of the operation frequency to the maximum possible frequency that a logic circuit can be employed or $\alpha = f_{op}/f_{Max}$ (see Chap.5).

A. Tajalli and Y. Leblebici, *Extreme Low-Power Mixed Signal IC Design: Subthreshold Source-Coupled Circuits*, DOI 10.1007/978-1-4419-6478-6_6,

To study the performance of STSCL topology and demonstrate the power efficiency of digital systems constructed based on this topology for low-activity-rate applications, a very low leakage (stand-by) static random access memory (SRAM) structure has been developed. In the proposed circuit, the tail bias current of each cell can be reduced down to a few pico-Amperes while the operation frequency can be kept as high as 2.1 MHz.

6.2 Power Efficiency in Low Activity Rates

It is already shown that SCL gates operating with small logic depth and high activity rate exhibit comparable or better power–delay product (PDP) with respect to the CMOS gates, mainly due to their lower output voltage swing [1,4].

For reduced activity rates, on the other hand, the power–delay product (PDP) or energy–delay product (EDP) advantage of SCL diminishes, since the static current consumption of the tail source tends to dominate the overall energy balance [4]. This observation is valid also for ultra-low-power SCL circuits operating in subthreshold regime [6]. Here, a more precise comparison including the leakage current of CMOS circuits is provided to make a more precise comparison between the two topologies.

6.2.1 STSCL Topology Performance

The total power consumption of a conceptual system constructed by N SCL gates is

$$P_{\text{diss,SCL}} = V_{\text{DD}} \sum_{i=1}^{N} I_{\text{SS}(i)} \tag{6.1}$$

where V_{DD} is the supply voltage of the system, N is the total number of gates, and $I_{\text{SS}(i)}$ is representing the bias current of ith gate. Here, it is assumed that all the cells are using the same supply voltage, V_{DD}. This assumption is generally correct since the gate delay in SCL topology does not depend on supply voltage. Hence, the supply voltage is generally set to the minimum possible value.

Based on (6.1), the power dissipation of a SCL-based circuit is constant and independent of the activity rate. Hence, this type of circuits are more power efficient when the circuit activity rate is maximized [7]. It is also possible to determine the bias current of each individual cell separately to optimize the power–delay trade-off as:

$$I_{\text{SS}(i)} = \frac{\ln 2 \cdot V_{\text{SW}} C_{L(i)}}{t_{d(i)}} \tag{6.2}$$

where V_{SW} is the voltage swing at the output of the proposed SCL gate, $C_{L(i)}$ is the capacitive load at the output of the gate, and $t_{d(i)}$ indicates the delay budget for the proposed gate. Since in STSCL circuits the NMOS differential pair transistors are in subthreshold regime, we can assume that V_{SW} is equal for all the gates and is independent of bias current ($V_{SW} \geq 4n_n U_T$ as discussed before in Chap. 3). To extract (6.2), delay of each gate has been estimated by:

$$t_{d(i)} \approx \ln 2 \times \tau_i = \ln 2 \cdot R_{L(i)} C_{L(i)} \tag{6.3}$$

where $R_{L(i)} \approx V_{SW}/I_{SS(i)}$ is the load resistance of the proposed gate. Regarding (6.2), it is also possible to scale the frequency of operation in a very wide range by scaling the tail bias current.

Finally, the relationship between the power consumption and the operating frequency (f_{op}) in a SCL-based digital system can be represented by

$$P_{diss,SCL} \approx \ln 2 V_{DD} V_{SW} f_{op} \cdot \sum_{i=1}^{N} C_{L(i)} N_{L(i)} \tag{6.4}$$

where $t_{d(i)}$ in (6.2) is replaced by

$$t_{d(i)} = 1/(N_{L(i)} f_{op}) \tag{6.5}$$

in which $N_{L(i)}$ stands for the logic depth of the block that the proposed gate is in it. Here, it is assumed that for a gate placed in a block with logic depth of $N_{L(i)}$, the delay of each gate needs to be $N_{L(i)}$ times smaller than the total clock period $(1/f_{op})$.

Assuming that $\overline{C_L}$ and $\overline{N_L}$ are the average values for the load capacitance and the logic depth in the proposed system, respectively, such that $\sum_{i=1}^{N}(C_{L(i)} N_{L(i)}) = N \overline{N_L C_L}$, then (6.4) can be more simplified to

$$P_{diss,SCL} \approx \ln 2 V_{DD} V_{SW} f_{op} N \cdot \overline{N_L} \cdot \overline{C_L} \tag{6.6}$$

which is proportional to $N \times \overline{N_L}$ and also the operating frequency. It linearly increases by f_{op} unlike CMOS topology which is proportional to $\sqrt{f_{op}}$ as will be discussed in the next section. It is noticeable that the power dissipation depends strongly on logic depth (N_L), and circuit complexity through N and C_L. To reduce the power consumption, it is desirable to reduce the voltage swing based on (6.6). However, as discussed in Chapter 3, voltage swing cannot be reduced very much due to degradation of NM.

The lower limit of power dissipation in SCL-based circuits is the minimum standby current of the SCL gates which can be as low as a few pico-Amperes [7] (also see Sect. 3.4.7). To have a good control on tail bias current in such low current levels, high threshold voltage (HVT) devices can be used. Since speed of operation in SCL topology does not depend on threshold voltage, using HVT for tail bias current does not affect the performance of the circuit.

6.2.2 CMOS Topology Performance

Conventional CMOS topology shows a very good power efficiency for a very wide range of applications and activity rates [8]. This is mainly due to its negligible static power consumption. The static power consumption of the CMOS circuits, however, is going to be more and more pronounced in modern nano-scale technologies. For nanometer-scale CMOS technologies where the off (subthreshold) leakage of each transistor can reach nA-levels, however, the SCL topology with its controllable tail bias current can offer reduced power consumption well below the leakage of CMOS, while maintaining a significant speed advantage over CMOS topologies.

Including leakage current, the total RMS power consumption of a digital CMOS system can be approximated by (see Sect. 5.2.3)

$$P_{\text{diss,CMOS}} \approx V_{\text{DD}} \sqrt{I_{\text{leak}}^2 + \zeta \cdot \alpha}. \tag{6.7}$$

Here, I_{leak} is the total leakage current consumption of the system, α represents the activity rate, and ζ is a proportionality factor representing the relationship between activity rate and dynamic current consumption of the system. Based on (6.7), as the activity rate grows, power dissipation increases proportional to $\sqrt{\alpha}$ when supply voltage is constant. However, by reducing the activity rate, the power consumption will be dominated by the leakage current as:

$$P_{\text{diss,CMOS}}|_{\alpha \to 0} \approx V_{\text{DD}} I_{\text{leak}} = V_{\text{DD}} \sum_{i=1}^{N} I_{\text{leak}(i)} \tag{6.8}$$

where N is the total number of gates in the system and $I_{\text{leak}(i)}$ is the leakage current of ith gate. It is also possible to present the system power consumption in this case as

$$P_{\text{diss,CMOS}}|_{\alpha \to 0} \approx N V_{\text{DD}} \overline{I_{\text{leak}}} \tag{6.9}$$

where $\overline{I_{\text{leak}}}$ is the average leakage current per cell in the proposed system: $\overline{I_{\text{leak}}} = \sum_{i=1}^{N} I_{\text{leak}(i)}/N$. As explained in Chap. 2, subthreshold channel residual (leakage) current can be represented by:

$$I_{\text{leak}} \approx I_{\text{subth}} \approx \mu C_{\text{ox}} \frac{W}{L_e} U_T^2 \cdot \cdot e^{-\frac{\Delta V_T}{n U_T}} \times e^{\frac{-V_{T0} + \eta V_{\text{DD}}}{n U_T}} \tag{6.10}$$

which implies that leakage current highly depends on temperature, variation on threshold voltage, and increases by V_{DD} due to the DIBL effect modeled by η in this equation.

6.2.3 Comparison

Comparing (6.6) and (6.9), when the activity rate of the circuit is low enough such that the stand-by current constructs the dominant part of the power consumption of the CMOS circuits, it is possible to use STSCL topology with a logic depth of not more than

$$\overline{N_L} > \frac{\overline{I_{\text{leak}}}}{\ln 2} \times \frac{1}{f_{\text{op}} V_{\text{SW}} \overline{C_L}} \tag{6.11}$$

to be able to reduce the power consumption. Based on (6.11), as the leakage current increases and load capacitance reduces by scaling down the technology feature size, the power efficiency of STSCL topology improves.

To derive (6.11), it is assumed that the proposed system will have the same number of gates (N) and the same supply voltage, implemented in either CMOS or SCL topologies which might not be always correct. Moreover, here the overhead of periphery circuitry has been neglected. The overhead of periphery circuit can be especially important in CMOS circuits where the supply voltage needs to be precisely controlled by precise voltage regulators [9].

Figure 6.1 shows the power dissipation of a chain of identical gates based on static CMOS and SCL topologies in 65-nm CMOS technology, both loaded with the same output capacitance and both operating in subthreshold regime. It can be seen that the overall dissipation of the CMOS chain at very low operating frequencies is limited by the leakage current which can be reduced by lowering the supply voltage, yet a dramatic reduction is not possible because the operational robustness diminishes as the current-drive capability of CMOS gates drops exponentially with the supply voltage [10, 11]. Meanwhile, the SCL topology with a constant tail bias current exhibits comparable operation speed at lower power dissipation, and much less dependence to process and supply voltage variations.

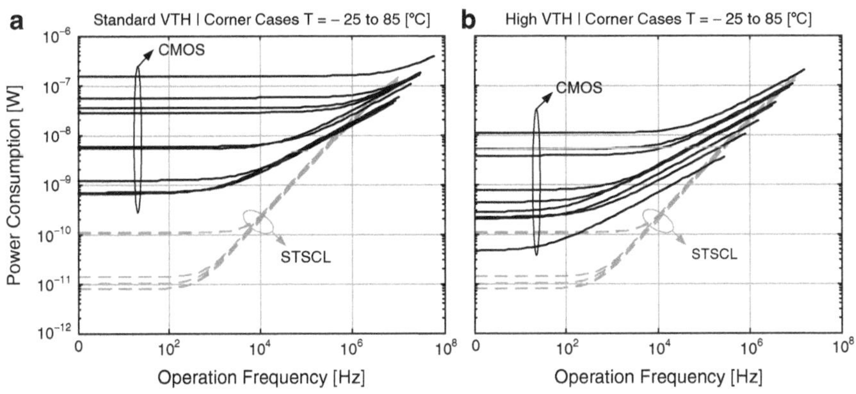

Fig. 6.1 Simulated power consumption of a chain of gates in 65-nm CMOS technology based on static CMOS (*solid line*) and STSCL topologies (*dashed line*). Variation of the power consumption due to the process corners and temperature variation is shown with standard-VT (**a**) and high-VT (**b**) CMOS. Operating conditions: $V_{\text{DD(CMOS)}} = 300\,\text{mV}$ and $V_{\text{DD(STSCL)}} = 400\,\text{mV}$

The leakage power dissipation in CMOS circuits can also be reduced significantly by using HVT transistors, which inevitably impacts the operation speed (Fig. 6.1b). The SCL topology, on the other hand, can be constructed using HVT transistors especially to control the tail bias current, without any detrimental effects on switching speed. This observation implies that subthreshold SCL circuits can offer significant advantages for very low activity rate applications where static CMOS circuits lose their effectiveness due to leakage and also the exponential dependence between operation frequency and supply voltage, such as in SRAM circuit operating in subthreshold regime.

The other important issue is the very wide variation of leakage and dynamic consumption in CMOS topology which can be as high as two orders of magnitude. This wide variation is mainly due to the exponential dependence of the subthreshold residual channel current in subthreshold regime on device V_T as depicted in (6.10).

It should be also mentioned that the superior power efficiency of the SCL topology compared to CMOS is not limited to only low-activity-rates. As illustrated in Fig. 6.1, the SCL topology exhibits less power consumption in higher activity-rates (operation frequencies) up to frequencies very close to the maximum operation frequency of CMOS circuit. The upper limit for activity rate in which SCL topology still exhibits a better performance can be estimated by comparing (6.1) and (6.7) for each specific systems.

6.3 Low-Leakage CMOS SRAMs

In ultra-low-power applications, the amount of power that each individual part of a system consumes is very important. One of the main building blocks in many modern integrated digital systems is the memory block. The continuous trend and demand for increasing the size of embedded memories on integrated systems for improving the performance, has made this type of circuits one of the key components in such systems. In many modern digital systems, static random-access memories (SRAMs) comprise a significant part of the total area and power consumption. For example, embedded cache memories are expected to occupy 90% of the total area in a system-on-a-chip (SoC) [17]. Therefore, it is necessary to reduce the static and dynamic power consumption of this type of circuits in addition to their area.

There are many challenges in design of low-voltage and low-power SRAM circuites. Although reducing the supply voltage in SRAM circuits helps to reduce their dynamic and static power consumption[2], however, this could not be done without special cares. This is mainly because static noise margin (SNM) of the SRAMs depends on supply voltage and degrades by supply reduction. Meanwhile, in lower supply voltages, SNM will be more sensitive to the process

[2] Reduction of leakage current is mainly due to reducing the drain-source voltage, and hence alleviating DIBL effect [13].

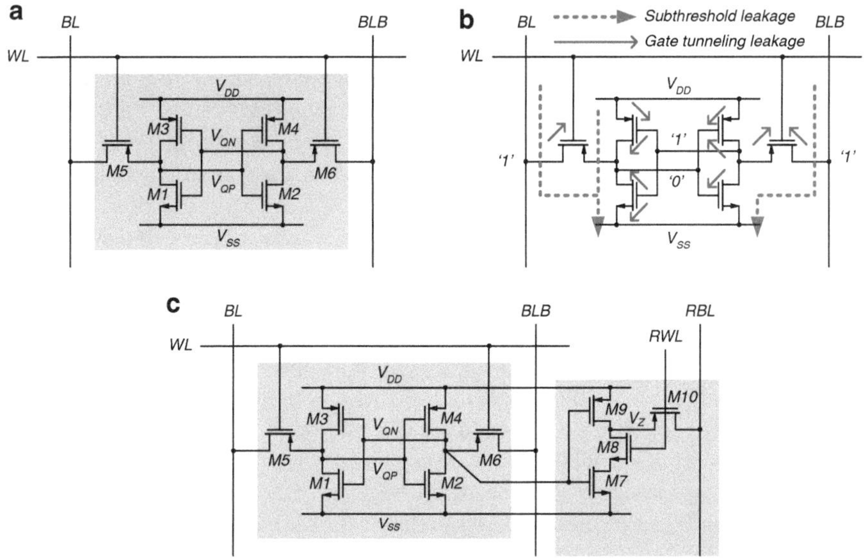

Fig. 6.2 (**a**) Conventional 6 transistor SRAM cell and (**b**) leakage paths in this configuration. (**c**) 10T SRAM for subthreshold operation [12]

variation [12]. Device mismatch[3] is the other main issue in design of SRAM cells. For example, proper write operation in conventional six-transistor (6T) SRAM circuits shown in Fig. 6.2a, depends on the ratio of transistor currents. The failure due to device mismatch could be observed not only on write mode, but also in read, hold, and access modes. Hence, any device mismatch can degrade the margin in different modes of operation. These effects could be more exacerbated in subthreshold region where device current exponentially depends on threshold voltage.

Figure 6.2b illustrates the different leakage paths in a conventional 6T SRAM bitcell. There are different paths for subthreshold leakage current. Transistors with $|V_{ds}| = V_{DD}$ where their gate-source voltage is zero are the main sources for subthreshold leakage current. Gate tunneling current can also be observed almost in all gate terminals.

One of the main issues for subthreshold operation, is the degradation of SNM in read mode. By reducing the supply voltage, read mode SNM is the main limiting factor against pushing the devices towards subthreshold regime. Therefore, the first step to design subthreshold SRAMs is mitigating this problem. Figure 6.2c shows a solution for implementing subthreshold SRAM cells [12] with improving the read mode SNM. In this configuration, an output buffer for read operation has

[3] Device mismatch is generally described by inter-die and intra-die process variations. Random dopant fluctuation (RDF) and line edge roughness (LER) are the main causes for intra-die variations which can result in threshold voltage mismatch (see Chap. 2).

been used. The buffering technique used here helps to improve the read SNM by isolating SRAM core and bit-lines. Therefore, it is possible to reduce the supply voltage to half of the supply voltage of conventional 6T structure with the same amount of SNM. In the 10T SRAM schematic shown in Fig. 6.2c, $M8$ is used to reduce the leakage current and hence be able to put more bit-cells on a bit-line (BL). As indicated in [12], this configuration can not hold the data for supply voltages less than $V_{DD} = 230\,\text{mV}$.

A more compact SRAM cell is introduced in [13], where each cell consists of 8 transistors (8T). Using this technique, the supply voltage can be reduced down to $V_{DD} = 350\,\text{mV}$ while SRAM operates at 25 kHz frequency.

A Schmitt trigger based 10T SRAM circuit introduced in [17] with improved read SNM and better process variation tolerance compared to the conventional 6T configuration (Fig. 6.3). Implemented in 0.13 μm, the supply voltage of circuit could be reduced down to 160 mV. The penalty that has been paid in this design for having a more robust operation in subthreshold region is 2.1× more cell area. Table 6.1 compares the performance of some of the recently reported low-leakage SRAM circuits. As can be seen, there is a tight relationship among supply voltage, speed

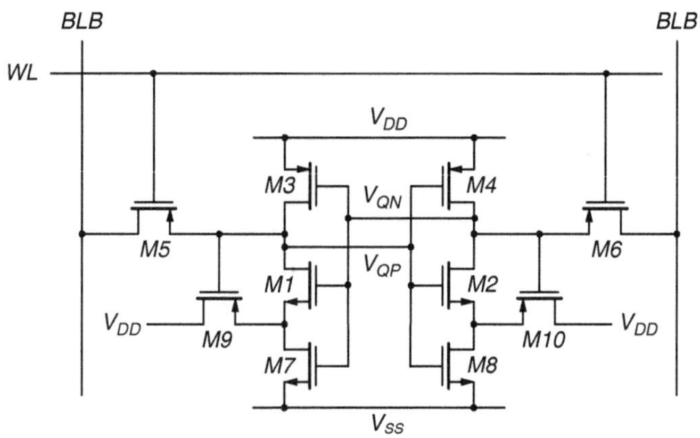

Fig. 6.3 Schmitt trigger based SRAM bitcell introduced in [17] operating at $V_{DD} = 160\,\text{mV}$

Table 6.1 Recently reported low-leakage SRAM cells

Reference	Year	Tech.	V_{DD} (V)	Leakage per cell (pA)	f_{CK} (kHz)	Memory size (kb)	Cell area (μm^2)
[12]	2007	CMOS 65 nm	0.4	11	500	256	
[13]	2008	CMOS 65 nm	0.35	8	25	256	
[16]	2008	CMOS 130 nm	0.2	120	100	480	2.68×2.80
[17]	2007	CMOS 0.13 μm	0.16		620@0.4 V	4	
[18]	2009	CMOS 90 nm	0.16	36	0.5	32	
[19]	2008	CMOS 65 nm	0.7	2	250	1,000	0.667
[20]	2009	CMOS 0.18 μm	0.4	10	2,100	1	

of operation, and leakage current. In the next section, an STSCL based SRAM cell is introduced that can reach very low leakage current and at the same time high operating frequency.

6.4 Low Stand-By Current STSCL Memory Cell

In this section, we are presenting an SRAM array which exhibits very low stand-by dissipation in idle state, and allows robust read and write operations at frequencies that are significantly higher than those achievable in CMOS-based topologies. This circuit can be embedded in a STSCL standard-cell library to improve the library capabilities.

6.4.1 Circuit Topology

The core of the proposed memory cell is based on a cross-coupled STSCL inverter to construct the positive feedback needed to store the data. The circuit schematic of an STSCL inverter and also the core of the proposed memory cell are shown in Fig. 6.4a, b, respectively.

In Fig. 6.4a, $M1$ and $M2$ construct the NMOS switching network, $M3$ and $M4$ are the load devices, and the tail bias current is controlled by $M5$ [7]. To construct the load resistances, $M3$ and $M4$ transistors with their bulk shorted to their drain terminals have been used. Using minimum size devices, this structure shows a very high resistivity in a wide voltage swing [6]. Due to the reverse subthreshold effect, the threshold voltage of $M5$ can be increase by selecting the length of this device slightly larger than the minimum size which helps to have a more precise current mirror [14, 15]. Transistors $M6$ and $M7$ in Fig. 6.4b are the access transistors.

The write operation is performed by pre-charging BL and BLB nodes to the desired voltage levels, and then turning on the access transistors M6-M7 in order to charge/discharge the output nodes QP and QN of the memory core (Fig. 6.4b). After turning off the access transistors, the positive feedback in the cell will preserve the new state. Since QP and QN have been already charged to the intended values, no extra settling time is required to accomplish the write operation of the cell. Therefore, the write operation is very fast.

To enable a fast read operation, as illustrated in Fig. 6.4c, an open-drain differential pair is formed by M8–M9, driven by the tail bias transistor M10 which is external to the cell and shared by the cells on a word-line. During the read cycle, M10 is turned on and conducts the current I_{READ}, which is steered to one of the output branches of BL/BLB depending on the stored data on the core. This output current is detected by a current-mode sense amplifier (SA) and will be converted to voltage. Therefore, the speed of the read operation is completely independent of the core tail bias current (I_{CORE}) and depends only on I_{READ} as well as the parasitic

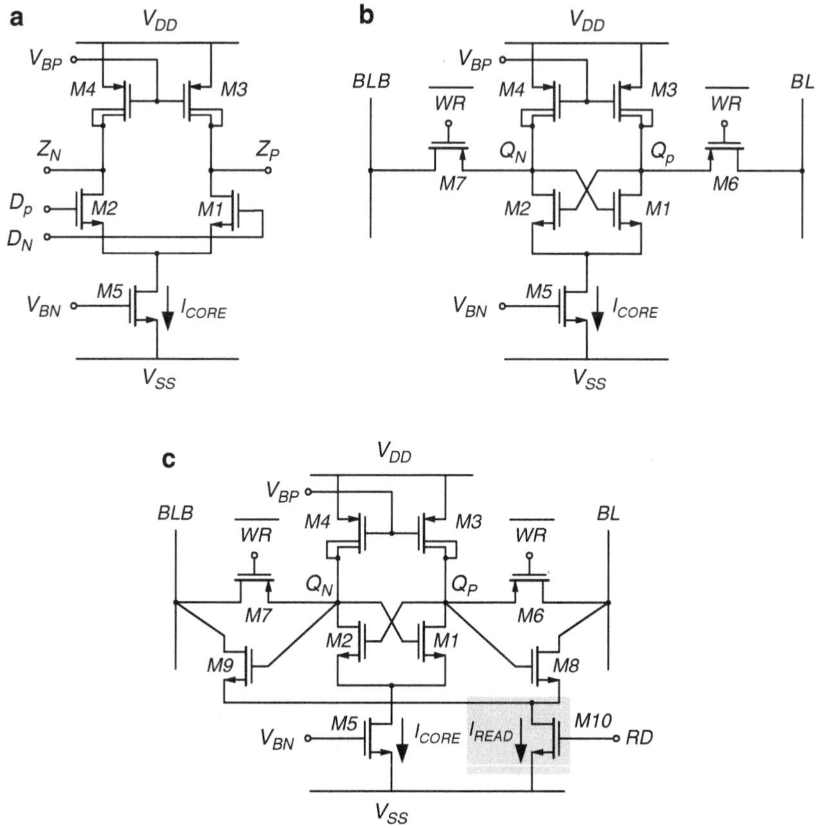

Fig. 6.4 (**a**) Schematic of a STSCL inverter. (**b**) The core of the proposed memory cell based on STSCL topology. (**c**) Completed memory cell. In this schematic, $M10$ is shared among all the memory cells on a word line to save area

capacitances of the nodes BL/BLB. In this work, a small aspect-ratio has been chosen for M10 to reduce the leakage current due to this device during the idle state. By setting RD = 0, the latch circuit will turn on and preserve the data.

Isolating the speed of RD/WR operation from the "hold" consumption in the proposed 9T memory cell permits the reduction of the core bias current down to leakage-current levels. The main limitation for further reducing the tail bias current below 10 pA is the turn-on current of the forward-biased source-bulk diode of the PMOS load devices. The forward voltage across this diode is equal to the voltage swing at the output of the core, which can be as low as $V_{SW} = 4nU_T \approx 140\,\text{mV}$ in room temperature (U_T is the thermal voltage) [7]. In this work, the tail bias current has been chosen to be twice the diode turn-on current.

6.4.2 Device Sizing

In contrast to conventional CMOS SRAM cells where the speed of operation depends on threshold voltages, HVT devices can be used throughout this cell to limit leakage without impacting speed. The length of MOS devices in Fig. 6.5a has been selected slightly larger than minimum feature size to increase the threshold voltage of devices. Since the tail bias current is very low, the NMOS differential pair devices are deeply in weak inversion, and hence:

$$V_{\text{GS}} \approx V_{T0} + n_n U_T \ln\left(\frac{I_{\text{CORE}}}{I_0}\right) \tag{6.12}$$

where V_{T0} is the threshold voltage of the device, and $I_0 = 2n_n(W/L_{\text{eff}})U_T^2$ [21]. To have a complete current switching in differential pair transistors, it is necessary that gate-source voltage of the turned on transistor remains larger than V_{SW} or $V_{\text{GS}} > V_{\text{SW}}$. Therefore, using a device with higher threshold voltage can help to satisfy this constraint. Assuming $V_{\text{GS}} \approx V_{\text{SW}}$, the minimum theoretical achievable supply voltage is:

$$V_{\text{DD,min}} \approx V_{\text{SW}} + V_{\text{CS}} \tag{6.13}$$

where V_{CS} is the headroom required to keep the tail bias transistor (M0) in saturation region. For very low bias currents, M0 is in subthreshold region, hence $V_{\text{CS}} > 4U_T$. Therefore, the minimum supply voltage is about $10U_T$. Measurements show that

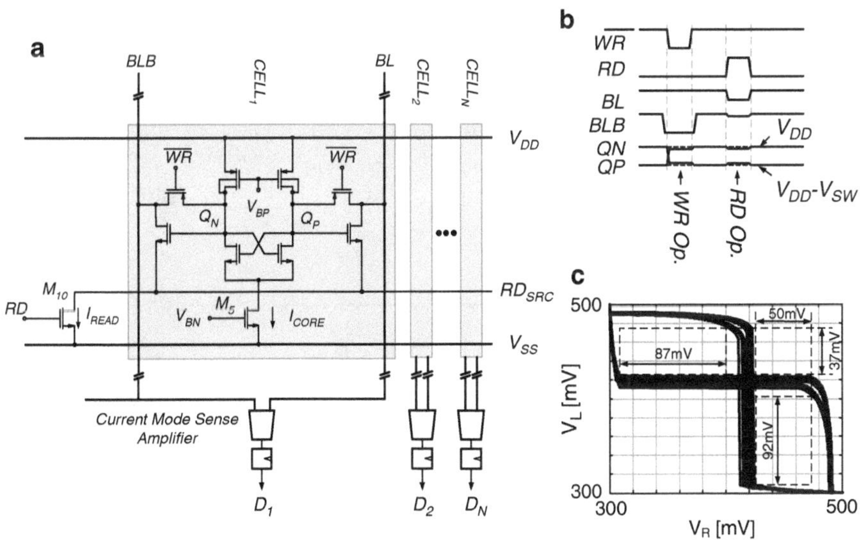

Fig. 6.5 (**a**) Circuit schematic, and (**b**) timing diagram of the STSCL-based SRAM cell. (**c**) Simulated butterfly curve of a cell in CMOS 65 nm (showing different corner cases) for $V_{\text{DD}} = 500$ mV and $V_{\text{SW}} = 200$ mV

the circuit supply voltage (including replica bias circuit and the amplifier used in replica bias) can be reduced to 350 mV for very low bias currents [7]. The minimum supply voltage will be higher when the bias current increases and the devices leave the weak inversion region.

With a static current consumption of 10 pA/cell, this SRAM core exhibits about three times smaller idle power dissipation compared to [13] while the RD/WR speed can be as high as 2.1 MHz (25 kHz for $V_{DD} = 350$ mV in 65-nm CMOS technology [13]).

Figure 6.5a, b depicts the topology and timing diagram of the proposed memory array. Figure 6.5c illustrates the Butterfly curves of the proposed memory cell in different process corners and temperatures. Here, the voltage swing is chosen to be 200 mV at the output of the SCL memory cell and supply voltage is 500 mV. Simulations show that the supply voltage can be reduced to 350 mV without degrading the static noise margin of the cell.

6.4.3 Sense Amplifier

The differential current generated during the read operation will be conducted to sense amplifier (SA) which is depicted in Fig. 6.6. During the hold or write modes (RD = 0), the SA is isolated from the memory. In this condition, M16 and M17 are off and SA operates as a latch and keeps the latest data has been read from the memory. The bias voltage of PMOS load devices, $V_{BP(SA)}$, is generated corresponding to the tail bias current of SA circuit (I_{SA}) to control the gain and output voltage swing

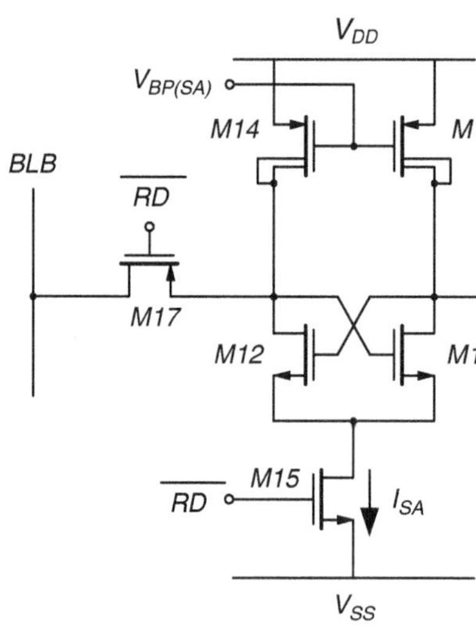

Fig. 6.6 Sense amplifier used to reconstruct the data at the output of memory cell

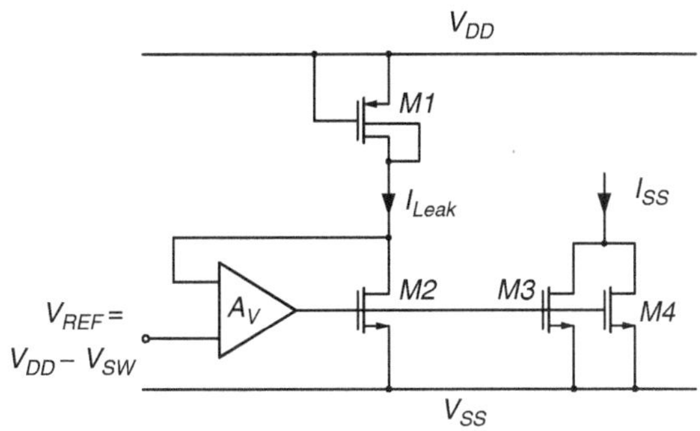

Fig. 6.7 Leakage detector and bias current generator circuit schematic

of SA. As the read signal is activated (RD = 1), tail bias current will be switched off and the load resistances and the read circuitry inside each memory cell (M8–M10 in Fig. 6.4c) will construct a single stage amplifier. Therefore, the circuit will amplify the output of the proposed memory cell.

6.4.4 *Leakage Current Detection*

The bias current of each memory cell, as discussed before, depends on leakage current due to the forward biased diode of the PMOS load devices. Hence, it is necessary to detect this current and adjust the bias current of the memory core with respect to that. Having an on-chip leakage current measurement circuit helps to track the PVT variations and hence compensate their effect.

Figure 6.7 illustrates a simple circuit that can be used for detecting the diode forward bias current called I_{Leak}. An amplifier is used to adjust the source-drain voltage of the PMOS transistors equal to the required V_{SW}. Then the leakage current is conducted to a current mirror and hence can be used to generate the tail bias current of memory cells. In this schematic, the leakage current is amplified to make sure that the memory core bias current, I_{SS}, is much larger than the leakage current.

6.5 Experimental Results

Test Setup: A 1-kb (8b × 256) SRAM array has been designed and fabricated using 0.18-μm CMOS technology, as a test vehicle to demonstrate the key principles discussed above. Supply voltage of the core memory circuit is directly accessible to measure the power consumption. To measure the supply current, an HP 4156A semiconductor parameter analyzer has been used. Also, a logic analyzer controls the write and read processes.

A single bit separated SRAM cell with buffers has been used to measure the butterfly curves. An internal replica bias circuit controls the voltage swing at the

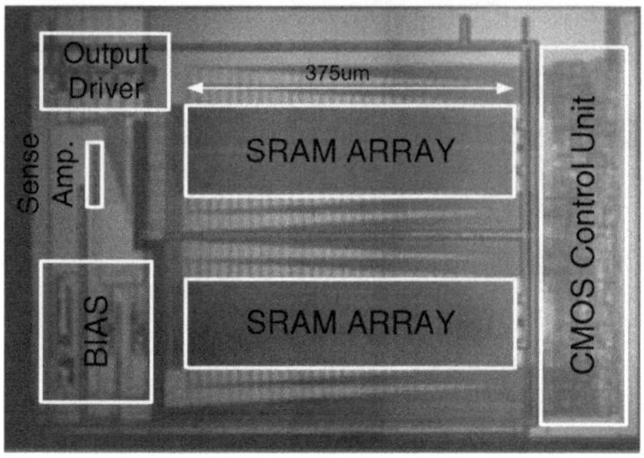

Fig. 6.8 The chip photomicrograph of the ultra low stand-by (leakage) current SRAM array (1 kb block) fabricated with conventional 0.18-μm CMOS technology

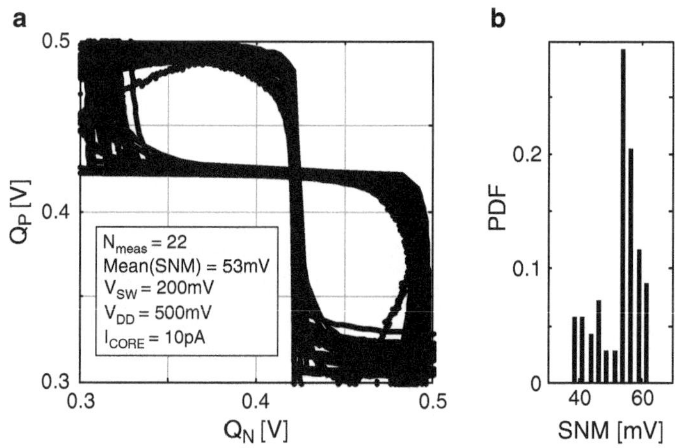

Fig. 6.9 Measured (**a**) butterfly curves and (**b**) statistical distribution of the SNM, for the proposed SRAM cell ($I_{CORE} = 10\,\text{pA}$, $V_{SW} = 200\,\text{mV}$, and $V_{DD} = 500\,\text{mV}$)

output of the memory cells [7]. The fabricated 1 kb SRAM array is shown in Fig. 6.8. The active area of the memory (including biasing and sense amplifiers) is $670\,\mu\text{m} \times 390\,\mu\text{m}$. The design has been done based on digital CMOS design rules.[4]

Noise Margin: Figure 6.9a shows the measured butterfly curves for the proposed SRAM circuit, where the static noise margin of the cell is not affected by the read operation. The average SNM (Fig. 6.9b) is measured to be 53 mV for $I_{CORE} = 10\,\text{pA}$ and $V_{SW} = 200\,\text{mV}$.

To investigate the influence of V_{SW} on SNM, measurements have been repeated for different output voltage swing values. Figure 6.10 shows that the SNM initially

[4] Generally, special design rules for layout of SRAM cells are applied to minimize the cell area.

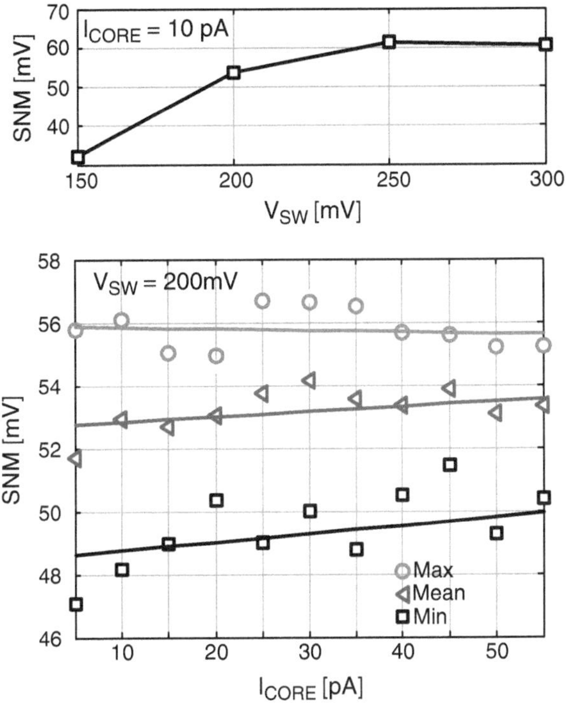

Fig. 6.10 Measured variation of the SNM versus V_{SW} (for $I_{CORE} = 10\,\text{pA}$) and variations of SNM versus tail bias current (I_{CORE}) for $V_{SW} = 200\,\text{mV}$

improves with increasing V_{SW}, and eventually saturates at $V_{SW} = 250\,\text{mV}$, mainly due to the saturation of the amplifier used in replica bias circuits. The dependence of SNM on the tail bias current is shown in Fig. 6.10, with average, minimum and maximum values for SNM plotted for different I_{CORE} levels. It can be seen that the SNM has only minor dependence on I_{CORE}. It remains very stable down to very low levels of bias current and that the variation on SNM is reduced by increasing I_{CORE}.

Speed of Operation: In the proposed memory, the main speed limiting factor is the read operation. To increase the speed of operation, it is necessary to increase I_{READ}, which can be achieved by increasing the voltage swing at the gate of M9 in Fig. 6.5a. Figure 6.11 shows the variation of the normalized power dissipation of the memory versus operating frequency.

Power Consumption: Measurements confirm that the total current consumption of the array is between 9.5 to 13 nA for different dies (corresponding to 9 to 12.5 pA per SRAM cell) at $V_{DD(SCL)} = 500\,\text{mV}$. At 10 pA core bias current and 1.5 MHz read/write clock frequency, fewer than 0.01% RD/WR errors were observed. The maximum clock frequency was found between 1.7 to 2.1 MHz for different dies. Table 6.2 summarizes the specifications of the proposed STSCL SRAM circuit.

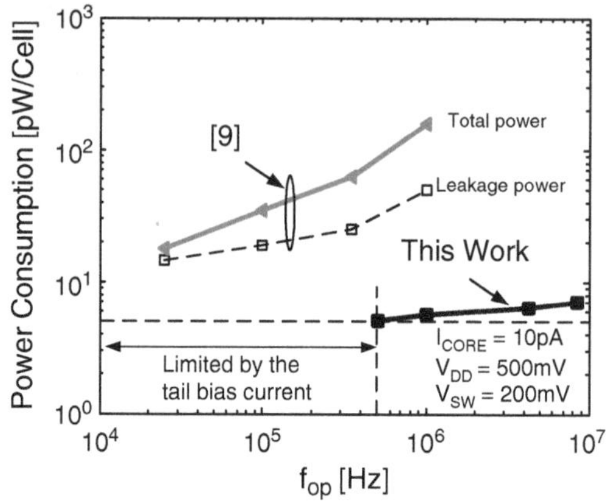

Fig. 6.11 Variation of the idle power consumption (per cell) versus operating frequency, comparing this work with the SRAM cell presented in [13]

Table 6.2 Performance summary for STSCL SRAM cell

Parameter	Value	Unit
Technology	0.18-μm CMOS	(-)
Supply voltage	>400	(mV)
Voltage swing	200	(mV)
Active area	670×390	(μm^2)
Stand-by current per cell	9–12.5	(pA)
Operating frequency	1.7–2.1	(MHz)
Static noise margin	53	(mV)

6.6 Observations and Discussion

CMOS circuits have been very widely used for implementing digital systems in different types of applications. Area and power efficiency of this type of circuits have made them very successful compared to many other types of circuits [8]. The tight tradeoff between power consumption, speed of operation, supply voltage, and device threshold voltage, however, has made the design of power efficient digital systems based on this topology and in modern nano-scale CMOS technologies very challenging.

In this work, a very low stand-by (leakage) memory cell based on STSCL topology has been designed and tested. Some very interesting observations can be made based on the results of this work:

Observation 1: The measurements in this work and also the results in [7] show that the power consumption of each STSCL cell can be reduced to few pico-Watts. Compared to the subthreshold leakage current of CMOS circuits that can be as high as nano-Amperes per cell, such a low leakage value can be critically important.

Observation 2: It is important to notice that in this type of circuits, the speed of operation depends on tail bias current of the cells and is independent of the threshold voltage of the MOS devices and also supply voltage[5].

In addition, as shown in (6.13) the minimum supply voltage when the devices are operating deeply in week inversion does not depend on threshold voltage of MOS devices. Therefore, the tight tradeoff that existed in CMOS topology among supply voltage, threshold voltage, power consumption, and speed of operation, is more relaxed in STSCL.

Observation 3: The other important observation is that STSCL topology can show comparable or even better power–delay performance compared to CMOS topology even in low activity rate circuits. This is contrary to the traditional observations that SCL circuits only have been used to implement high activity systems [4].

The main reason is that the static power consumption of the CMOS circuits can not be ignored in very low power circuits. Therefore, the possibility of reducing the bias current of STSCL circuits below the subthreshold leakage current of CMOS circuits will make the power–delay performance of this type of circuits comparable to CMOS circuits.

Observation 4: The main issue associated with STSCL topology is its larger area occupation in comparison to the CMOS topology. Increased number of transistors as well as need to two separate n-well regions to put the PMOS load devices are the main reason for having a larger area. Larger area is the price paid to have a simpler power management system and also lower power consumption.

References

1. S. Badel, "MOS current-mode logic standard cells for high-speed low-noise applications," *PhD Dissertation*, Ecole Polytechnique Fédérale de Lausanne (EPFL), Switzerland, 2008
2. P. Heydari and R. Mohanavelu, "Design of ultrahigh-speed low-voltage CMOS CML buffers and latches," in *IEEE Tranactions on Very Large Scale Integration (VLSI) Systems*, vol. 12, no. 10, pp. 1081–1093, Oct. 2004
3. A. Tajalli, P. Muller, and Y. Leblebici, "A power-efficient clock and data recovery circuit in 0.18-μm CMOS technology for multi-channel short-haul optical data communication," *IEEE J. Solid-State Circuits*, vol. 42, no. 10, pp. 2235–2244, Oct. 2007
4. J. M. Musicer and J. Rabaey, "MOS current mode logic for low power, low noise CORDIC computation in mixed-signal environment," in *Proceedings of International Symposium on Low Power Electronics and Design (ISLPED)*, pp. 102–107, 2000
5. M. Pedram and J. Rabaey, *Power Aware Design Methodologies*, Kluwer, 2002
6. A. Tajalli, E. Vittoz, Y. Leblebici, and E. J. Brauer, "Ultra low power subthreshold MOS current mode logic circuits using a novel load device concept," in *Proceedings of European Solid-State Ciruits Conference (ESSCIRC)*, Munich, Germany, pp. 281–284, Sep. 2007

[5] In the proposed SRAM topology, speed of READ operation depends on I_{READ} and hence the threshold voltage of M10. This is a specific case and in general speed of operation does not depend on device threshold voltage in STSCL topology.

7. A. Tajalli, E. J. Brauer, Y. Leblebici, and E. Vittoz, “Sub-threshold source-coupled logic circuit design for ultra low power applications,” *IEEE J. Solid-State Circuits*, vol. 43, no. 7, pp. 1699–1710, Jul. 2008
8. S. -M. Kang and Y. Leblebici, *CMOS Digital Integrated Circuits*, McGraw-Hill, 2003
9. B. Nikolič, “Design in the power-limited scaling regime,” in *IEEE Transactions on Electron Devices*, vol. 55, no. 1, pp. 71–83, Jan. 2008
10. B. H. Calhoun, and A. Chandrakasan, “Ultra-dynamic voltage scaling (UDVS) using sub-threshold operation and local voltage dithering,” *IEEE J. Solid-State Circuits*, vol. 41, pp. 238–245, Jan. 2006
11. B. Zhai, S. Hanson, D. Blaauw, and D. Sylvester, “A variation-tolerant sub-200 mV 6-T sub-threshold SRAM,” *IEEE J. Solid-State Circuits*, vol. 43, no. 10, pp. 2338–2348, Oct. 2008
12. B. H. Calhoun and A. P. Chandrakasan, “A 256-kb 65-nm sub-threshold SRAM design for ultra-low-voltage operation,” *IEEE J. Solid-State Circuits*, vol. 42, no. 3, pp. 680–688, Mar. 2007
13. N. Verma and A. P. Chandrakasan, “A 256 kb 65 nm 8T subthreshold SRAM employing sense-amplifier redundancy,” *J. Solid-State Circuits*, vol. 43, no. 1, pp. 141–149, Jan. 2008
14. C. Y. Lu and J. M. Sung, “Reverse short-channel effects on threshold voltage in submicrometer salicide devices,” in *IEEE Electron Device Letters*, vol. 10, no. 10, pp. 446–448, Oct. 1989
15. C. Subramanian, “Reverse short channel effect and channel length dependence of boron penetration in PMOSFETs,” in *International Electron Device Meeting*, pp. 423–426, Dec. 1995
16. T. -H. Kim, J. Liu, J. Keane, and C. H. Kim, “A 0.2 V, 480 kb subthreshold SRAM with 1 k cells per bitline for ultra-low-voltagre computating,” *IEEE J. Solid-State Circuits*, vol. 43, no. 3, pp. 518–529, Feb. 2008
17. J. P. Kulkarni, K. Kim, K. Roy, “A 160 mV robust Schmitt triger based subthreshold SRAM,” *IEEE J. Solid-State Circuits*, vol. 42, no. 10, pp. 2303–2313, Oct. 2007
18. I. J. Chang, J. -J. Kim, S. P. Park, and K. Roy, “A 32 kb 10 T sub-threshold SRAM array with bit-inteleaved and differential read scheme in 90 nm CMOS,” *IEEE J. Solid-State Circuits*, vol. 44, no. 2, pp. 650–658, Feb. 2009
19. Y. Wang, et al., “A 1.1 GHz 12 μA/Mb-leakage SRAM design in 65 nm ultra-low-power CMOS technology with integrated leakage reduction for mobile applications,” *IEEE J. Solid-State Circuits*, vol. 43, no. 1, pp. 172–179, Jan. 2008
20. A. Tajalli and Y. Leblebici, “Subthreshold SCL for ultra-low-power SRAM and low-activity-rate digital systems,” to appear in *European Solid-State Circuits Conference (ESSCIRC)*, Sep. 2009
21. C. C. Enz and E. A. Vittoz, *Charge-based MOS Transistor Modeling*, Wiley, 2006

Part II
Scalable and Ultra-Low-Power Analog Integrated Circuits

Chapter 7
Widely Adjustable Continuous-Time Filter Design

7.1 Introduction

In most of the integrated systems, analog part acts as an interface between the real world and the internal processing system. Thus, to realize a specific high performance integrated system, characteristics of the analog part becomes critically important.

In this work, several techniques for implementing high-performance and widely adjustable analog circuits have been developed. This concept is explained in more details in Fig. 7.1. The heart of this system is a digital unit that is used to do the required processing job. The operation frequency of this part can be adjusted with respect to the work load and other higher level issues such as power optimization and battery life time. These adjustments can be done using a phase-locked loop (PLL) and consequently an appropriate biasing circuits. The proposed PLL provides the internal clock as well as the required bias current for the STSCL gates in the digital signal processing unit.

In this system, the analog input signal will be converted to digital signal by an ADC circuit. In front of this ADC, a low-pass filter for anti-aliasing purpose and also for removing the high frequency noise, is employed. It might be necessary to use a low noise amplifier at the front end in order to increase the input signal level and at the same time relax the noise requirements of the following stages. In addition to the wide tuning range, these blocks need to consume a very low amount of power.

In the following, some techniques for implementing widely adjustable continuous-time filters will be described. First, a very short review on design of subthreshold transconductance operational amplifiers (OTAs) is provided. Then, the design of a power scalable transconductor-C (g_m-C) filter with improved linearity performance is explained. It is shown that using some simple modifications, considerable improve in the linearity performance in biquadratic transconductor-C filters can be achieved. In addition, a very low frequency MOSFET-C filter with scalable power-frequency characteristics is described. This circuit employs the floating resistance which has been developed in Chap. 3. Finally, measurement results are provided to be compared with the expected performance.

A. Tajalli and Y. Leblebici, *Extreme Low-Power Mixed Signal IC Design: Subthreshold Source-Coupled Circuits*, DOI 10.1007/978-1-4419-6478-6_7,

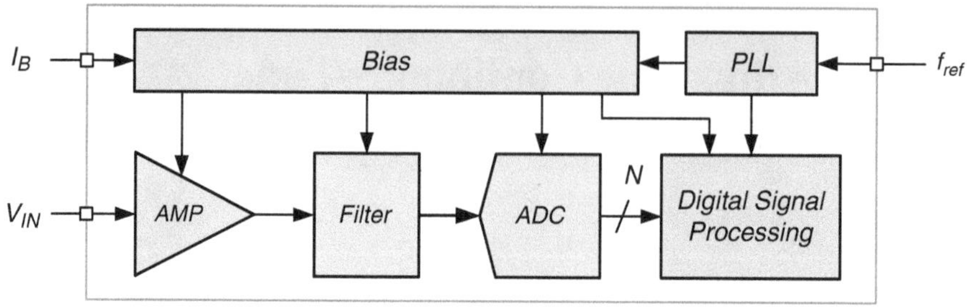

Fig. 7.1 A conceptual block diagram of a widely adjustable mixed-mode integrated circuit

7.2 Amplifier Design

Amplifiers are probably the most critical building blocks in the field of analog circuit design. Here, a simple approach for implementing two different amplifiers with scalable power dissipation with respect to the operation frequency (or unity gain bandwidth), will be presented. The amplifiers are aimed to be used in replica bias circuit (see Chap. 3), and also MOSFET-C filter will be detailed in this chapter.

7.2.1 Low Power Folded-Cascode Amplifier

To implement a stable, low-power, high-gain, and power-scalable amplifier for low frequency applications (such as biasing circuits described in Sects. 3.4.3 and 3.4.4), folded-cascode topology can be a proper choice.

In a replica bias circuit, where the output swing of a STSCL circuit needs to be controlled, generally there is a very large loading capacitance and this capacitance is directly appears at the output of amplifier, and hence creates a very low frequency pole. A simplified schematic is shown in Fig. 7.2a. This low frequency pole at V_{BP} in addition to the second pole at V_P can cause some stability issues. For this reason, a single pole amplifier such as folded-cascode topology which is shown in Fig. 7.2b can be used to relax the stability issue.

Illustrated in Fig. 7.2b, the amplifier exhibits a unity gain bandwidth (UGBW) which is proportional to the transconductance of the input differential devices (g_m) as well as inversely proportional to the load capacitance (C_L):

$$\mathrm{UGBW} = \frac{g_m}{2\pi C_L}. \tag{7.1}$$

Biased in subthreshold regime, then: $g_m = I_B/(2nU_T)$ (n is the subthreshold slope factor of the input NMOS devices and U_T stands for the thermal voltage). Therefore:

$$\mathrm{UGBW} = \frac{1}{2nU_T} \cdot \frac{I_B}{C_L} \cdot \frac{1}{2\pi} \tag{7.2}$$

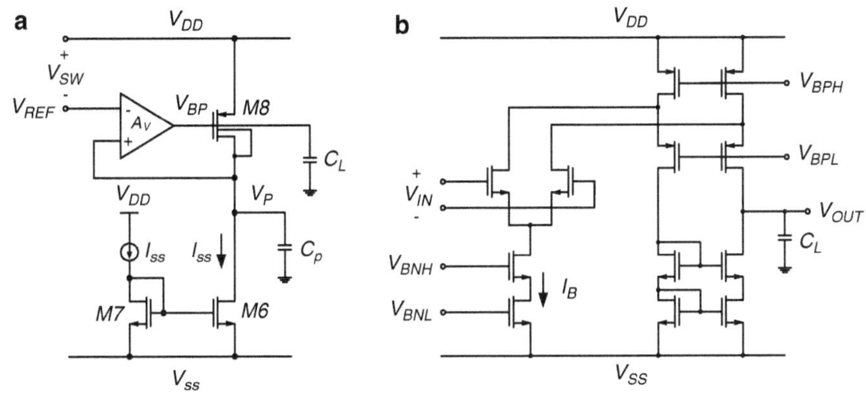

Fig. 7.2 (**a**) Simplified replica bias circuit. (**b**) Conventional folded cascode amplifier circuit topology

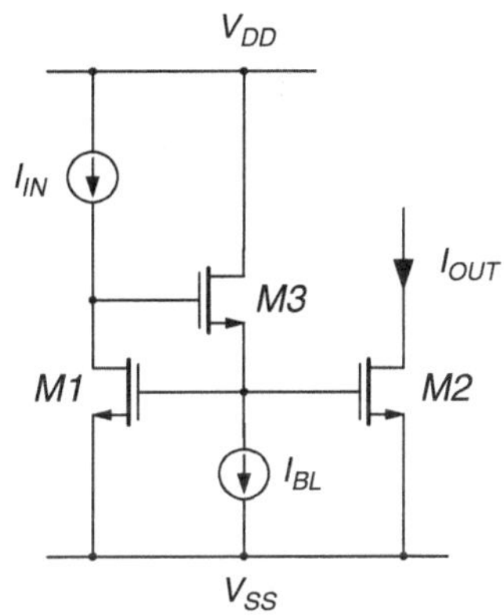

Fig. 7.3 Modified current mirror schematic to be used in very low bias current levels

which is proportional to the input bias current. It can be also shown that in the first order approximation, the gain and the phase margin of the folded-cascode amplifier are independent of the tail bias current. Therefore, as far as the circuit can be biased properly in subthreshold, the amplifier can be employed in different tail bias currents and hence different UGBW frequencies.

For bias currents below 100 pA, the current mirrors used in Fig. 7.2b start to enter linear region. This is mainly due to the shorted drain-gate voltages. As the gate voltage reduces due to reduction of the bias current, the drain voltage also reduces and hence pushes the transistor toward linear region. To overcome this problem, either the aspect ratio of the devices in current mirror should be reduced or the technique shown in Fig. 7.3 can be used in order to keep the drain voltage high enough to be in saturation region. In this schematic, a level shifter constructed by $M3$ and I_{BL} is used to increase the V_{DS} voltage of the current mirror devices ($M1$ and $M2$), and hence avoid operating in triode region.

The loop gain of the replica bias system shown in Fig. 7.2a can be calculated by:

$$\mathrm{LG}(s) = -\frac{A_V \cdot \frac{1}{n_p - 1}}{(1 - s/p_1) \cdot (1 - s/p_2)} \tag{7.3}$$

where n_p is the subthreshold slope factor of $M8$ in Fig. 7.2a, and gate-to-drain gain of $M8$ is: $-1/(n_p - 1)$. Indeed, in the replica bias circuit shown in Fig. 7.2a, there are two dominant poles at nodes V_{BP} and V_P:

$$p_1 = \frac{-1}{R_{\mathrm{OUT}} C_L} \tag{7.4}$$

where R_{OUT} is the equivalent output resistance of the OTA, and

$$p_2 = \frac{-1}{R_L C_P} \tag{7.5}$$

where $R_L \approx V_{\mathrm{SW}}/I_{\mathrm{SS}}$ is the equivalent resistance of the PMOS load (transistor $M8$). Since $C_L >> C_P$ and $R_{\mathrm{OUT}} >> R_L$, therefore, $| p_1 | << | p_2 |$ and the dominant pole of system is at the node V_{BP}. To have an acceptable phase margin (PM), the nondominant pole of this system, i.e., p_2, should be larger than the loop unity gain bandwidth. In order to have a phase margin of 60°, $| p_2 | \geq 3 \times \mathrm{UGBW}$, hence, using (7.2):

$$\frac{I_B}{I_{\mathrm{SS}}} \leq \frac{2n}{3} \cdot \frac{U_T}{V_{\mathrm{SW}}} \cdot \frac{C_L}{C_P} \cdot (n_p - 1). \tag{7.6}$$

It is very important to notice that by changing the bias current of STSCL circuit, I_{SS}, the bias current of amplifier, I_B, should also be scaled proportionally. If I_B does not scale proportional to I_{SS}, then under certain conditions, the nondominant pole of the system gets close to the dominant pole and pushes the system toward instability.

7.2.2 *Widely Adjustable Two-Stage Amplifier*

As will be explained later, one of the critical blocks for designing a power-performance scalable MOSFET-C filter is the widely adjustable amplifier that is required in this topology. To implement a filter with scalable power consumption proportional to its cutoff frequency, it is necessary to design a scalable power amplifier, as well. As illustrated in Fig. 7.4a, in this work a two stage amplifier topology has been utilized for this purpose. It can be shown that the UGBW of this amplifier is also proportional to the bias current of the input stage as presented in (7.2). Meanwhile, to have a phase margin of at least 60°, it is necessary to have [1]:

$$\frac{1}{3} \cdot \frac{C_C}{C_L + C_C} \geq \frac{g_{m1}}{G_L}. \tag{7.7}$$

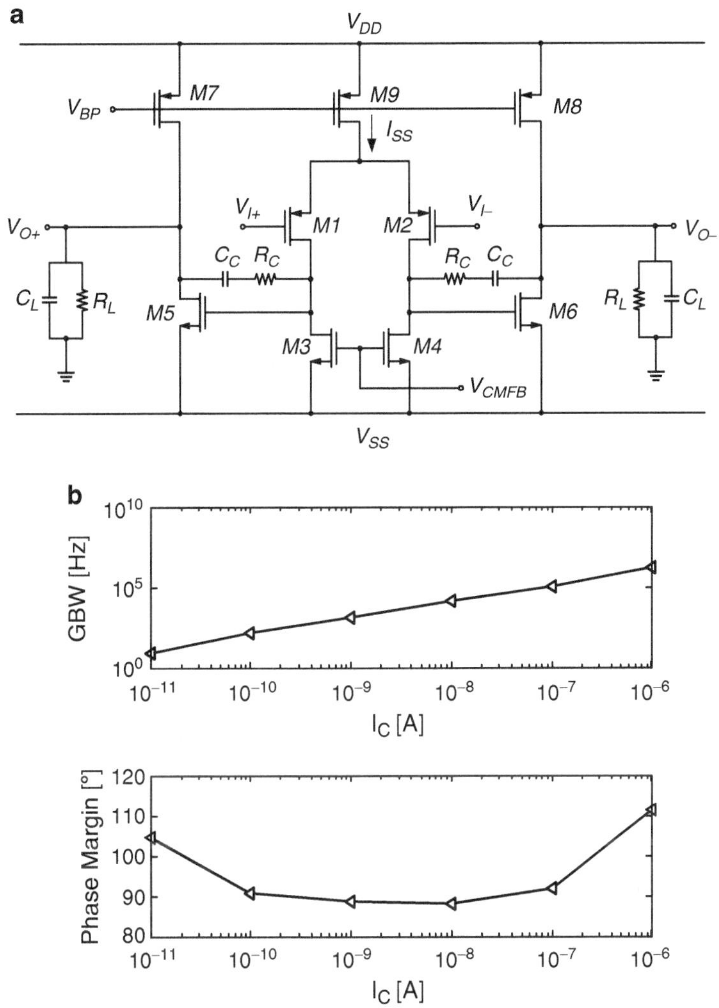

Fig. 7.4 (**a**) Circuit schematic of the amplifier. (**b**) Simulated unity gain bandwidth (UGBW) and phase margin of the amplifier for different current bias values. In this plot, I_C is the reference current value used to change the filter cutoff frequency

Since the value of g_{m1} and $G_L = 1/R_L$[1] are both proportional to the bias current, by proper choosing the size of the devices, the right hand side of (7.7) can be made bias current independent. Therefore, while the devices are in subthreshold regime, the stability of the circuit can be guaranteed. In this figure, R_C is implemented using NMOS devices to follow the variations of the bias current. Figure 7.4b shows the simulated gain and phase margin of the proposed amplifier in different bias currents.

[1] R_L is the variable resistors used to construct the MOSFET-C filters will be described in Sect. 7.4.

7.3 Transconductor-C Filter Design

Transconductor-C or g_m-C topology is very suitable for implementing very high [2] or very low frequency ([3] and [4]) filters. The main issue associated with this type of filters is their poor linearity performance. In this type of filters, transconductors are the critical components and directly affect the linearity of the circuit. To reach the desired linearity performance, it is necessary that all the transconductors remain linear for their entire input differential voltage swing. This requirement calls for some complicated techniques to improve the linearity of the transconductor circuit. Employing complicated circuit techniques is generally associated with some degradation in frequency and noise performance of the filter. This problem becomes more evident in widely adjustable filters, where the transconductance needs to be varied in a very wide range.

In the following, a very simple approach for improving the linearity of the filter is proposed which reveals the demand for having linear transconductor circuits. Using very simple circuit topologies for transconductors helps to achieve the desired tuning range with a good linearity and noise performance, simultaneously.

7.3.1 Proposed Biquadratic Filter Topology

Single stage differential pair operational transconductance amplifier (OTA), as illustrated in Fig. 7.5a, is one of the simplest transconductor topologies that can be used for implementing g_m-C filters. Transconductance of this OTA can also be tuned over a very wide range; from weak to strong inversion. The input transistors can be biased in weak, medium, and strong inversion regimes. This property makes this topology very suitable for widely tunable filters. However, the main drawback of this topology is its very limited linearity range. Indeed, the linear input voltage swing of this OTA is limited to few U_Ts (independent of the bias current) in weak-inversion (subthreshold) regime [5] and to about $V_{\mathrm{DS,sat}}$ (proportional to $\sqrt{I_{\mathrm{SS}}}$) in strong inversion [6], which is not sufficient for most of the applications.

Figure 7.5b depicts the maximum input voltage swing to have a nonlinearity less than 5% at the output current of a differential pair circuit versus bias current and at different device aspect rations. Depicted in this figure, as long as the devices are in subthreshold regime, this voltage swing is almost constant and is a fraction of U_T. However, by increasing the bias current and entering into the strong inversion, this range increases. Biased in strong inversion, linearity can be improved by reducing the device aspect ratio and hence increasing the V_{DSsat}. As depicted in Fig. 7.5b, the linearity performance of the simple transconductor shown in Fig. 7.5a is very poor and the linearity of a g_m-C filter uses this block will be even worse. Therefore, it is necessary to employ special linearity improvement techniques to achieve the desired linearity performance. The approach which is proposed here is based on canceling the nonlinearity effect of the transconductor in topology level, and hence make it

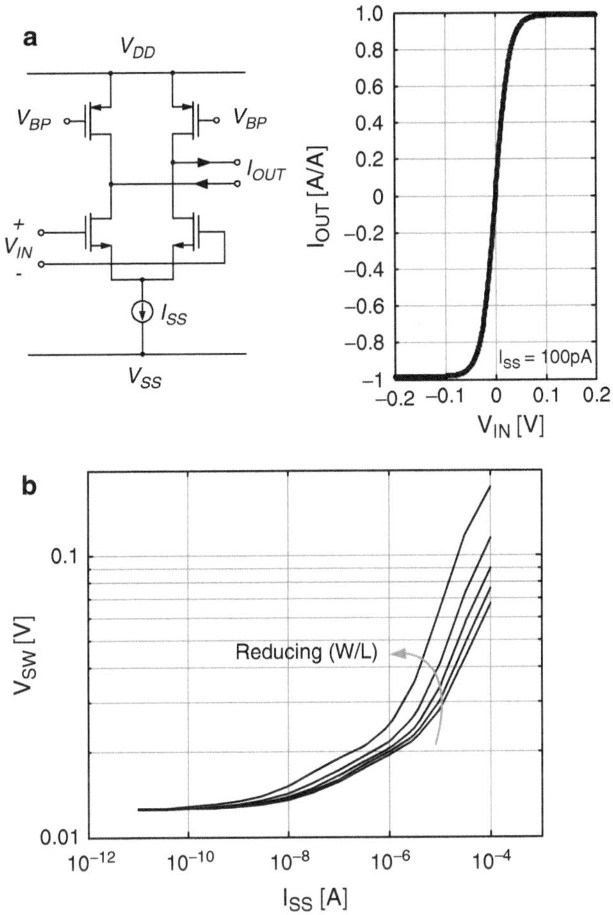

Fig. 7.5 (**a**) Single stage differential operational transconductance amplifier (OTA) can be used as a widely adjustable transconductor. Typical I/V characteristics of the differential pair OTA also is shown. (**b**) Maximum voltage swing at the input of differential pair OTA to have a nonlinearity less than 5% at the output current (nominal $(W/L) = 1.0\,\mu\text{m}/0.4\,\mu\text{m}$)

possible to use simple and power efficient transconductors. It is obvious that using transconductors with better linearity performance in this approach will result in even less circuit nonlinearity.

7.3.1.1 Proposed Circuit Topology

Figure 7.6a shows a conventional second order biquadratic g_m-C filter. In this simplified circuit diagram, there are two transconductors that convert the voltage to current as the following:

$$I_M = G_{m1} \cdot [(V_{\text{IP}} - V_{\text{IN}}) + (V_{\text{OP}} - V_{\text{ON}})] \tag{7.8}$$

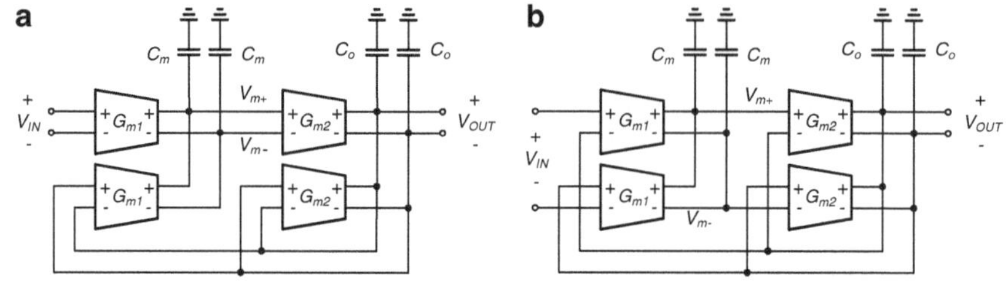

Fig. 7.6 Biquadratic g_m-C filter: (**a**) conventional topology and (**b**) modified topology with improved linearity performance

and:

$$I_O = G_{m2} \cdot [(V_{\mathrm{MP}} - V_{\mathrm{MN}}) + (V_{\mathrm{OP}} - V_{\mathrm{ON}})] \tag{7.9}$$

while the frequency characteristic of the filter is:

$$H(s) = \frac{\omega_0^2}{s^2 + (\omega_0/Q)s + \omega_0^2} \tag{7.10}$$

in which the cutoff frequency of the system is given by:

$$\omega_0 = \frac{G_m}{\sqrt{C_M C_O}} \tag{7.11}$$

and the quality factor of the filter is:

$$Q = \sqrt{\frac{C_O}{C_M}} \tag{7.12}$$

Based on (7.8) and (7.9), each transconductor converts two differential voltages to currents, and then the currents will be summed up together. Based on the configuration shown in Fig. 7.6a, the main problem arises when a transconductor should convert a differential voltage (such as $V_{\mathrm{IP}} - V_{\mathrm{IN}}$) to current. In this case, the transconductor needs to be very linear for the entire input voltage swing. To alleviate this requirement, (7.8) and (7.9) can be rewritten as the following:

$$I_M = G_{m1} \cdot [(V_{\mathrm{IP}} - V_{\mathrm{OP}}) + (V_{\mathrm{ON}} - V_{\mathrm{IN}})] \tag{7.13}$$

and:

$$I_O = G_{m2} \cdot [(V_{\mathrm{MP}} - V_{\mathrm{OP}}) + (V_{\mathrm{ON}} - V_{\mathrm{MN}})]. \tag{7.14}$$

In this way, the total current at the output of each transconductor, and hence the filter transfer function calculated in (7.10) remain unchanged. The only difference is that each transconductor needs to convert the difference of the two signals that

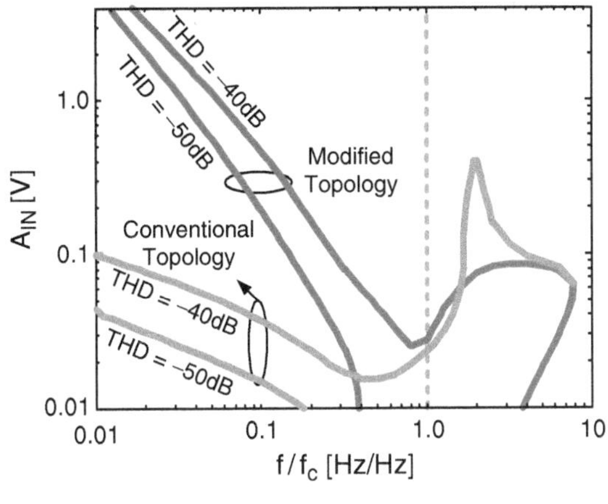

Fig. 7.7 Comparing the linearity performance of the two biquadratic filters shown in Fig. 7.6 based on behavioral modeling. Here, it is assumed that the input differential pair transistors are biased in subthreshold regime and transconductance can be calculated using (7.15)

are in phase together (or have a phase difference smaller than 90° for the in-band frequencies). Therefore, it is expected that the linearity performance of the filter improves considerably.

Implementations of the biquadratic filter which is based on (7.13) and (7.14) are shown in Fig. 7.6b. Figure 7.7 compares the linearity performance of the two filters shown in Fig. 7.6 based on behavioral modeling. In this figure, the input signal swing to have a THD (total harmonic distortion) of −40 dB and −50 dB are plotted for both topologies. As can be seen, for in band signal frequencies, the voltage swing can be much higher for the modified topologies shown in Fig. 7.6b.

In the proposed model, it is assumed that the input devices are biased in subthreshold regime and the frequency is normalized to the cutoff frequency of the filter. In very low input frequencies, the phase different between V_I and V_O, and also between V_M and V_O are very small. Hence, the linearity improvement is considerable. By increasing the frequency, and hence increasing the phase shift between the proposed signals, linearity enhancement decreases; however, the linearity performance is still much better for the modified topology shown in Fig. 7.6b.

In very high frequencies (in Fig. 7.7: $f \approx 2f_c$), the linearity performance of the two filters becomes comparable. As long as the input differential pair devices are in subthreshold regime, the linearity performance depicted in Fig. 7.7 remains valid. Therefore, this approach is very suitable for implementing widely tunable g_m-C filters with a good linearity performance. Similar improvement can be achieved for the devices in strong inversion.

It should be mentioned that this technique is applicable to other types of g_m-C filters, such as gyrator-based topologies. Meanwhile, it is possible to improve the linearity of the filter even more by employing some linearizing technique, such as the one shown in Fig. 7.8. The floating resistance needed in this transconductor can be implemented using the resistor shown in Fig. 7.11a to achieve the linearity and wide tuning range, simultaneously.

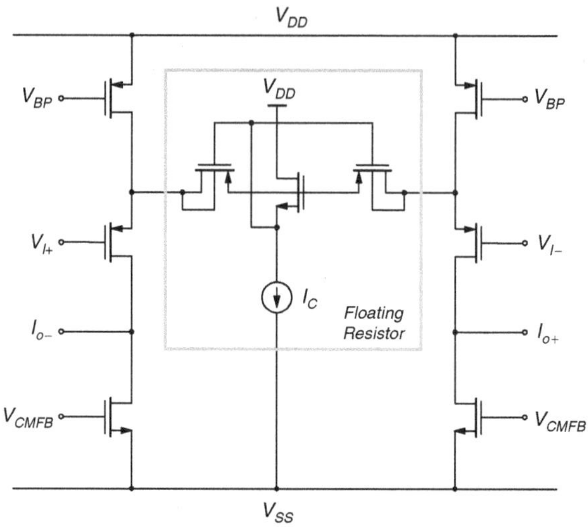

Fig. 7.8 Linearized transconductance suitable for wide tuning range applications

7.3.2 *Dynamic Range*

As long as the input differential pair transistors of the proposed OTA are in subthreshold regime, the large signal transconductance can be expressed by:

$$G_m = \frac{\partial I_{\text{OUT}}}{\partial V_{\text{IN}}} = \left(\frac{I_{\text{SS}}}{2nU_T}\right) \cdot \frac{1}{\cosh^2\left(V_{\text{IN}}/(2n_n U_T)\right)}. \tag{7.15}$$

In this case, the linearity performance does not depend on bias current of the OTA. Hence, it is expected that the filter exhibits a relatively constant linearity performance as depicted in Fig. 7.5 and can also be deduced from (7.15). By entering into the strong inversion in large current values, however, the linearity starts to improve.

On the other hand, it is expected that the total output rms (root mean square) noise remains independent to the cutoff frequency of the filter. Assuming that $G_{m1} = G_{m2} = G_m$ in Fig. 7.6, it can be concluded that the output noise power density ($v_{n,\text{out}}^2$) is:

$$v_{n,\text{out}}^2 = \mid H(j\omega) \mid^2 \cdot \left(1 + \frac{\omega^2}{\omega_0^2 Q^2}\right) \cdot \frac{i_{n,G_m}^2}{G_m^2} \tag{7.16}$$

where $i_{n,G_m}^2 = 4kT\gamma_{G_m} \cdot G_m$ is the current noise power corresponding to each transconductor (γ_{G_m} is the noise excess factor for the proposed transconductor). Therefore, based on (7.16), the output noise power density is inversely proportional to the G_m value. On the other hand, since the filter bandwidth is proportional to the

G_m value, the total output noise power which is proportional to the filter bandwidth, will remain unchanged with scaling the bias current (I_{SS}) or equivalently G_m.

Having a constant total rms noise in addition to the relatively constant linearity performance means that the dynamic range of the proposed filter remains constant as long as the differential pair devices in OTA are in subthreshold. By entering into the strong inversion, the DR will improve slightly proportional to the linearity improvement. This property is especially important when it is required to have a constant DR over the entire tuning range. Using programmable capacitance array or transconductor cell, generally this property could not be achieved.

7.3.3 Sixth Order g_m-C Filter

A simple OTA with folded cascode topology has been used to implement two sixth order Butterworth g_m-C filters based on Fig. 7.6a, b. Folded cascode OTAs consume more power compared to the simple differential pair OTAs; however, they can provide much more input common-mode range which can improve the linearity performance in both topologies shown in Fig. 7.6.

To reduce the chip area, a single filter which is switchable between the two topologies has been designed. For this purpose, each biquadratic stage uses two CMOS transmission gate switches to deliver the input signals to the transconductors according to Fig. 7.6a or b. Meanwhile, MOS type capacitors have been used in order to reduce the required Si area. Simulations show that two filters exhibit similar frequency responses and input referred noise values for the entire tuning range. Achievable cutoff frequency tuning range is $f_c = 20\,\text{Hz}$ to $10\,\text{MHz}$, corresponding to the control current values of $I_C = 10\,\text{pA}$ to $10\,\mu\text{A}$. In Sect. 7.5, extensive comparison between the simulation and measurement results of the proposed filters are provided.

7.4 MOSFET-C Filter Design

Cutoff frequency of a MOSFET-C filter can be adjusted by changing the size of capacitors or resistors. Using triode MOS based resistors, it is possible to have enough flexibility to compensate for process and environmental variations and tune the filter cutoff frequency on the desired value [7]. There are also some reports using varactors to provide the desired tuning range [8]. However, to have a very wide tuning range, generally programmable capacitor or resistor banks are needed. In this approach, the size of capacitors or resistors can be changed in a very wide range, and hence provide the desired adjustability range. The complexity and extra area due to the switchable components reduces the power and area efficiency of this approach.

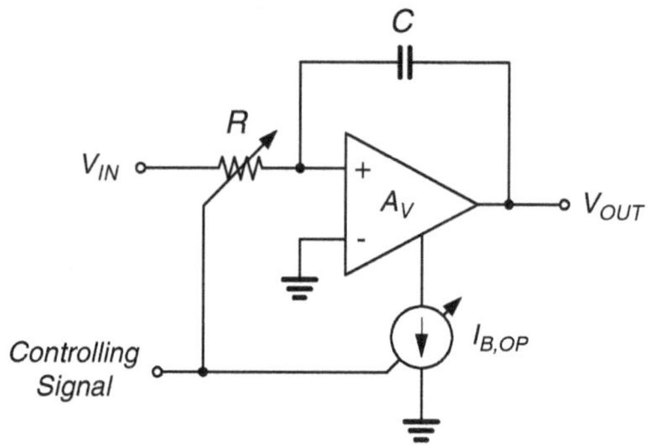

Fig. 7.9 Tunable active-RC (MOSFET-C) filter using a variable resistor. The power consumption of the amplifier is scalable with respect to the filter cutoff frequency

7.4.1 Circuit Topology

Figure 7.9 proposes a first order MOSFET-C filter that uses a variable resistance for adjusting its cutoff frequency. Here, a widely tunable resistor in addition to a high-gain and robust OTA with scalable power consumption are the main building blocks to implement a wide tuning range MOSFET-C filter. To scale the circuit power consumption with respect to the filter cutoff frequency (f_c), it is necessary to be able to change the power consumption of the amplifier (through $I_{B,\mathrm{OP}}$) proportional to the f_c (or inversely proportional to R).

7.4.2 High-Valued Pseudo-Resistance

To have a very wide tuning range as well as low power consumption, MOS devices biased in subthreshold regime can be employed. The exponential I/V characteristics of MOS devices in subthreshold makes the wide variation range for biasing condition possible. However, MOS devices in subthreshold regime have very poor linearity performance. As depicted in Figs. 7.10a and c, a PMOS device shows a medium linearity for voltage swing of in the order of U_T. To extend the linearity range of the device without the need for very large size devices, the configuration of Fig. 7.10b can be utilized [9]. In this configuration, the bulk terminal of the PMOS device is connected to its drain; hence, based on EKV model [5], and as explained in Chap. 3, the equivalent resistance of this device is:

$$R_{\mathrm{SD}} = \left(\frac{\partial I_{\mathrm{SD}}}{\partial V_{\mathrm{SD}}}\right)^{-1} = \left(\frac{n_p U_T}{I_{\mathrm{SD}}}\right) \cdot \left(\frac{\mathrm{e}^{V_{\mathrm{SD}}/U_T} - 1}{(n_p - 1)\mathrm{e}^{V_{\mathrm{SD}}/U_T} + 1}\right). \tag{7.17}$$

Based on (7.17), this device can be used as a resistor with medium linearity in a wider voltage swing range compared to the conventional configuration shown in Fig. 7.10a. The maximum voltage swing in this configuration is limited to about 500 mV, where the source-bulk diode starts to conduct a current comparable to the current of PMOS device.

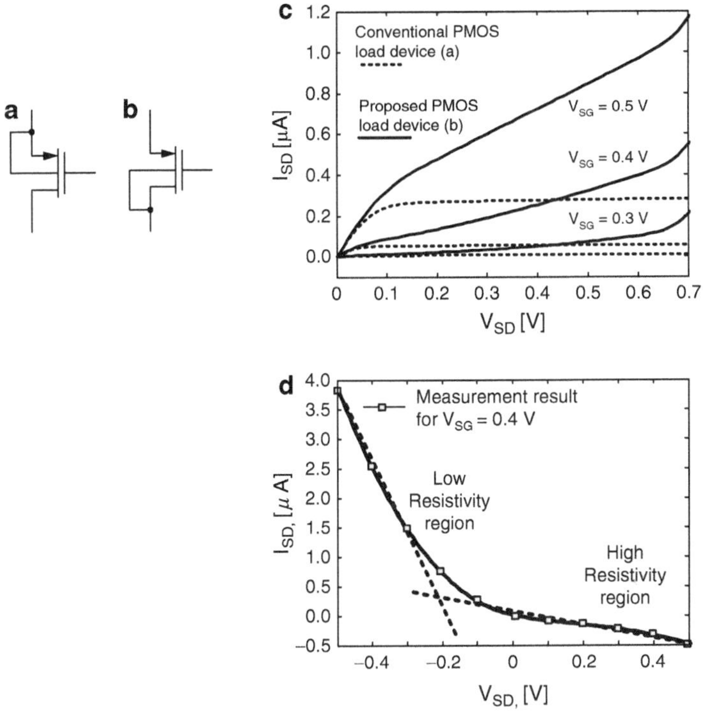

Fig. 7.10 High-valued resistance implementation based on subthreshold PMOS device: (**a**) conventional PMOS device and its I/V characteristics, (**b**) proposed PMOS device and its I/V characteristics with extended linearity range [9], (**c**) I/V characteristics of the devices shown in (**a**) and (**b**). (**d**) Measured I/V characteristics of the proposed floating resistor for $V_{SD} < 0$ V, and $V_{SD} > 0$ V

As explained in Chap. 3, when V_{SD} becomes negative, the current direction reverses, and the device switches to conventional configuration in which the bulk is connected to the source (Fig. 7.10d). In this case, the drain current will increase rapidly. This property can help to implement high valued *floating resistors* with a very wide adjusting range by connecting two back to back PMOS transistors as shown in Fig. 7.11a. The measured I/V characteristics of this floating resistance show moderate linearity in a very wide voltage range. Based on measurement results shown in Fig. 7.11b, this floating resistance exhibits medium linearity performance, and can be used to implement a widely tunable MOSFET-C filter. The adjustability range of the proposed floating resistance is shown in Fig. 7.11c.

Analysis: Here, a short analysis on behavior of the proposed floating resistance is provided. Regarding Fig. 7.11a, since the two transistors in series are not linear devices, $V_A \neq (V_{IN+} + V_{IN-})/2$. Using EKV model, it can be shown that:

$$V_A = V_0 - U_T \ln\left(\frac{\cosh \frac{\Delta V}{2n_p U_T}(n_p - 1)}{\cosh \frac{\Delta V}{2n_p U_T}}\right) \tag{7.18}$$

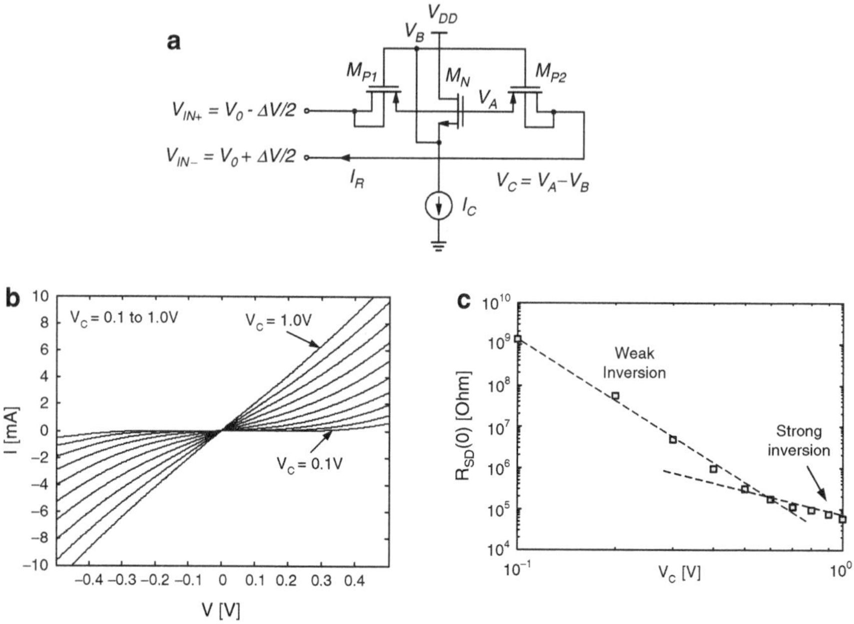

Fig. 7.11 Proposed floating resistance: (**a**) circuit schematic, (**b**) measured I/V characteristics of the proposed configuration for different V_C values, and (**c**) measured resistance of the proposed floating resistor with respect to the gate-source voltage of MN ($V_C = V_{GS,MN} = V_{SG,MP1,2}$). Here, $(W/L)_{pMOS} = 0.24\,\mu m/0.40\,\mu m$ and $(W/L)_{nMOS} = 1.0\,\mu m/0.40\,\mu m$

Therefore, V_A depends on input voltage swing. The voltage at this node has a "V" shape with respect to ΔV. The minimum occurs at $\Delta V = 0$, and it increases by increasing $\mid \Delta V \mid$. Having the value of ΔV, it is possible to calculated the current flow through MP1 and MP2:

$$I_R = I_0 \cdot e^{\frac{V_0 - V_B}{n_p U_T}} \cdot e^{-\frac{\Delta V}{2n_p U_T}} \cdot \left(1 - e^{\frac{\Delta V}{2U_T}} \cdot \frac{\cosh \frac{\Delta V}{2n_p U_T}}{\cosh \frac{\Delta V}{2n_p U_T}(n_p - 1)}\right). \quad (7.19)$$

Based on this, the circuit shown in Fig. 7.11 achieves its maximum resistivity when $\Delta V = 0\,V$, and then the resistivity drops when $\mid \Delta V \mid$ increases. If we generate V_B from V_A using a level shifter as shown in Fig. 7.11a, the linearity can be improved slightly. The reason is that when V_A increases by increase of $\mid \Delta V \mid$ as can be deduced from (7.18), the drop in resistivity will be canceled out partially by reduction of V_{SG} of the PMOS devices. In this case, the maximum value of the resistance will be no more at $\mid \Delta V \mid = 0$, but in two different points symmetrically placed with respect to the $\mid \Delta V \mid = 0$.

As a summary, when V_B has a constant value, the maximum resistance occurs at $\mid \Delta V \mid = 0$, while if V_B generated from V_A, the maximum resistivity will occur

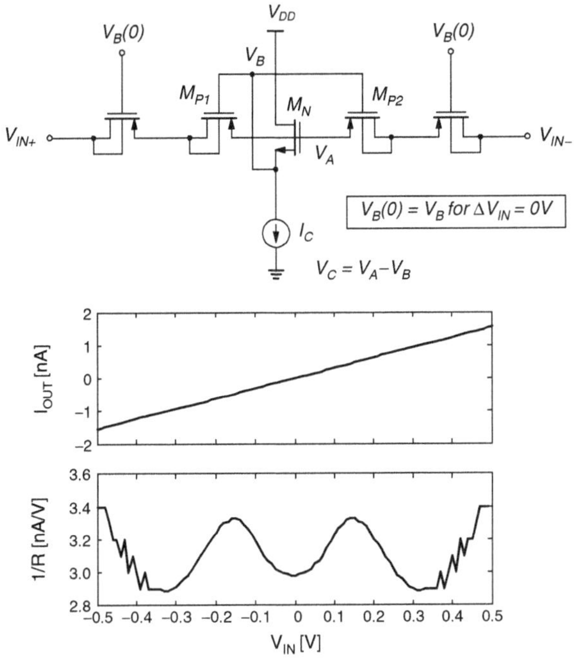

Fig. 7.12 High-valued floating resistance with improved linearity

at two different symmetric points. One can quickly propose using a combination of the two possible topologies in order to improve the linearity. For example, a series of the two circuits like the one shown in Fig. 7.12 could be used to improve the linearity. As can be seen, using a very simple technique linearity has been improved considerably.

It is interesting to study the performance of this circuit when V_{SG} becomes negative or the device will be in accumulation mode. In this situation, the equivalent resistance of the device will become very high with little dependence on the gate voltage. The behavior of the circuit for two different possible topologies are shown in Fig. 7.13. Both circuit topologies show a very large resistivity with a relatively good linearity performance. As depicted in this figure, the resistance is in the order of 50–500 GΩ. Monte Carlo simulations show very little variation on absolute resistance value and the linearity of this resistance.

7.4.3 *Dynamic Range*

The topology of the MOSFET-C shown in Fig. 7.9 is well suited for implementing constant dynamic range (DR) widely adjustable filters. This property is mainly due

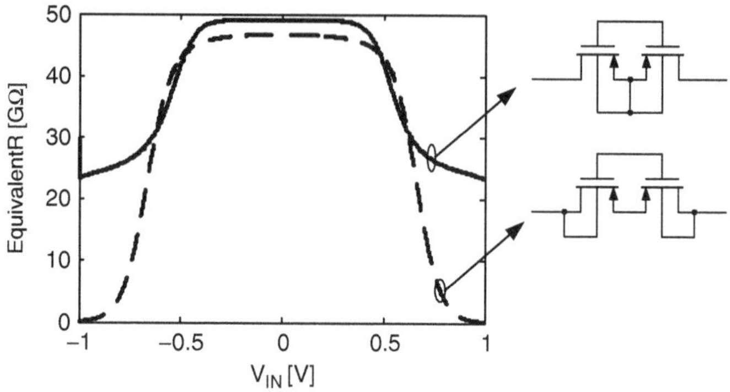

Fig. 7.13 Extreme high-valued resistance using negative V_{SG} values

to the almost constant noise and linearity performance of the filter over its tuning range. The total rms (root-mean square) input referred noise of the filter shown in Fig. 7.9 is:

$$v^2_{n,\text{rms,in}} = \gamma_F \cdot (k \cdot T/C) \tag{7.20}$$

in which γ_F indicates the circuit excess noise factor and depends on topology of the amplifier, resistors, and filter frequency transfer function especially filter quality factor (Q), k is Boltzmann's constant, and T is the junction temperature in Kelvin. Regarding (7.20) and assuming that the noise of amplifier scales with its power consumption (or equivalently, assuming that γ_F is bias independent), constant capacitor size in addition to the scalable amplifier power consumption results in constant rms filter noise in different cutoff frequencies.

On the other hand, based on (7.17) the linearity of the resistor introduced in Fig. 7.10b is independent of the bias current or V_{SG}, and the dependence on V_{SD} is the same for different bias currents. In other words, the nonlinear part of I/V characteristics of the device only depends on V_{SD}/U_T, and hence the nonlinear component remains the same for different values of bias current or V_{SG} value. Using Taylor expansion for (7.17):

$$R = R_0 \cdot \mathrm{e}^{-\frac{V_{SD}}{U_T} \cdot \frac{n_p - 1}{n_p}} \approx R_0 \cdot \left(1 - \alpha \cdot V_{SD} + \frac{\alpha}{2} \cdot V^2_{SD}\right) \tag{7.21}$$

where:

$$\alpha = \frac{1}{U_T} \times \frac{n_p}{n_p - 1} \tag{7.22}$$

Therefore, nonlinearity of the proposed resistance depends weakly on the biasing condition through n_p, and hence can be assumed to be approximately independent of the biasing conditions. Based on this, as long as the devices are in subthreshold regime, the linearity performance of the resistance shown in Fig. 7.11 remains unchanged. The linearity improves by entering into the medium and the strong inversion regions.

As the noise and linearity performance of the proposed filter remain relatively constant with respect to the change in biasing condition, and hence the cutoff frequency of a filter, it can be concluded that the dynamic range of the filter remains almost constant over its tuning range.

7.4.4 Second Order MOSFET-C Filter

Using the proposed floating resistor topology, a second order MOSFET-C filter has been designed. As illustrated in Fig. 7.14, the cutoff frequency and the quality factor (Q) of this filter can be tuned independently by adjusting the value of the resistors [10].

Simulations show that the cutoff frequency of the filter can be adjusted from 10 Hz to 200 kHz. The linearity performance of the filter remains almost constant as while as the devices are in subthreshold. For high bias currents, when the devices are entering into medium and strong inversion, linearity slightly improves. In low input frequencies ($f_{\text{in}} << f_c$), the circuit transfer characteristic depends on the ratio of resistors. Therefore, the nonlinearity of the resistors is not very much important as far as they are well matched. In higher frequencies, when both capacitors and resistors are participating in constructing the output signal, then nonlinearity of the resistors become important.

In Sect. 7.5.1, the linearity performance of this filter based on measurement results have been studied extensively.

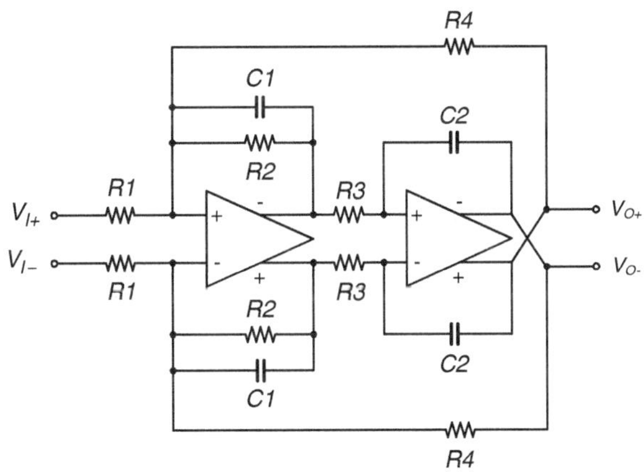

Fig. 7.14 A second order MOSFET-C filter. All the resistors are implemented using the proposed floating resistor shown in Fig. 7.11a. Quality factor of this filter can be tuned through $R2$ independent to the cutoff frequency. In this design, $R1 = R3 = R4$

7.5 Experimental Results

A second order MOSFET-C filter and a sixth order g_m-C filter based on the topologies introduces in this chapter have been implemented in 0.18-μm CMOS technology. The chip photomicrograph of the filters is shown in Fig. 7.15. In the following, the measurement results on these two test chips will be explained. The most important parameters for this study are tuning range, Δf_C, power efficiency over the entire tuning range P_{diss}/f_C, linearity, noise, and dynamic range behavior.

Internal output buffers have been used to isolate the outputs of each filter from external loading effects. The cutoff frequency of both filters can be adjusted using external bias currents. Measurements have been done using chip-on-board test setup.

7.5.1 *MOSFET-C Filter*

The proposed second order MOSFET-C filter occupies a silicon area of 420 μm× 210 μm while uses MiM capacitors. As depicted in Fig. 7.15, it is possible to compare the area of the proposed floating resistors (8 μm × 10 μm) with the other components in this filter.

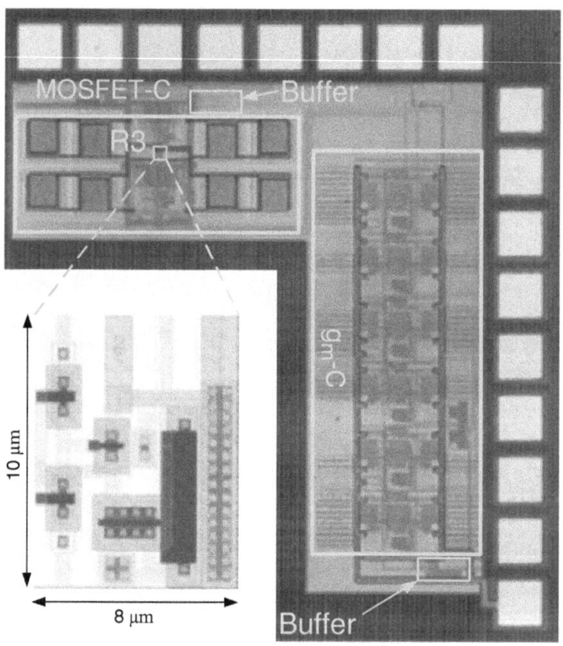

Fig. 7.15 Chip photomicrograph of the proposed filters implemented in 0.18 μm CMOS technology

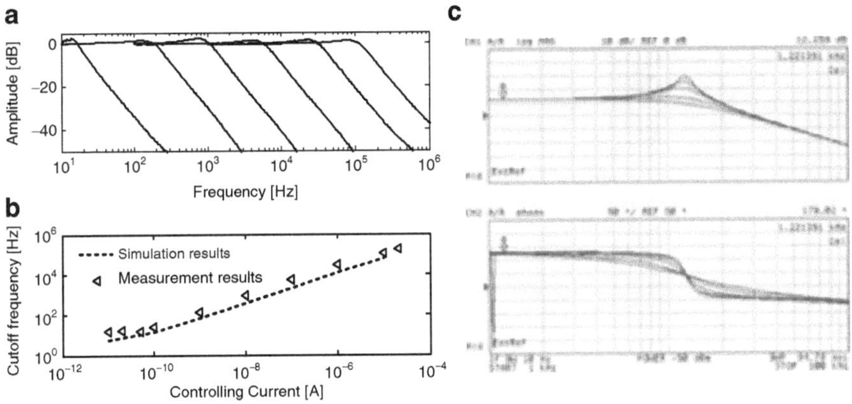

Fig. 7.16 Measured MOSFET-C filter characteristics: (**a**) frequency transfer characteristics. (**b**) cutoff frequency versus tuning current in comparison to the simulation results, and (**c**) Q tuning by changing $R2$ value at $I_C = 1$ nA

Frequency Response: The measured frequency response of the filter versus input controlling current (I_C) is shown in Fig. 7.16a. In this measurement, bias current of all the resistors as well as the bias current of the amplifiers are scaling with respect to I_C. As can be seen in this figure, the controlling current can be as low as I_C =100 pA for $f_C \simeq 20$ Hz. This low cutoff frequency has been achieved using 2 pF filter capacitors. Based on this, it can be seen that this topology can be very suitable for implementing very low frequency filters.

Figure 7.16b compares the tunability of this filter in comparison to the simulation results. The measured cutoff frequency of the filter is $f_C = 20$ to 184 Hz which is slightly less than five decades. The measured frequency response shows a very good agreement with the simulation results. The small difference that can be seen between measurement and simulation results which is a relatively constant ratio over the entire range is mainly due to the difference between capacitor values in the simulations and measurements. The normalized power consumption of the proposed second order filter is 1,080 pW/Hz.

As depicted in Fig. 7.16c, it is possible to adjust the Q of the filter independent to the cutoff frequency through changing $R2$ in Fig. 7.14. Measured output phase of the proposed second order MOSFET-C filter shows that there is a negligible variation on the filter cutoff frequency when the quality factor of the filter is changing. On the other hand, based on Fig. 7.16a changing the cutoff frequency does not change the quality factor of the filter.

Dynamic Range: The measured third intercept point (IP3) and noise of this filter is shown in Fig. 7.17a, b. By changing the controlling current, I_C, the cutoff frequency has been changed, and in each point noise and IP3 of the filter are measured.

As expected, the total noise power remains fairly constant over the entire tuning range. The total noise power for this filter is in the range of 45–55 μV_{rms} when the cutoff frequency is changing from about 80 Hz to 184 kHz.

Meanwhile, while the devices in the proposed floating resistors shown in Fig. 7.11a are in subthreshold regime, the filter exhibits a constant IP3. When the

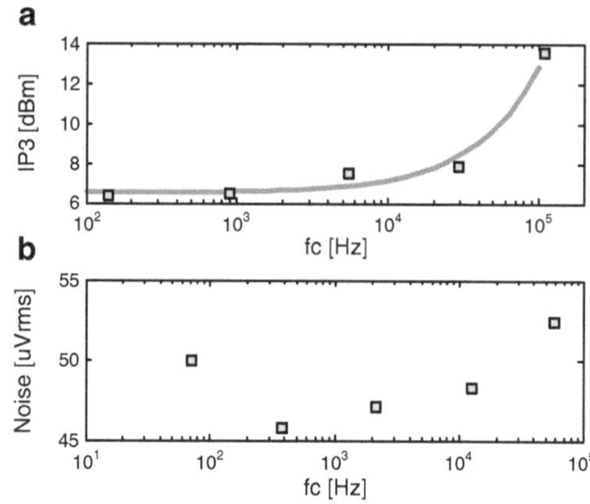

Fig. 7.17 Measured (**a**) third order intermodulation intercept point and (**b**) noise of the proposed MOSFET-C filter

devices enter strong inversion, IP3 improves by increasing the controlling current. This behavior can be seen in the measurements as depicted in Fig. 7.17a. The IP3 of the filter is slightly less than 8 dBm for low cutoff frequencies and starts to increase for frequencies above 10 kHz and finally reaches 14 dBm for $f_C = 100$ kHz.

7.5.2 g_m-C Filter

This section explains the measurement results for a g_m-C filter prototype fabricated with 0.18- μm technology. The proposed sixth order g_m-C filter occupies a silicon area of 620 μm × 250 μm while uses MOS capacitors in order to reduce the chip area. The transconductor used for implementing this filter are based on simple differential pair topology shown in Fig. 7.5a. A folded-cascode topology is employed to increase the input common mode range of the transconductors.

The fabricated filter is based on a configurable topology that can be switched between conventional and modified biquadratic g_m-C topologies shown in Figs. 7.6a and b. In this way it is possible to measure and compare the performance of both topologies. This has been done using a simple switching network constructed of transmission gates at the input of G_{m1} and G_{m2} in Fig. 7.6.

Frequency Response: The measured frequency response of the filter versus input frequency controlling current (I_C) is shown in Fig. 7.18a. Based on measurement results, the controlling current can be as low as $I_C = 10$ pA for $f_C \simeq 100$ Hz. The upper cutoff frequency limit is $f_C \simeq 10$ MHz which corresponds to 10 μA controlling current.

The controlling current (I_C) directly is applied to the tail of differential pair transistors of each transconductor shown in Fig. 7.5a. As illustrated in Fig. 7.18, for cutoff frequencies less than 1MHz there is no need for adjustment of Q. At this range of frequencies, the quality factor of the filter depends on the ratio of capacitors and transconductors as depicted by (7.12).

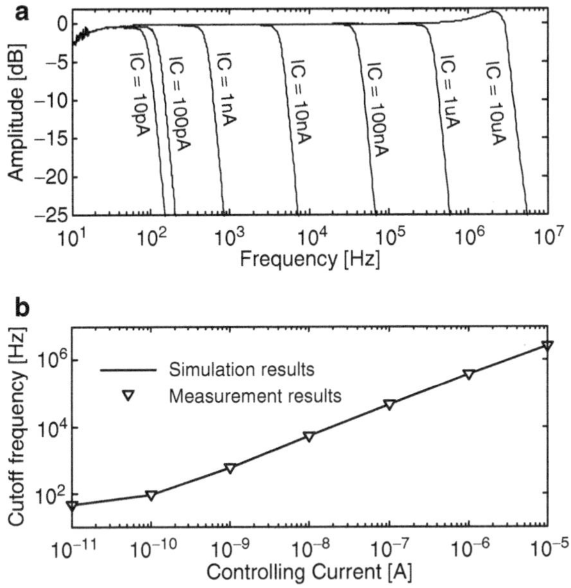

Fig. 7.18 Measured g_m-C filter characteristics: (**a**) frequency transfer characteristics and (**b**) cutoff frequency versus tuning current in comparison to the simulation results

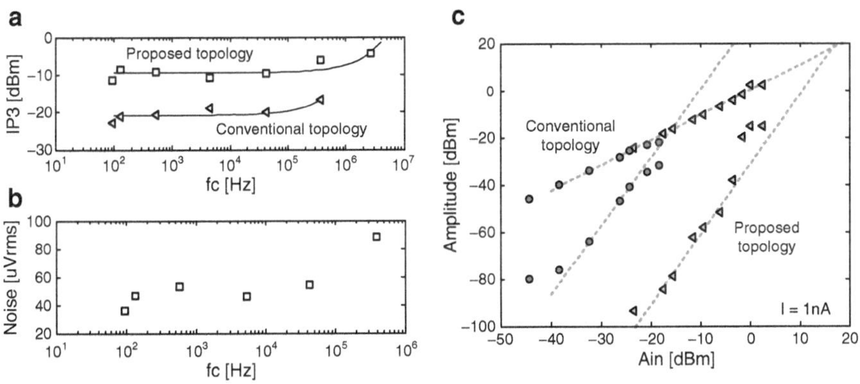

Fig. 7.19 Measured: (**a**) third order intermodulation intercept point (IP3) and (**b**) noise of the proposed g_m-C, for different filter cutoff frequencies. (**c**) Third order harmonic distortion (HD3) of the proposed g_m-C filter in comparison the conventional topology when $I_C = 1\,\text{nA}$, and $f_{\text{in}} = f_c/4$

The normalized power consumption of the proposed sixth order filter is 344 pW/Hz. Figure 7.18b compares the tunability of this filter in comparison to the simulation results which are in a very good agreement.

Dynamic Range: The measured noise and third intercept point (IP3) of this filter are shown in Fig. 7.19a, b. As expected, the total noise power remains relatively constant for the entire tuning range. In high frequencies, the noise power has increased due to the increase of quality factor of the filter. Meanwhile, as long as the input differential pair devices are in subthreshold regime, filter exhibits a constant IP3. When devices enter strong inversion, IP3 improves by increasing the controlling current.

Compared to the conventional topology, the IP3 of the filter has been improved by about 10 dB. This improvement is about 30 dB for total harmonic distortion (THD). The measurement results show that the in-band harmonic distortion is 25–35 dB less for the modified biquadratic filter. Figure 7.19c shows the measured third harmonic distortion (HD3) results for these two topologies at $I_C = 1$ nA. More than 10 dB improvement in IP3 has been achieved using very simple modification in the filter topology. It is clear that using transconductors with better linearity performance can result in even better IP3 values.

7.5.3 *Figure of Merit*

Table 7.1 summarizes the specifications of the two filters designed in this work. While g_m-C filter consumes 60 pW/Hz/pole and occupies and area of 0.027 mm^2/pole, the MOSFET-C filter consumes 540 pW/Hz/pole occupying 0.045 mm^2/pole. The extra normalized power consumption and silicon area in MOSFET-C filter are the costs paid for achieving better linearity performance. The designed MOSFET-C filter exhibits four orders of magnitude tuning range while the adjustability range of the g_m-C filter is about five decades.

Figure 7.20 compares the figure of merit (FOM) of this work with some previously published reports based on the figure of merit (FOM) introduced in [11]:

$$\text{FOM} = 10\log\left(\frac{\text{IMFDR}_{\text{lin.}} \times \Gamma_f}{P_{\text{diss}}/(N \times f_c)}\right) \tag{7.23}$$

Table 7.1 Specifications of the Filters

Parameter	MOSFET-C	g_m-C	Unit
V_{DD}	1.8	1.8	(V)
Technology	0.18-μm CMOS	0.18-μm CMOS	[-]
Order	2	6	[-]
$f_{c,\text{min}}$	20	100	(Hz)
$f_{c,\text{Max}}$	184 k	10 M	(Hz)
$f_{c,\text{Max}}/f_{c,\text{min}}$	9,200	100,000	(Hz/Hz)
Normalized P_{diss}	540	60	(pW/Hz/pole)
Area	0.09	0.16	(mm^2)
Normalized Area	0.045	0.027	(mm^2/pole)
Noise	50	60	(μV$_{\text{rms}}$)
IP3	7	−8	(dBm)
IMFDR	70	55	(dB)
FOM	202	197	[-]

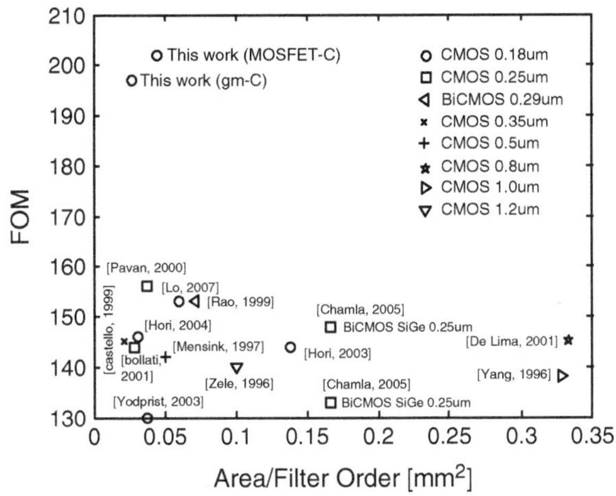

Fig. 7.20 FOM comparison to some other reports versus normalized filter area (area is normalized to the order of the filter). The data points used in this figure are extracted from [11] and [12]

in which $IMFDR_{lin.}$ stands for intermodulation free dynamic range (without unit) and Γ_f indicates the ratio of the maximum to minimum filter cutoff frequencies.

Figure 7.20 shows that the proposed filters exhibit much better FOM compared to the other already published reports. This improvement is mainly due to the simple topologies used to implement the circuits which has also concluded in a very area efficient implementation as illustrated in Fig. 7.20.

7.6 Conclusion

In this chapter, we introduced two techniques for implementing continuous-time filters (one MOSFET-C filter and one g_m-C filter) with very wide tuning range.

The proposed MOSFET-C filter uses a compact floating resistor implemented by subthreshold pMOS devices that can be adjusted in a very wide range. This technique is especially suitable for implementing very low frequency filters with a good linearity and dynamic range.

The g_m-C filter applies simple differential pair transconductors in balanced configuration to improve the linearity of the filter. This structure makes it possible to change the transconductance of the cells in a very wide range and have a good linearity performance.

Both filters are employing constant filter capacitances which implies constant rms noise level for the entire tuning range. Measurements show that the linearity also remains almost constant for both topologies and for their entire tuning range. In both filters, power consumption scales proportional to the cutoff frequency which makes these topologies very power efficient. Implements in 0.18-μm CMOS technology, the area of MOSFET-C and g_m-C filters are 0.09 mm^2 and 0.16 mm^2, respectively.

References

1. W. Sansen, *Analog Desing Essentials*, Springer, May 2006
2. S. Pavan, Y. Tsividis, and K. Nagaraj, "Widely programmable high-frequency continuous-time filters in digital CMOS technology," *IEEE J. Solid-State Circuits*, vol. 35, no. 7, pp. 503–511, Apr. 2000
3. A. Arnaud, R. Fiorelli, and C. Galup-Montoro, "Nanowatt, sub-nS OTAs, with sub-10-mV input offset, using series-parallel current mirrors," *IEEE J. Solid-State Circuits*, vol. 41, no. 9, pp. 1–10, Sep. 2006
4. P. Bruschi, N. Nizza, F. Pieri, M. Schipani, and D. Cardisciani, "Fully integrated single-ended 1.5-15-Hz low-pass filter with linear tuning law," *IEEE J. Solid-State Circuits*, vol. 42, no. 7, pp. 1522–1528, Jul. 2007
5. C. Enz, F. Krummenacher, and E. Vittoz, "An analytical MOS transistor model valid in all regions of operation and dedicated to low-voltage and low-current applications," *Analog Int. Circ. Signal Proc. J.*, vol. 8, pp. 83–114, Jun. 1995
6. P. R. Gray, P. J. Hurst, S. H. Lewis, and R. G. Meyer, *Analysis and Design of Analog Integrated Circuits*, Wiely, Fourth Ed., 2000
7. M. Banu, and Y. Tsividis, "An elliptic continuous-time CMOS filter with on-chip automatic tuning," *IEEE J. Solid-State Circuits*, vol. 20, no. 6, pp. 1114–1121, Dec. 1985
8. S. Chattarjee, Y. Tsividis, and P. Kinget, "0.5-V analog circuit techniques and their application in OTA and filter design," *IEEE J. Solid-State Circuits*, vol. 40, no. 12, pp. 2373–2387, Dec. 2005
9. A. Tajalli, E. Vittoz, Y. Leblebici, and E. J. Brauer, "Ultra low power subthreshold MOS current mode logic circuits using a novel load device concept," in *Proceedings of European Solid-State Ciruits Conference (ESSCIRC)*, Munich, Germany, pp. 281–284, Sep. 2007
10. M. Banu, and Y. Tsividis, "Fully integrated active RC filters in MOS technology," *IEEE J. Solid-State Circuits*, vol. 18, no. 6, pp. 644–651, Dec. 1983
11. D. Chamla, A. Kaiser, A. Cathelin, and D. Belot "A $G_m - C$ low-pass filter for zero-IF mobile applications with a very wide tuning range," *IEEE J. Solid-State Circuits*, vol. 40, no. 7, pp. 1143–1150, Jul. 2005
12. T.-Y. Lo, and C.-C. Hung, "A wide tuning range $G_m - C$ continuous-time analog filter," in *IEEE Transactions on Circuits and Systems-I: Fundamental Theory and Applications*, vol. 54, no. 4, pp. 713–722, Apr. 2007
13. C. Enz, M. Punzenberger, and D. Python, "Low-voltage log-domain signal processing in CMOS and BiCMOS," in *IEEE Transactions on Circuits and Systems-II: Analog and Digital Signal Processing*, vol. 46, no. 3, pp. 279–289, Mar. 1999
14. G. Bollati, S. Marchese, M. Dimecheli, and R. Castello, "An eight-order CMOS low-pass filter with 30-120 MHz tuning range and programmable boost," *IEEE J. Solid-State Circuits*, vol. 36, no. 7, pp. 1056–1066, Jul. 2001
15. J.-M. Stevenson, et al., "A multi-standard analog and ddigital TV tuner for cable and terrestrial applications," in *Proceedings of IEEE International Solid-State Circuits Conference (ISSCC) Digest of Technical Papers*, pp. 210–211, Feb. 2007
16. J. Fields, et al., "A 200 Mb/s CMOS EPRML channel with integrated servo demodulator for magnetic hard disks," in *Proceedings of IEEE International Solid-State Circuits Conference (ISSCC) Digest of Technical Papers*, pp. 314 315, Feb. 1997
17. E. Vittoz, "Weak Inversion for Ultimate Low-Power Logic," in *Low-Power Electronics Design*, Editor C. Piguet, CRC, 2005
18. A. Chandrakasam, and R. Brodersen, "Minimizing power consumption in digital CMOS circuits," in *Proceedings of the IEEE*, vol. 83, no. 4, pp. 498–523, Apr. 1995
19. P. Bruschi, F. Sebastiabo, and N. Nizza, "CMOS transconductors with nearly constant input ranges over wide tuning range," in *IEEE Transactions on Circuits and Systems-II: Analog and Digital Signal Processing*, vol. 53, no. 10, pp. 1002–1006, Oct. 2006
20. C. Enz, and E. Vittoz, *Charge-Based MOS Transistor Modeling: The EKV Model for Low-Power and RF IC Design*, Wiley, 2006

21. B. Pankiewicz, M. Wojcikowski, S. Szczepanski, and Y. Sun, "A field programmable analog array for CMOS continuous-time OTA-C filter applications," *IEEE J. Solid-State Circuits*, vol. 37, no. 2, pp. 125–136, Feb. 2002
22. A. Vasilopoulos, G. Vitzilaios, G. Theodoratos, and Y. Papananos "A low-power wideband reconfigurable integrated active-RC filter with 73 dB SFDR," *IEEE J. Solid-State Circuits*, vol. 41, no. 9, pp. 1997–2008, Sep. 2006
23. N. Rao, V. Balan, and R. Contreras, "A 3-V 10100-MHz continuoustime seventh-order 0.05 equiripple linear phase filter," *IEEE J. Solid- State Circuits*, vol. 34, pp. 1676–1682, Nov. 1999
24. J. A. De Lima and C. Dualibe, "A linearly tunable low voltage CMOS transconductor with improved common-mode stability and its application togm-C filters," in *IEEE Transactions on Circuits Systems-II: Analog Digital Signal Processings*, vol. 48, no. 7, pp. 649–660, Jul. 2001
25. S. Hori, T. Maeda, N. Matsuno, and H. Hida, "Low-power widely tunable Gm-C filter with an adaptive dc-blocking, triode-biased MOSFET transconductor," in *Proceedings of European Solid-State Ciruits Conference (ESSCIRC)*, Leuven, Belgium, pp. 99–102, 2004
26. S. Hori, T. Maeda, H. Yano, N. Matsuno, K. Numata, N. Yoshida, Y. Takahashi, T. Yamase, R. Walkington, and H. Hida, "A widely tunable CMOS Gm-C filter with a negative source degeneration resistor transconductor," in *Proceedings of European Solid-State Ciruits Conference (ESSCIRC)*, Estoril, Portugal, pp. 449–452, 2003
27. R. H. Zele and D. J. Allstot, "Low-power CMOS continuous-time filters," *IEEE J. Solid-State Circuits*, vol. 31, no. 12, pp. 157–168, Dec. 1996
28. C. H. J. Mensink, B. Nauta, and H.Wallinga, "A CMOS Soft-Switched transconductor and its application in gain control and filters," *IEEE J. Solid-State Circuits*, vol. 32, no. 7, pp. 989–998, Jul. 1997
29. R. Castello, I. Bietti, and F. Svelto, "High-frequency analog filters in deep-submicron CMOS technology," in *IEEE International Solid-State Circuits Conference (ISSCC) Digest of Technical Papers*, pp. 74–75, Feb. 1999
30. U. Yodprasit and C. Enz, "A 1.5-V 75-dB dynamic range third-order $G_m - C$ filter integrated in a 0.18 μm standard digital CMOS process," *IEEE J. Solid-State Circuits*, vol. 38, no. 7, pp. 1189–1197, Jul. 2003
31. F. Yang and C. Enz, "A low-distortion BiCMOS seventh-order bessel filter operating at 2.5 V supply," *IEEE J. Solid-State Circuits*, vol. 31, no. 3, pp. 321–330, Mar. 1996

Chapter 8
Scalable Folding and Interpolating ADC Design

8.1 Introduction

Analog-to-digital converters (ADCs) are one of the most critical building blocks in mixed-signal integrated circuits. The signals in analog domain are generally required to be converted to digital signals with enough resolution for further processing in the digital part of a system. For this purpose, after amplification and filtering, input signal will be digitized by an ADC block. As the dynamic range and speed of operation in this block are both very critical, generally this part of circuit consumes a considerable amount of power. Therefore, design of ultra-low power ADC circuits is very demanding.

In the next two chapters, some techniques for implementing ultra-low power and scalable ADC data converters are proposed.

8.2 Previous Art

Most of the reported ULP ADCs are based on the successive approximation register (SAR) topology [1–6]. The simple topology of this type of ADCs which consists of sampling switches, a comparator, a charge redistribution digital-to-analog (DAC) converter, and a simple digital block, makes it very suitable for medium resolution and low frequency applications (see Fig. 8.1). While the power consumption of logic part is fairy negligible, the two main sources of energy dissipation in this topology are (a) charging the binary weighted capacitors to reference voltages and (b) the comparison process. Since this topology needs two reference voltages, generally V_{DD} and V_{SS} are used for this purpose. In this way, the power consumption associated with the reference buffers can be eliminated [1], while the sensitivity to supply voltage variations increases.

As shown in Table 8.1, it is possible to reduce the power consumption of the data converter to only a few micro-Watts and still maintain a high resolution. The power

A. Tajalli and Y. Leblebici, *Extreme Low-Power Mixed Signal IC Design: Subthreshold Source-Coupled Circuits*, DOI 10.1007/978-1-4419-6478-6_8,

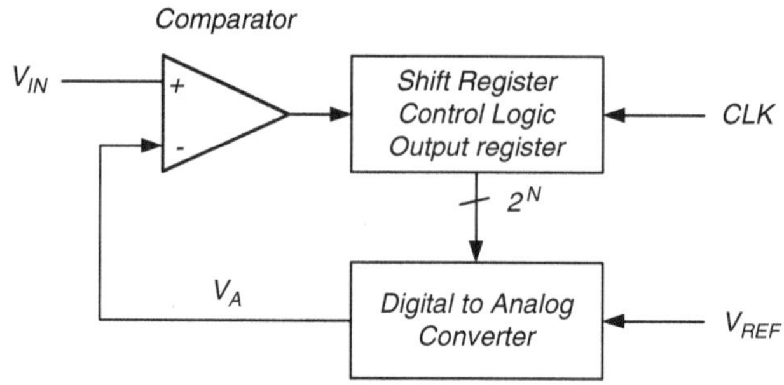

Fig. 8.1 Topology of a SAR ADC

Table 8.1 Reported ultra low power ADCs

Year	Reference	V_{DD} (V)	P_{diss} (μW)	Resolution (b)	f_s (kHz)	Technology	Area (mm^2)	Stand-by P_{diss} (nW)
2002	Scott [1]	1.0	3.1	8	100	0.25 μm	0.053	0.041
2003	Sauerbrey [2]	0.5	1.0	9	4.1	0.18 μm	0.11	
2004	Bonfini [3]	2.8	17.35	10	2.9	0.8 μm	0.8	
2007	Verma [4]	1.0	25	8–12	100–200	0.18 μm	0.63	
2007	Hong [5]	0.9	2.47	8	1,800	0.18 μm	0.7	
2007	Gambini [6]	0.45	7	6	1,500	90 nm	0.12	
2008	van Elzakker [7]	1.0	1.9	10	1,000	65 nm		
2008	Daly [8]	0.4	2.84	6	400	0.18	2	

consumption per conversion of this type of the reported ADCs is generally well below 1 pJ/conversion-step based on the figure of merit defined by:

$$\mathrm{FOM} = \frac{P_{\mathrm{diss}}}{2^{\mathrm{ENOB@DC}} \cdot \mathrm{ERBW}} \tag{8.1}$$

where ENOB and ERBW are effective number of bits and effective resolution bandwidth of the ADC, respectively [5].

The ADC reported in [1] uses a conventional SAR topology with a differential pair based comparator. The static bias current of the differential pair circuit is chosen such that it exhibits a negligible amount of noise at the input. Speed of operation of this comparator is determined by its static power consumption. Two NOT gate based buffers have been also used in this comparator whose power consumption depends on operation frequency. Therefore, the proposed ADC has a power consumption is composed of two parts: the first part is a static power which does not depend on sampling frequency, and the second part is dynamic power which depends on sampling frequency.

This comparator topology (regenerative resetable comparator based on differential pair) has also been used in [2]. In this work, a modified topology for SAR ADC has been proposed that separates the capacitive DAC and sample and hold (S & H) circuits. In this way, the input capacitance of the circuit does not depend on DAC capacitor array. The proposed ADC circuit can be operated with a supply voltage of as low as 0.5 V without using low V_T devices.

In [4], a rate and resolution scalable SAR ADC for micro-sensor networks has been reported. The supply voltage in this design has been chosen to be $V_{DD} = 1.0\,V$, very close to the optimum point where there is a balance between analog and digital part power consumption. The comparator circuit is constructed by few pre-amplification stages followed by a latch circuit. Both pre-amplifier and latch circuits are using auto-zeroing technique for cancelation of offset. To improve the input common-mode range, a comparator with combination of NMOS and PMOS input differential pairs has been introduced in [5]. The SAR ADC introduced in this work can be operated in a relatively wide range of sampling frequencies. While the power consumption of digital part and reference voltage part of this ADC are both scaling with operating frequency, the analog part exhibit a constant power dissipation. Therefore, this ADC is more suitable for higher frequencies.

The minimum reported power consumption per conversion-step is as low as 4.4 fJ/conversion-step in [7]. In this work, a new approach for charging and discharging the weighted capacitor array circuit is described to minimize the required energy for this process. Based on the proposed approach, some of the large capacitors are charging (or discharging) in multiple steps to reduce the energy dissipation which is proportional to $C \times \Delta V^2/2$.

While the power consumption of the ADCs reported in Table 8.1 are nominally more than 1 μW, in this chapter some techniques for reducing the power consumption below this limit will be proposed. A power-scalable folding and interpolating (FAI) ADC is proposed that is very suitable for ULP and medium resolution applications. Regular structure of the current-mode FAI ADCs provides this opportunity to change the power consumption and speed of operation in a very wide range.

8.3 Folding and Interpolating Analog-to-Digital Converter

Folding and interpolating ADCs have been already used widely for digitizing high bandwidth signals with medium range of resolution. This type of ADCs need fewer number of comparators and hence consume less power and silicon area compared to the flash architecture [9, 10]. A flash ADC can be considered as a fully parallel architecture which is very suitable for high frequency applications. However, as the number of comparators needed to implement the ADC increases rapidly with the number of bits ($N_{comp} = 2^{N_b} - 1$, N_b is the resolution bits of the ADC); hence, this technique is generally used for very high speed and low resolution applications ($N_b \leq 6$). Due to the less number of comparators needed in FAI ADCs, this architecture is more suitable for low power implementation of medium resolution (6–10 bits) data converters.

8.3.1 Basics

There are two main reasons for using FAI architecture for ultra-low-power ADC implementation with scalable sampling frequency. The first reason is that this

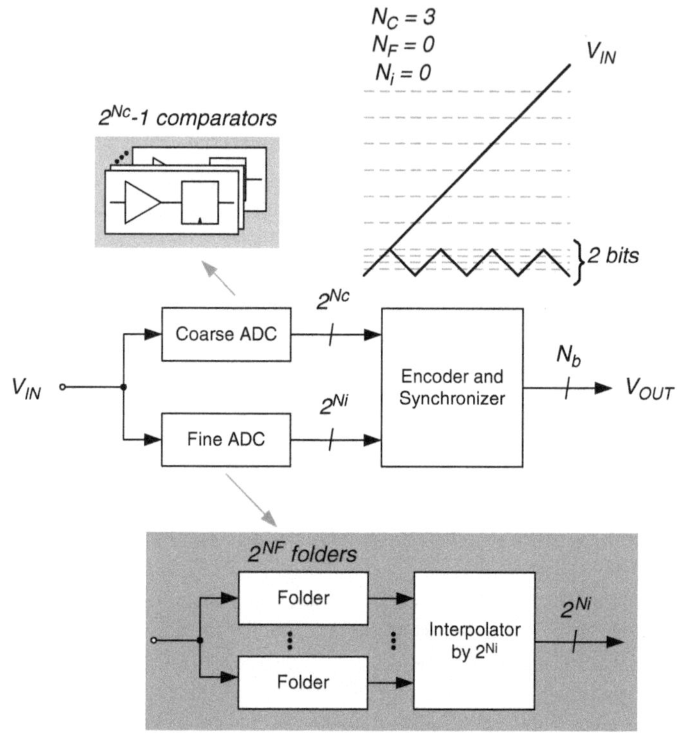

Fig. 8.2 Topology of a FAI ADC

topology can lead to very good power efficiency (below 1 pJ/con.-step). Specially using current-mode techniques, it is possible to reduce the power consumption of the circuit considerably. The second reason is that due to the regular structure of current-mode approach, this topology is suitable for scalable sampling frequency structures. These aspects are exploited in more details in the following.

As illustrated in Fig. 8.2, a FAI ADC consists of two parts: a coarse ADC with N_C bits of resolution and a folding and interpolating part with $N_{fi} = N_F + N_i$ bits of resolution. Coarse ADC is a flash ADC that extracts the highest N_C most significant bits. The rest of bits, i.e., $N_{fi} = N_b - N_C$ bits, will be extracted by the folding and interpolating part.

In the fine quantizer part, the input analog signal is folded by 2^{N_C} times. The folded signal, then will be converted to digital by a second flash ADC. Using this technique, the number of comparators will be reduced to $2^{N_C} + 2^{N_{fi}} - 2$ which is much smaller than 2^{N_b} in flash topology. In the last step, the digitized outputs of coarse and fine ADC parts need to be encoded, synchronized, and combined [10].

Using interpolation technique, it is possible to simplify the comparators to zero-cross detector circuits. Interpolator can be realized using resistors [21] or current mirrors [13]. Meanwhile, using more than one folder stage can help to simplify the design even more. For example, consider an 8B ADC: using a 3B coarse ADC, 4

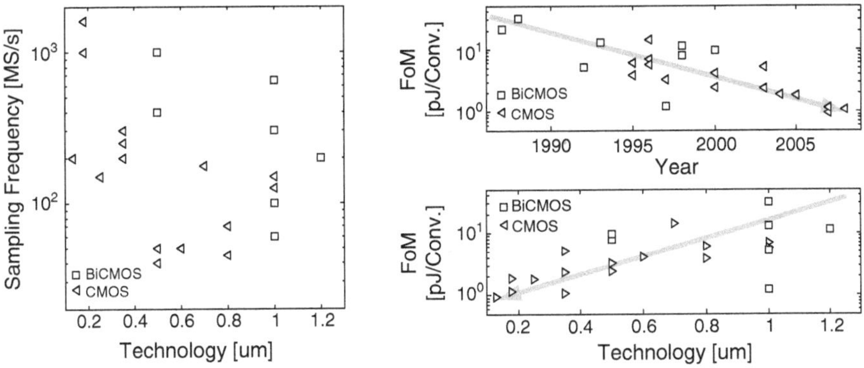

Fig. 8.3 Performance improvement of the reported FAI ADCs versus time and technology nodes

folder stages ($N_F = 2$), and interpolation factor of 8 ($N_i = 3$), the total number of comparators will be $(2^3 - 1) + (2^5 - 1) = 38$, instead of $2^{N_b} = 255$ for a full parallel flash ADC. This reduction in the number of comparators leads to proportional reduction in area and power consumption.

As shown in Fig. 8.2, for each full swing transition at the input, the output of the folder stage show 2^{N_C} transitions. The main issue with this behavior is the need for more bandwidth at the output of each folding stage. Therefore, a careful design is required to make sure that the limited bandwidth at the output of the folding stage will not degrade the general performance of the fine ADC in Fig. 8.2 [11].

Figure 8.3 shows the evolution of the reported FAI ADCs. While most of the early FAI data converters have been designed in BiCMOS or bipolar technologies, the speed improvement in the modern sub-micron CMOS technologies has enabled the designers to implement very high speed ADCs in this technology, as well. This figure also depicts how technology scaling has led to performance improvement in FAI ADCs. By technology scaling, it has been possible to implement GS/s range CMOS FAI ADCs. Meanwhile, figure of merit of the reported FAI ADCs has been improved using more advanced CMOS technologies. Most of the reported FAI ADCs are operating with a sampling frequency above 10 MS/s.

8.3.1.1 Nonideality Effects in FAI ADCs

One of the main issues in FAI topology is the effect of bandwidth limitation at the output of folder stage. Regarding the transfer characteristics of folder circuit, the signal at the output of folder stage have higher frequency components in comparison to the input signal. Therefore, the circuit bandwidth needs to be much higher than the input signal frequency (f_{in}). It is shown that the instantaneous output frequency (f_{out}) can be as high as:

$$f_{\text{out}} = K_F \times 2^{N_F} f_{\text{in}} \tag{8.2}$$

where K_F is a constant number ($K_F = \sqrt{2}$ in [11] and $K_F = \pi/2$ in [20]). Therefore, the performance of ADC can be degraded if the folder circuit does not have high enough bandwidth at the output. The bandwidth limitation can cause signal attenuation at the output, create some group delay, and even alter the zero cross points [11]. This problem can be mitigated using a front-end track-and-hold (TAH) circuit which determines the overall analog bandwidth of the system [20]. In this case, the main limiting factor is the settling behavior of the TAH circuit during the hold mode. The TAH circuit can be placed in front-end, which in this case will be very power hungry, or it can be distributed among folding stages as reported in [20]. In this case, the performance of FAI ADCs which are not using a frond end TAH is limited by distortion associated with the nonlinear folder stage [11, 13]. By increasing the input signal frequency the distortion due to displacement of zero crossing points increases even more.

In addition, mismatch among differential pair devices construct the folder circuit also causes some distortion [11, 13, 18].

8.3.2 Building Blocks and Design Tradeoffs

In this section, the general performance of flash ADCs as the basic building block for constructing a FAI ADC will be analyzed.

8.3.2.1 Resistor Ladder

A key component in a flash or FAI type of ADC is resistor ladder which is used to generate reference voltage levels. The value of resistors should be selected small enough to exhibit a very small time constant. The small time constant helps to have a fast settling time after each sampling. Regarding Fig. 8.4, the time constant in each node can be calculated by

$$\tau_j = R_{\mathrm{Lad}} C_{\mathrm{Lad}} \cdot \frac{j \cdot (2^{N_b} - j)}{2^{N_b}} \tag{8.3}$$

where N_b is the total number of resistors in the ladder and it is assumed that all the resistors in the ladder are equal to R_{Lad} and are connected to a parasitic capacitance of C_{Lad} as shown in Fig. 8.4. The maximum time constant occurs at the node $j = 2^{N_b} - 1$ which is equal to $\tau_{\mathrm{Max}} = R_{\mathrm{Lad}} C_{\mathrm{Lad}}/2$. To have a fast enough settling time and negligible error at the reference voltage of an ADC with resolution of N_b bits, it can be shown that the unit resistance should be smaller than:

$$R_{\mathrm{Lad}} < \frac{2}{C_{\mathrm{Lad}} f_s \ln (2^{N_b+1})} \tag{8.4}$$

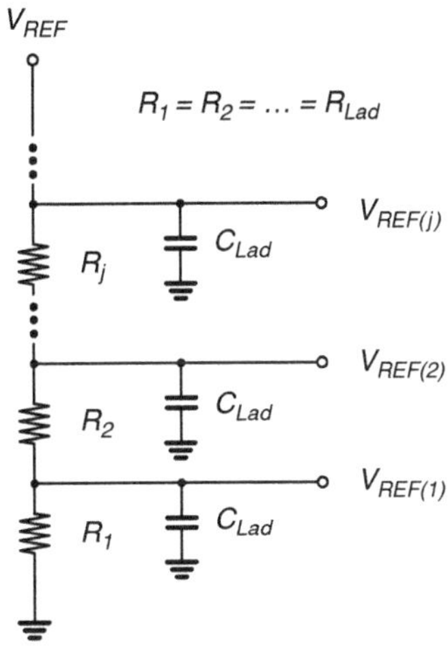

Fig. 8.4 Ideal resistor ladder to generate reference voltages

which indicates that the unit resistance depends on sampling frequency, f_s, load capacitance, C_{Lad}, and the resolution through N_b. The power consumption of the ladder circuit will also depend on the unit resistance value and can be calculated from (8.4):

$$P_{\mathrm{Lad}} \approx \frac{V_{\mathrm{REF}}^2 C_{\mathrm{Lad}} f_s \ln\left(2^{N_b+1}\right)}{2^{N_b+1}}. \tag{8.5}$$

The other important issue with the resistor ladder is matching properties of the components. Any mismatch among the resistors causes some variation on the reference values and hence reduces the circuit resolution. The variation on a reference value respect to its ideal values is

$$\frac{\Delta V_{\mathrm{Ref}(i)}}{V_{\mathrm{Ref}(i)}} \approx \frac{1}{n R_{\mathrm{Lad}}} \sum_{j=1}^{i} \Delta R_j \tag{8.6}$$

where $\sum_{j=1}^{2^{N_b}} \Delta R_j = 0$. Based on (8.6), integral nonlinearity (INL) of a flash ADC with ideal comparators will be limited by the matching property in resistor ladder. Thus, the limitation on INL due to the resistor mismatch can be indicated by:

$$\mathrm{INL}_{\mathrm{Ladder}} \approx \alpha_{\mathrm{Ladder}} \cdot \frac{\Delta R}{R_{\mathrm{Lad}}} \tag{8.7}$$

as it is shown in Figs. 8.5a and b. These figures are based on behavioral modeling in MATLAB. The behavioral modeling depicts that:

$$\alpha_{\mathrm{Ladder}} \approx 0.65 \times 2^{0.5381 N_b}. \tag{8.8}$$

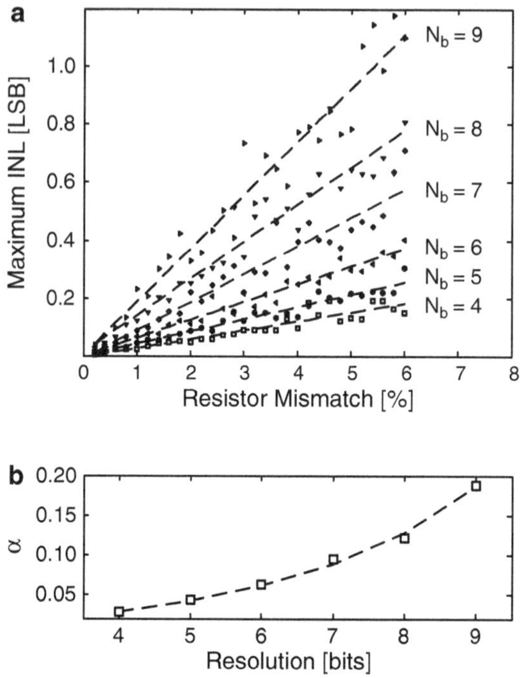

Fig. 8.5 (**a**) INL degradation due to the mismatch on resistors of reference voltage ladder simulated in MATLAB. (**b**) α_{Ladder} as a function of ADC resolution

Now, one can calculate the maximum acceptable mismatch of the resistors to have a integral nonlinearity error not more than $\text{INL}_{\text{Ladder}}$:

$$\frac{\Delta R}{R_{\text{Lad}}} < \frac{\text{INL}_{\text{Ladder}}}{\alpha_{\text{Ladder}}} \tag{8.9}$$

which practically puts a lower limit on area of resistor ladder. Indeed, variation on resistor value (ΔR), has a normal distribution with mean value of zero and standard deviation of:

$$\sigma_R \approx \frac{A_R}{\sqrt{W_R L_R}} \tag{8.10}$$

where A_R is a process dependent parameter and W_R and L_R are the width and the length of the proposed resistance and its area is $S_R = W_R \times L_R$. With this in mind, the lower limit on resistor area will be:

$$S_R > A_R^2 \cdot \frac{\alpha}{\text{INL}_{\text{Ladder}}}. \tag{8.11}$$

8.3.2.2 Offset Effect on Linearity

The other important factor that limits the performance of a flash ADC is the offset of comparators and pre-amplifiers which are shown in Fig. 8.6.

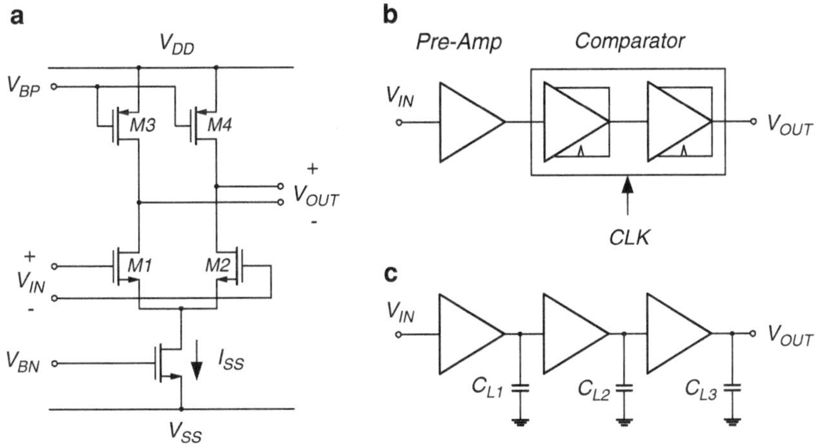

Fig. 8.6 Differential pair based pre-amplifier and comparator: (**a**) pre-amplifier, (**b**) a comparator consisting of pre-amplification and latch stages, and (**c**) a simple model for the proposed three stage circuit

The offset of the single stage amplifier shown in Fig. 8.6a can be estimated by:

$$\sigma_{\mathrm{OS}}^2 \approx \frac{A_{\mathrm{VTN}}^2}{W_N L_N} + \frac{A_{\mathrm{VTP}}^2}{W_P L_P} \frac{1}{A_V^2} \tag{8.12}$$

where A_{VTN} and A_{VTP} are process dependent parameters representing the matching properties of the threshold voltage of the MOS transistors [12], W and L are standing for width and length of NMOS and PMOS transistors, and A_V is the gain of the differential stage. Here, we have neglected the mismatch on $\beta = \mu C_{\mathrm{ox}} W/L$ [12].

As (8.12) implies, offset puts a lower limit on the size of transistors. The total input referred offset of pre-amplifier and comparator circuits should be small enough to have a negligible effect on the resolution or linearity performance of the ADC. Figure 8.7 depicts the effect of input referred offset on performance of ADC. In this figure we can see that as the input referred noise increases, the INL will degrade and the degradation is linearly proportional to the offset voltage. In this figure, input referred offset voltage V_{OS} is normalized to the corresponding LSB voltage, V_{LSB}. Meanwhile, INL is more sensitive to the offset when the bit resolution of ADC (N_b) is higher.

Indeed, the input referred offset can be modeled by a Normal random number with mean value of zero and RMS value of $\sigma_{\mathrm{OS,in}}$. In the presence of offset, the difference of the two consecutive transition points will be $V_{T(i)} = V_{\mathrm{REF}(i)} + V_{\mathrm{OS}(i)}$ and $V_{T(i+1)} = V_{\mathrm{REF}(i+1)} + V_{\mathrm{OS}(i+1)}$. Hence:

$$\Delta V_{\mathrm{REF}(i)} = \frac{V_{\mathrm{REF}}}{2^{N_b}} + V_{\mathrm{OS}(i+1)} - V_{\mathrm{OS}(i)} \tag{8.13}$$

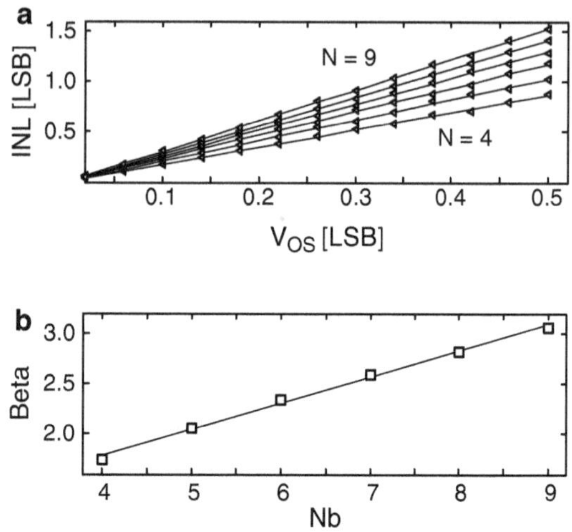

Fig. 8.7 Comparator offset effect on INL of the ADC deduced from MATLAB behavioral modeling

and then differential nonlinearity (DNL) will be:

$$\mathrm{DNL}_i = V_{\mathrm{OS}(i+1)} - V_{\mathrm{OS}(i)}. \tag{8.14}$$

Since V_{OS} has a normal distribution, then using (8.15) DNL would have a normal distribution with rms value of $\sqrt{2} \times \sigma_{\mathrm{OS,in}}$ or:

$$\sigma_{\mathrm{DNL}} = \sqrt{2} \times \sigma_{\mathrm{OS,in}}. \tag{8.15}$$

Therefore, to make sure that the DNL of the proposed ADC will not exceed $\mathrm{DNL}_{\mathrm{Max}}$:

$$\sigma_{\mathrm{OS,in}} < \frac{\mathrm{DNL}_{\mathrm{Max}}}{3\sqrt{2}}. \tag{8.16}$$

Here, it is assumed that $\mathrm{DNL}_{\mathrm{Max}} \approx 3\sigma_{\mathrm{DNL}}$. Using (8.16) and (8.12), now it is possible to size the devices and hence design the front end circuit.

It is also possible to estimate the acceptable offset voltage based on INL. Regarding Fig. 8.7 one can show that:

$$\mathrm{INL} = \beta \cdot \frac{\sigma_{\mathrm{OS,in}}}{V_{\mathrm{LSB}}} \tag{8.17}$$

in which $\beta \approx 0.255N_b + 0.7765$ as extracted from behavioral modeling shown in Fig. 8.7b.

8.3.2.3 Offset Effect on Speed and Power

Assuming $A_{\mathrm{VTN}} \approx A_{\mathrm{VTP}}$, only for simplicity, then the total input referred offset of the circuit shown in Fig. 8.6b can be estimated by:

$$\sigma_{\mathrm{OS,in}}^2 \approx A_{\mathrm{VT}}^2 \times \sum_{i=1}^{3} \left(\frac{1}{W_{n(i)} L_{n(i)} A_V^{2(i-1)}} + \frac{1}{W_{p(i)} L_{p(i)} A_V^{2i}} \right). \tag{8.18}$$

Based on (8.18), for a high enough stage gain, the offset of the second and the third stages become quickly negligible in comparison to the offset of the first stage. Therefore, it is possible to simplify this expression to:

$$\sigma_{\mathrm{OS,in}}^2 \approx A_{\mathrm{VT}}^2 \times \left(\frac{1}{W_{n1} L_{n1}} + \frac{1}{A_V^2 W_{p1} L_{p1}} + \frac{1}{A_V^2 W_{n2} L_{n2}} \right). \tag{8.19}$$

To simplify (8.19) for design purpose, one can assume that all the three terms are participating equally on the offset voltage, hence:

$$\begin{cases} S_{n1} = W_{n1} L_{n1} = 3A_{\mathrm{VT}}^2 / \sigma_{\mathrm{OS,in}}^2, \\ S_{p1} = W_{p1} L_{p1} = 3A_{\mathrm{VT}}^2 / (\sigma_{\mathrm{OS,in}}^2 A_V^2), \\ S_{n2} = W_{n2} L_{n2} = 3A_{\mathrm{VT}}^2 / (\sigma_{\mathrm{OS,in}}^2 A_V^2). \end{cases} \tag{8.20}$$

Using (8.20) it is possible to calculate the total input capacitance of the ADC by:

$$C_{\mathrm{IN}} = (2^{N_b} - 1) \times C_{\mathrm{ox}} \times \left(\frac{2}{3} W_{n1} L_{n1} + 2A_M W_{n1} L_{\mathrm{ov}} \right) \tag{8.21}$$

where the term $2W_{n1}L_{n1}/3$ represents the effect of C_{GS}, the term $2A_M W_{n1} L_{\mathrm{ov}}$ represents the effect of C_{GD}, A_M is the Miller effect ($A_M \approx 2$), and L_{ov} is the gate-drain overlap length.

Having the parasitic capacitance of the intermediate nodes, it is also possible to estimate the power consumption of the pre-amplifier and comparator to operate at f_s. Considering that the voltage swing at the output of single stage amplifier is $V_{\mathrm{SW}} = R_L I_{\mathrm{SS}}$ (R_L is the equivalent output resistance of the PMOS load devices and I_{SS} is the tail bias current) and controlled to be constant through V_{BP} (see Fig. 8.6a). To operate at f_s, it is necessary that $\tau_L = R_L C_L$ (C_L is the loading capacitance at the output of differential stage) be much smaller than $T_s = 1/f_s$. To have a proper transient response, let us assume that:

$$\tau_L = \frac{1}{4} \cdot \frac{T_s}{2}. \tag{8.22}$$

Thus:

$$I_{\mathrm{SS}} \geq \frac{8 f_s}{V_{\mathrm{SW}} C_L} \tag{8.23}$$

and the total current consumption of the pre-amplifier and comparator stages will be:

$$I_{\text{DD}} \geq (2^{N_b} - 1) \times \frac{8f_s}{V_{\text{SW}}} \cdot \sum_{i=1}^{3} C_{L(i)} \tag{8.24}$$

where

$$\begin{cases} C_{L1} = C_{\text{ox}}(W_{n1}L_{\text{ov}} + W_{p1}L_{\text{ov}} + A_M W_{n2}L_{\text{ov}} + W_{n2}L_{n2}), \\ C_{L2} = C_{\text{ox}}(W_{n2}L_{\text{ov}} + W_{p2}L_{\text{ov}} + A_M W_{n3}L_{\text{ov}} + W_{n3}L_{n3}), \\ C_{L3} = C_{\text{ox}}(W_{n3}L_{\text{ov}} + W_{p3}L_{\text{ov}} + A_M W_{n3}L_{\text{ov}} + W_{n3}L_{n3}). \end{cases} \tag{8.25}$$

where C_L as shown in Fig. 8.6c representing the parasitic output capacitances. To calculate C_{L3} it is assumed that the following stages have the same size as the third stage in Fig. 8.6 since there is no limitation due to the offset. Also, notice that (8.24) does not include the power consumption of the digital encoder circuit.

Using (8.16), (8.24), and (8.25), it is possible to estimate the power-speed trade-off in a fully parallel flash ADC and base on (8.1) show that:

$$\text{FOM}_{\text{Flash}} \approx 24 \times \beta \times 2^{N_b} C_{\text{ox}} \cdot \frac{V_{\text{DD}} A_{\text{VT}}^2}{V_{\text{SW}} V_{\text{REF}}} \cdot \frac{1}{\text{INL}_{\text{Max}}} \tag{8.26}$$

where β introduced in (8.17). It should be mentioned that to drive (8.26), the power consumption of the resistor ladder have been ignored and also simplified values for load capacitances has been used. Therefore, the figure of merit derived in (8.26) only represents the lower achievable limit. This equation also illustrates the effect of device mismatch and C_{ox} on overall performance of the ADC. In more advanced technologies where $C_{\text{ox}} \propto 1/t_{\text{ox}}$ increases and device matching improves ($A_{\text{VT}} \propto t_{\text{ox}}^2$), it is expected that the power efficiency of this topology improves.

One of the main issues with the flash topology is its high input capacitance. Depicted in Fig. 8.8, the input capacitance of a flash ADC can increase rapidly by improving the resolution. Figure 8.8 shows the estimated total power consumption, input capacitance, and FOM of a flash ADC as a function of resolution and operation frequency based on behavioral modeling. As this figure implies, the minimum achievable FOM for an 8B flash ADC is about 2 pJ/conv.-step which is very high. As it will be shown later, it is possible to reduce the FOM considerably using FAI topology.

8.4 Design of FAI ADC

This section describes the topology and the main building blocks used for implementing the proposed 8-bit FAI ADC.

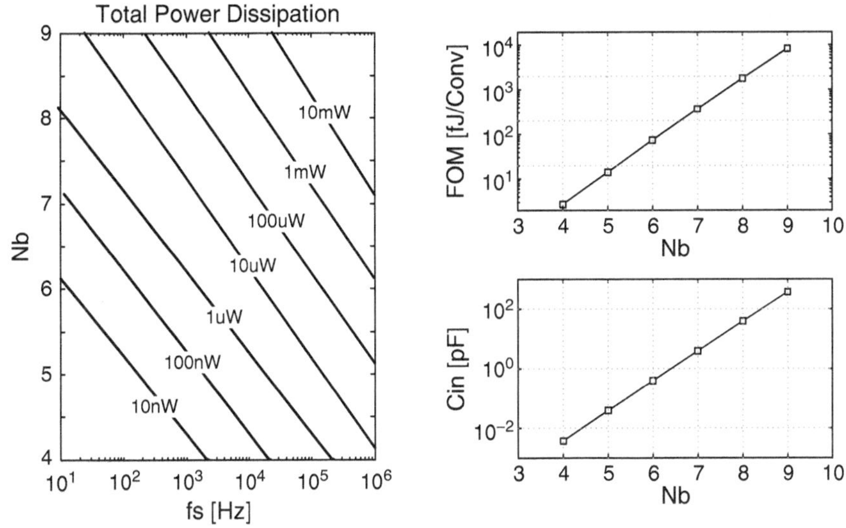

Fig. 8.8 Minimum achievable FOM using flash topology for ADC based on behavioral modeling. This figure also shows the power consumption (excluding encoder part) and the total input capacitance of the ADC as a function of N_b

8.4.1 *Circuit Topology*

As discussed in the previous section, using FAI topology it is possible to reduce the area and power consumption of the flash ADCs considerably. Figure 8.9 shows a possible folding scheme. Assume that the folding and interpolating part needs to extract the five LSB bits in the proposed FAI ADC. In this approach, the input analog signal is folded by four folder stages. These four signals can be delivered to four zero-crossing detectors to extract the two LSB bits. To extract the rest of the three bits, it is possible to use interpolating technique and generate the eight intermediate signals between each two consecutive folding signals. Therefore, the entire input signal range will be divided by $4 \times 8 = 32$ sections, and hence it is possible to extract 5 LSB bits. Since there are 32 folded and interpolated signals, each comparator needs to detect the zero crossing points.

Figure 8.10 shows one of the common circuits that is generally used for implementing folder circuit. Each transconductor (G_m) in this schematic is constructed based on a simple differential pair with a nonlinear transfer characteristics. Operating in subthreshold regime, as it is intended in this work, the output current of the transconductor shown in Fig. 8.10 can be expressed by:

$$I_{\text{OUT}} = I_{\text{SS}} \tanh\left(\frac{V_{\text{IN}}}{2nU_T}\right) \tag{8.27}$$

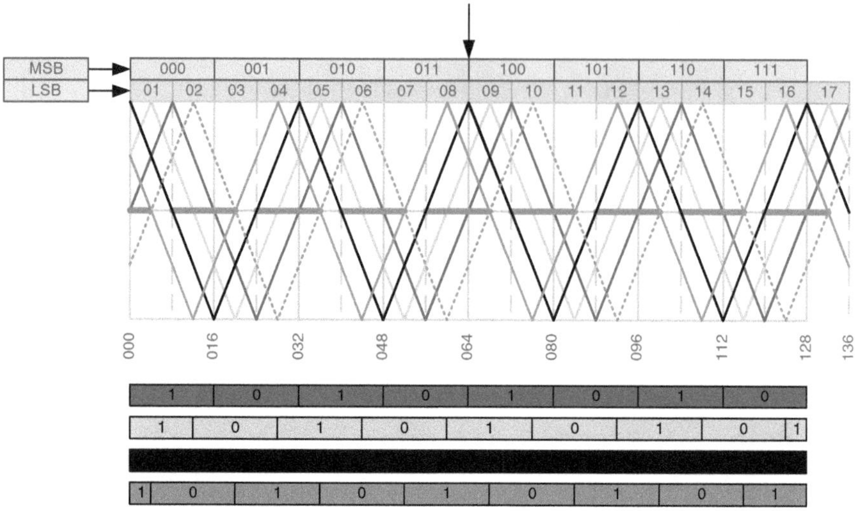

Fig. 8.9 Folding scheme: four folders are used to generate four folded signals. Each two consecutive folded signals can be used to generate interpolated signals

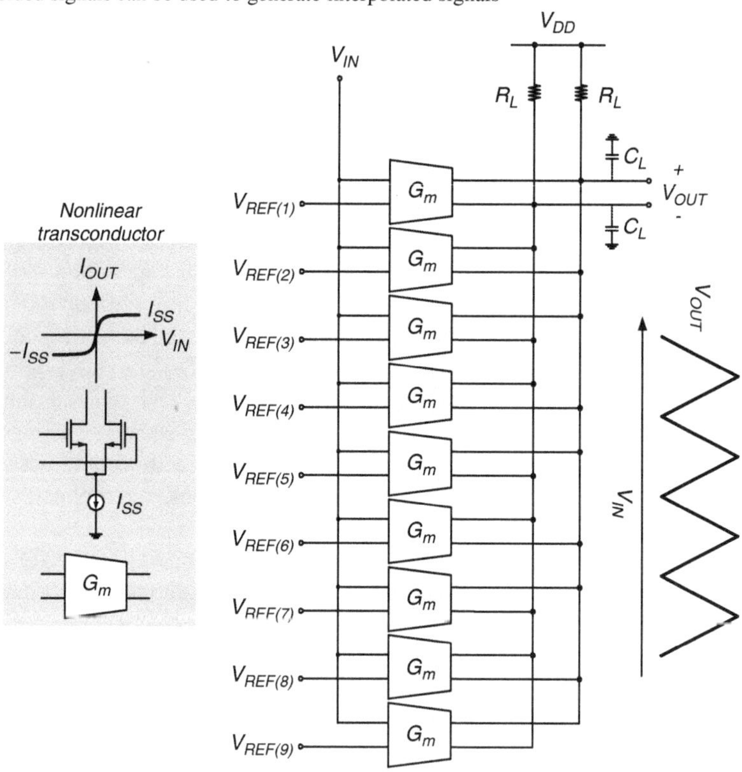

Fig. 8.10 Sample folder circuit ($N_F = 3$) uses nonlinear transconductors

where I_{SS} is the total tail bias current of each differential pair. Considering Fig. 8.10, the output current of each transconductor in the folder circuit will be

$$I_{OUT} = I_{SS} \tanh\left(\frac{V_{IN} - V_{REF(i)}}{2nU_T}\right) \times (-1)^{i+1}, \quad i = 1, ..., 2^{N_C} + 1 \tag{8.28}$$

and hence the output voltage will be

$$V_{OUT} = R_L I_{SS} \sum_{i=1}^{2^{N_C}+1} \left(\tanh\left(\frac{V_{IN} - V_{REF(i)}}{2nU_T}\right)(-1)^{i+1}\right). \tag{8.29}$$

To have more than one folded signal (such as Fig. 8.9), it is simply possible to shift the $V_{REF(i)}$ values. To have four folded signals as depicted in Fig. 8.9:

$$V_{OUT(j)} = R_L I_{SS} \sum_{i=1}^{2^{N_C}+1} \left(\tanh\left(\frac{V_{IN} - V_{REF(j,i)}}{2nU_T}\right) \times (-1)^{i+1}\right), j = 1, ..., 4. \tag{8.30}$$

where the difference between the two consecutive reference voltages is

$$\Delta V_{REF} = \frac{V_{REF}}{2^{N_b - N_C}}. \tag{8.31}$$

In the next step, interpolation will take place between all the two consecutive folded signals. This can be done by weighted sum of the output voltages as:

$$V_{OUT(k)} = \alpha V_{OUT(j)} + (1-\alpha) V_{OUT(j+1)} \tag{8.32}$$

where:

$$\alpha = \frac{k}{2^{N_b - N_C - N_F}}, k = 1, ..., 2^{N_i} - 1.$$

It is also possible to do the interpolation in current domain and among $I_{OUT(j)}$ signals. For this purpose, current mode interpolators can be used on top of the differential pair stage as depicted in Fig. 8.11a. The circuit can be even more simplified by merging the folder and interpolator circuits as shown in Fig. 8.11b [13]. Since current-mode interpolation eliminates the need for one additional stage, it can help to reduce the folding and interpolating circuit power consumption.

Operating in subthreshold regime, it is possible to calculate the inherent nonlinearity of a current-mode interpolator. Rewriting (8.32) in current domain:

$$\begin{aligned} I_{OUT(\alpha)} &= \alpha I_{OUT(j)} + (1-\alpha) I_{OUT(j+1)} \\ &= I_{SS}\left(\alpha \tanh \tfrac{V_{IN} - \Delta V_{REF}/2}{2n\mathrm{UT}} + (1-\alpha) \tanh \tfrac{V_{IN} - \Delta V_{REF}/2}{2n\mathrm{UT}}\right) \end{aligned} \tag{8.33}$$

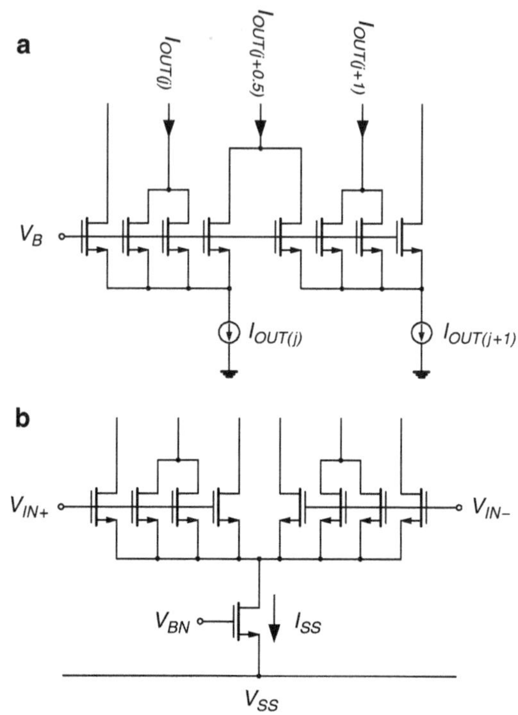

Fig. 8.11 (**a**) Current mode interpolator. (**b**) Merged folder and interpolator stage

The ideal zero cross point of $I_{OUT(\alpha)}$ needs to be between $V_{REF(j)}$ and $V_{REF(j+1)}$ with a distance of $z_i = \alpha\Delta V_{REF}$ from $V_{REF(j)}$. However, the zero cross point of $I_{OUT(\alpha)}$ calculated in (8.33) can be different from this value [14]. It can be shown that the real zero cross point will be:

$$z_r = nU_T \ln \frac{d + \sqrt{d^2 + 4}}{2} \tag{8.34}$$

where d is defined as:

$$d = (2\alpha - 1) \times \left(e^{\frac{\Delta V_{REF}}{2nU_T}} - e^{-\frac{\Delta V_{REF}}{2nU_T}} \right).$$

Therefore, the inherent INL of a current-mode interpolator which is biased in subthreshold regime is not zero and depends on ΔV_{REF} value. As depicted in Fig. 8.12, this error can be very small while ΔV_{REF} is small.

8.4.2 *Ultra Low Power Resistor Ladder*

To design a low-power FAI ADC with scalable power-frequency, it is necessary to implement a very-low-power and precise resistor ladder with scalable equivalent

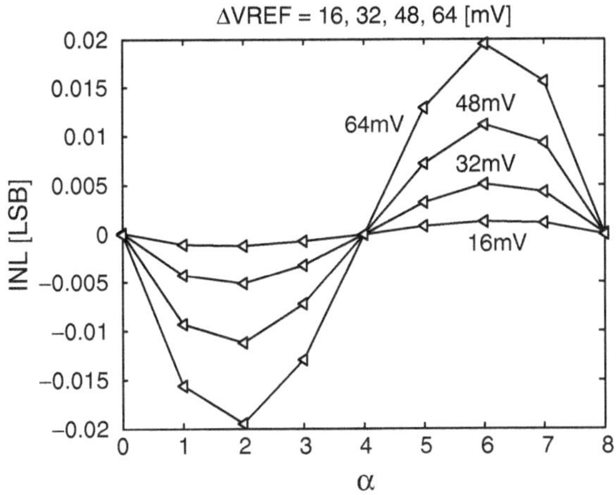

Fig. 8.12 Inherent INL of a current-mode interpolator biased in subthreshold regime

resistivity of the components. Scalability of the resistors helps to adjust the power consumption of this part of circuit with respect to the ADC sampling frequency. Indeed, the time constant in each node of a resistor ladder should be small enough to have a fast settling after each sampling in the ADC, as discussed in (8.4).

Meanwhile, it has been shown that the power dissipation of the resistor ladder which depends on sampling frequency, f_s, load capacitance of each resistor, C_{Lad}, and also resolution, N_b can be calculated by:

$$P_{\mathrm{Lad}} > \frac{\ln\left(2^{N_B+1}\right)}{2^{N_B+1}} \cdot C_{\mathrm{Lad}} f_s V_{\mathrm{REF}}^2. \tag{8.35}$$

Using conventional techniques, it is not possible to reduce the power consumption of this part below a few μW since the required resistance will be very large. Meanwhile, the resistivity of the ladder should be adjustable with respect to the sampling frequency. To implement a high valued resistance for resistor ladder, the topology shown in Fig. 8.13b can be used [15, 16]. In this topology, MR exhibits a very high resistivity which can be controlled over a very wide range by adjusting the source-gate voltage (V_{SG}) of the device. In Fig. 8.13c, MLS is used to adjust the V_{SG} of MR by tuning I_{RES}. Therefore, each resistance is constructed by two MOS devices and a current source. When the number of resistors in the ladder is high (e.g., 256 for an 8-bit flash ADC), then the power consumption due to the controlling part (MLS and I_{RES}) can be significant. Figure 8.13d shows a remedy to reduce the number of the devices in the required controlling part by sharing MLS and I_{RES} among more than one resistance. Since the total number of resistances in the ladder of the proposed FAI ADC is not high, the resistance of Fig. 8.13c has been used in this work.

To ensure that in Fig. 8.13d all the resistors observe the same value of source-drain voltage, i.e,:

$$V_{\mathrm{SD}(i)} = \frac{V_{\mathrm{AB}}}{n}, \quad i = 1, \ldots, n \tag{8.36}$$

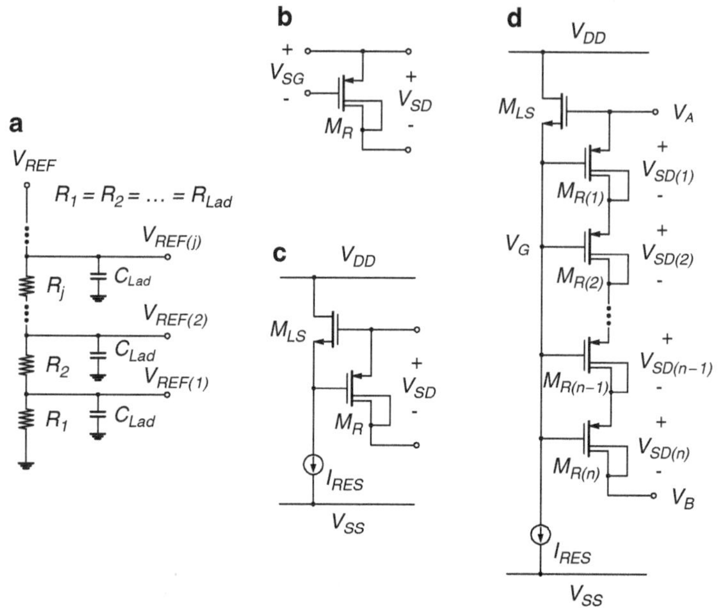

Fig. 8.13 Low power resistor ladder implementation: (**a**) ideal resistor ladder used to generate reference voltages, (**b**) high-value resistance based on subthreshold PMOS device, (**c**) biasing the proposed high-value resistance where the resistivity can be adjusted through I_{RES}, and (**d**) compact resistor ladder sharing the same biasing circuitry for more than one resistance

the devices should be sized very carefully. For a sample PMOS device MR(i) , one can show that:

$$\frac{V_{SD(i)}}{U_T} = \ln\left(\frac{I_{SD}}{2n_p\mu C_{ox}U_T^2} \cdot \frac{L_{(i)}}{W_{(i)}} \cdot e^{\frac{V_G - V_A}{n_p U_T}} \cdot e^{\frac{i \times V_{SD(1)}}{n_p U_T}}\right) + 1 \tag{8.37}$$

Therefore, it is possible to change the size of transistors to have a constant source-drain voltage as (8.36). To properly size the devices:

$$\frac{W_{(i)}}{L_{(i)}} = \frac{W_{(1)}}{L_{(1)}} \times e^{(i-1)\frac{V_{SD}}{n_p U_T}} \tag{8.38}$$

With this approach, the voltage drop across source-drain of all the transistors will be equal.

8.4.3 Comparator Circuit

Comparators are critical components in design of FAI ADCs. As discussed in Sect. 8.3.2, the performance of comparators can directly affect the performance

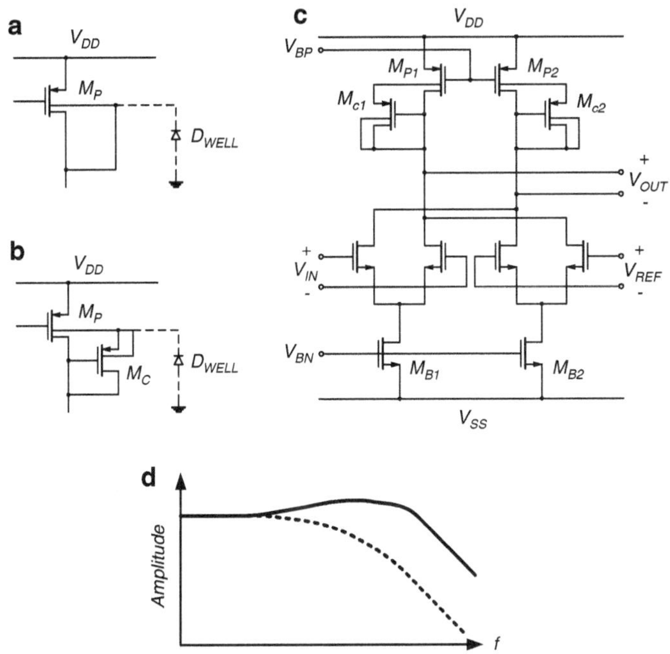

Fig. 8.14 (**a**) High valued load resistance. (**b**) Decoupling the parasitic capacitance of the well-substrate from output node. (**c**) Subthreshold pre-amplifier stage. (**d**) Improvement of frequency response through parasitic capacitance decoupling

of ADC. To reduce the sensitivity of the circuit to offset of comparators, a low gain pre-amplifier stage is used in front of each comparator.

The pre-amplifier used in this work is based on a single stage double differential amplifier as shown in Fig. 8.14. As the tail bias current reduces, a very high-valued load resistance is required to get enough gain from this stage. PMOS devices (MP1 and MP2) with their bulk connected to their drains are used to construct the required high-valued load resistances as explained in [15]. A replica bias circuit controls the voltage swing (V_{SW}) at the output of pre-amplifiers through V_{BP}. Using a replica bias, the gain of thc prc-amplifier stage is:

$$A_{V0} \approx \frac{n_p}{n_n(n_p - 1)} \tag{8.39}$$

where n_p and n_n are subthreshold slope factors of the PMOS load devices and NMOS differential pair devices, respectively. The gain predicted in (8.39) is about 3.2 for the proposed technology.

Figure 8.14a illustrates that the reverse biased diode of the nwell-to-substrate PN junction (D_{WELL}) appears directly at the output of the pre-amplifier and hence reduces the circuit bandwidth. To decouple this capacitance from the output node, a very high value load resistance has been added in series to the bulk-drain of the load devices. This resistance is implemented by MC as illustrated in Fig. 8.14b. Using this technique, the double difference preamplifiers (Fig. 8.14c) and comparator stages have been implemented. In each transition, the parasitic capacitance due to D_{WELL} charges and discharges with a delay due to the RC delay constructed by resistance of MC and capacitance of D_{WELL}. Therefore, this structure acts as a zero in pre-amplifier transfer function and can improve the speed of circuit response as illustrated in Fig. 8.14d.

8.4.4 Encoder

Topology: The outputs of the fine and coarse sub-ADCs are needed to be merged to get the final outputs [13]. For this purpose, the outputs of the coarse sub-ADC needs to be synchronized with the outputs of the fine sub-ADC after error correction as illustrated in Fig. 8.15 [17]. In the first step, at the output of the coarse sub-ADC, the majority detector circuits are used to remove the possible bubbles at the output thermal code. Then the thermal code is converted to Gray code and finally to binary codes. This coding scheme helps to have less bit error rate (BER).

At the output of fine sub-ADC, the cyclical code to binary code converter first converts the fine bits to Gray code to correct the bubble errors in the fine path and then converts the Gray code to binary code. Hence, there is no need to a separate error correction in the fine path.

Bit Synchronization: Synchronizing coarse and fine bits is an important issue in the design of an FAI ADC due to different paths for the fine and course bits. A small timing mismatch between the coarse and fine quantizers can cause non-linearity error. The bit synchronization block in Fig. 8.15 uses 8 cycle pointers

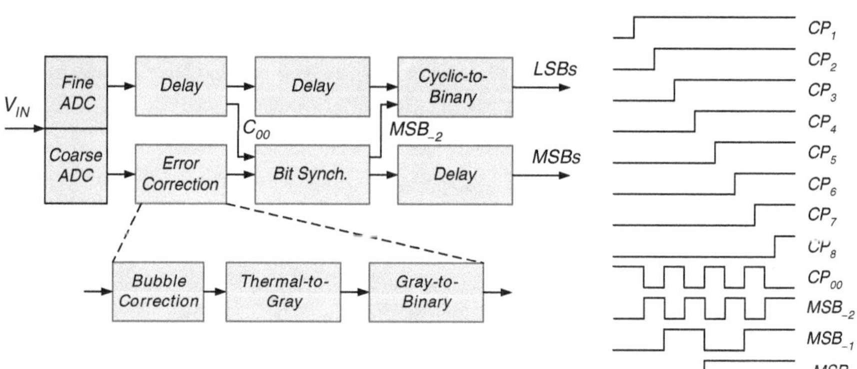

Fig. 8.15 Error correction and encoder using pipelined STSCL topology. Waveforms of the bit synchronization block. MSB, MSB_{-1}, and MSB_{-2} are the outputs. C_{00} is the synchronization bit and CP_1–CP_8 are cycle pointers

(CP_1–CP_8) and the synchronization bit (C_{00}) to generate 3 MSBs [19]. Cycle pointers are basically the outputs of the flash quantizer that after error correction are fed into the bit synchronization block. Figure 8.15 shows the waveforms of the bit synchronization block. The equations for generating MSB, MSB_{-1} and MSB_{-2} are:

$$MSB = CP_5 + CP_4 \cdot C_{00}$$
$$MSB_{-1} = CP_1 + CP_6 \cdot C_{00} + \overline{CP_5} \cdot \overline{C_{00}} \cdot CP_3 + \overline{CP_4} \cdot C_{00} \cdot CP_2$$
$$MSB_{-2} = CP_8 + CP_1 \cdot C'_{00}$$

Bubble Correction: Comparator metastability, threshold voltage variations, device mismatches, and other interference may cause unwanted transitions at the output of the comparators which are called bubble errors. An error correction circuitry is used in the course path to reject the bubble errors. The error correction block consists of majority detection cells in a pipelined structure. The output of a majority cell is at logic "1" when at least two out of the three inputs are at logic "1". Figure 8.16 shows the schematic of the latched majority cell.

Cyclical Code to Binary Code Conversion: The output of the fine quantizer is called cyclical code which can be easily converted to binary code using XOR operators. The fine bits (31 bits) and the LSB output of the bit synchronization block (MSB_{-2}) are the inputs of the code conversion block. Generation of the fine and coarse binary outputs remains synchronized by using the output of the bit synchronization block, MSB_{-2}, as an input of the converter.

Figure 8.17 shows the schematic of the circuit for cyclical code to binary code conversion. In fact, the cyclical code is first converted to Gray code which can eliminate the bubble errors and then converted to binary code using sequential XOR

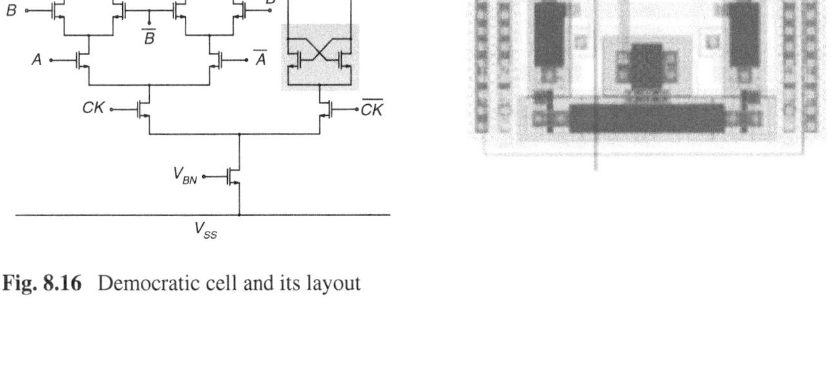

Fig. 8.16 Democratic cell and its layout

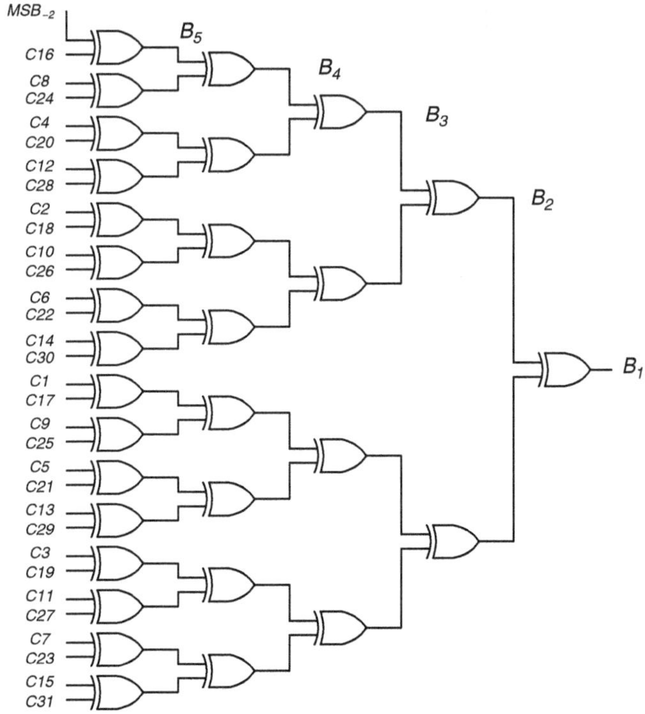

Fig. 8.17 Cyclical code to binary code converter circuit

operation. Sequential Gray code to binary code conversion uses the minimum number of XOR gates which is efficient from both power consumption and the area of the circuit points of views, but obviously generates the outputs with a high latency [19].

The outputs of the bit synchronization block should be latched during conversion because of the pipelined operation of the encoder (not shown in Fig. 8.17). These bits and the outputs of the conversion block form the outputs of the encoder.

Circuit Implementation: The digital part of the ADC is designed based on pipelined STSCL topology including totally 196 gates [16]. While in digital CMOS circuits, a precise control on supply voltage is required to adjust the power consumption of the circuit with respect to the operation frequency, in STSCL circuits the power control can be achieved simply by adjusting the tail bias current of the gates.

Here, STSCL topology has been used to implement the digital encoder circuit. To improve the power efficiency of STSCL digital part, two techniques have been employed:

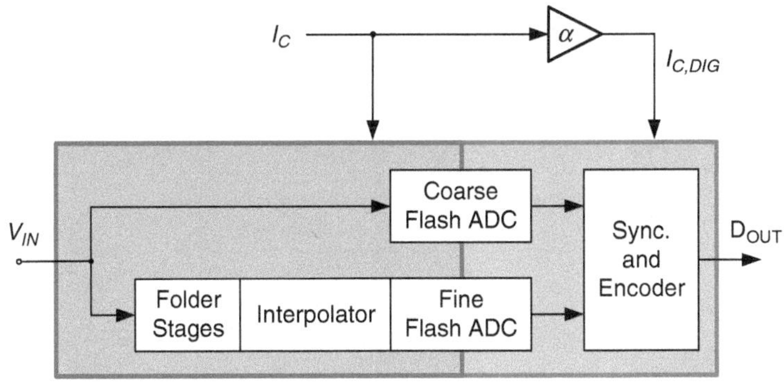

Fig. 8.18 Control of power consumption with respect to the operating frequency in the proposed subthreshold source-coupled FAI ADC

- Using stacked NMOS differential pairs in the switching network to construct compound logic operations. In this way, it is possible to merge the functionality of two or more STSCL gates in only one gate and reduce the power dissipation and area, simultaneously (see Chap. 5).
- Using pipelining technique that reduces the logic depth to practically one gate as described in Chap. 5.

Figure 8.16 illustrates how these two techniques have been employed to design an STSCL majority detector cell. Stacking of three layers of NMOS differential pairs help to do the desired complicated logic operation in only one stage. Meanwhile, a latch has been used at the output of the majority cell for implementing pipelining technique. When the clock signal is high, the logic circuit is in evaluation phase and when clock goes low, the evaluated value will be kept at the output node for the rest of the clock period. Therefore, the next stage can start its evaluation phase. As mentioned before, pipelining can help to reduce considerably the power dissipation of STSCL circuits when logic depth is deep.

Bias current of the digital circuit is a fraction of bias current of the analog part, and hence the same controlling system could be used for both parts (Fig. 8.18). This scheme simplifies considerably the control of power consumption in digital part. In addition, using large enough transistor sizes can minimize the effect of current mismatch in both analog and digital parts.

8.5 Simulation and Experimental Results

8.5.1 Encoder

Simulation results show that the encoder can operate in a wide range of frequencies by adjusting the bias current of the gates. Figure 8.19 shows the maximum

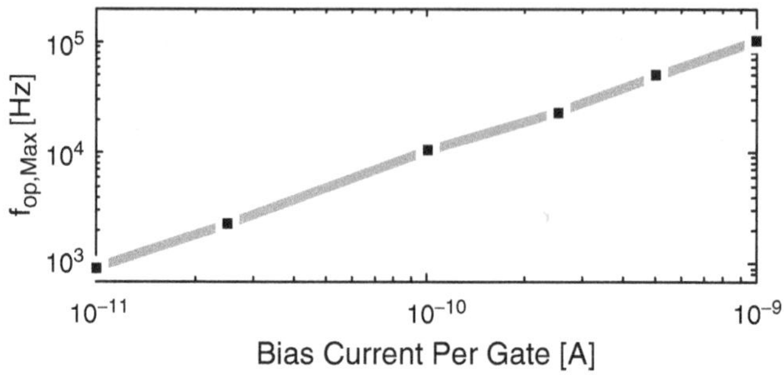

Fig. 8.19 Maximum operation frequency of the digital section as a function of tail biascurrent

frequency of operation of the encoder as a function of the tail bias current of STSCL gates. Pipelining has helped to improve the power-delay performance of the circuit as explained in [16]. The bias current of the digital circuit is set to be a fraction of the bias current of the analog section (Fig. 8.18); therefore, a separate controlling unit is avoided. In this experiment, the supply and the desired swing voltage were 1.0 V and 0.2 V, respectively [22].

The maximum frequency of operation shows a linear behavior with respect to the bias current per gate in the range of 250 pA–50 nA. Further increasing the bias current brings the differential pair transistors from weak inversion to medium inversion which degrades the linear dependency of the frequency of operation to the bias current.

At a tail bias current of 1 nA and supply voltage of 1 V per gate, the encoder consisting of a total of 196 gates is shown to operate at 100 kHz clock frequency with near-perfect eye opening.

8.5.2 *FAI ADC Performance*

Figure 8.20 shows the photomicrograph of the prototype test chip fabricated in 0.18-μm CMOS technology. The total active area of the circuit is 0.6 mm^2. The bias current of the analog and digital parts are controlled externally with respect to the sampling frequency. The sampling frequency of the proposed ADC can be adjusted from 800 S/s to 80 kS/s where the power consumption is scaling proportional to the sampling frequency from 17 nW to 1.9 μW with ENOB of 6.5.

Figure 8.21 shows the measured integral non-linearity (INL) and differential non-linearity (DNL) of the proposed FAI ADC which are 1.0 LSB and 0.4 LSB, respectively.

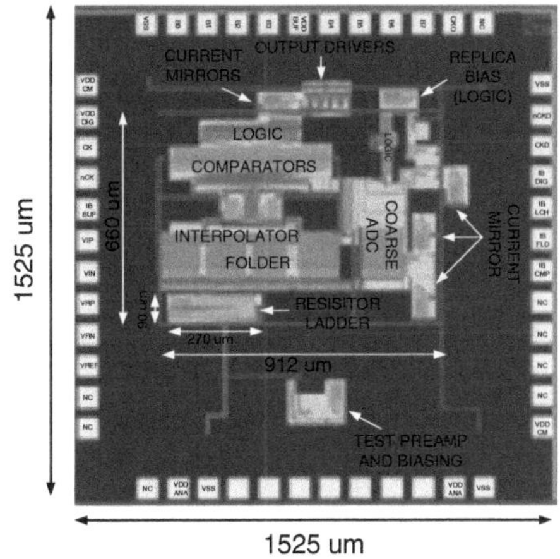

Fig. 8.20 Photomicrograph of the proposed chip implemented in 0.18-μm CMOS technology

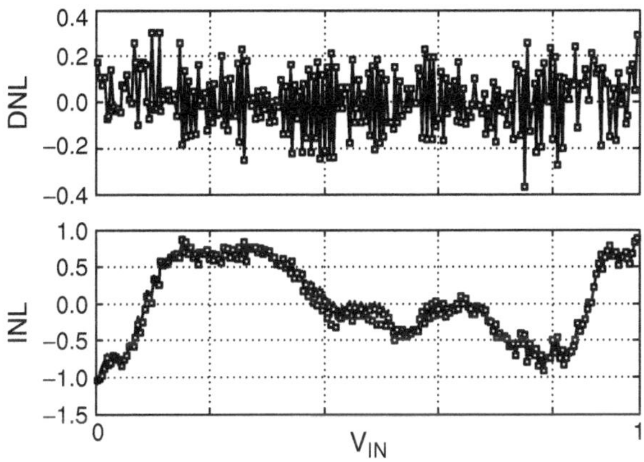

Fig. 8.21 Measured differential non-linearity (DNL) and integral non-linearity (INL)

8.6 Conclusion

An ultra-low-power folding and interpolating ADC with scalable sampling frequency operating in subthreshold region has been introduced. Using current-mode approach, it is possible to have a wide operating range (800 S/s to 80 kS/s) while the power consumption scales linearly proportional to it (17 nW to 1.9 μW from 1.2 V supply voltage). Completely novel circuit techniques for improving the speed of operation and also reducing the power consumption in comparator circuit and resistor ladder have been developed. The active area of the ADC is 0.6 mm^2 and is implemented in 0.18-μm CMOS technology. Measurements show that the INL and DNL of the ADC are 1.0 LSB and 0.4 LSB, respectively.

Also, a pipelined encoder for the proposed 8 bits FAI ADC has been designed using subthreshold SCL technique. Simulation results show that the encoder can operate over a wide frequency range between 10 kHz and 50 MHz. The speed and power consumption of the circuit are bias dependent and can be simply adjusted by tuning the bias currents of the STSCL gates. For this range of operating frequencies, the power consumption of encoder varies between 20 nW and 200 μW. The supply voltage can be lowered until the swing voltage at the output reaches to its minimum allowed value. The circuit also generates low amplitude current spikes which does not affect the supply voltage of the circuit significantly.

References

1. M. D. Scott, B. E. Boser, and K. S. J. Pister, "An ultra-low power ADC for distributed sensor networks," in *Proceedings of European Solid-State Circuits Conference (ESSCIRC)*, pp. 255–258, Sep. 2002
2. J. Sauerbrey, D. Schmitt-Landseidel, and R. Thewes, "A 0.5-V 1-μW successive approximation ADC," *IEEE J. Solid-State Circuits*, vol. 38, no. 7, pp. 1251–1265, Jul. 2003
3. G. Bonfini, and et al., "An ultralow-power switched opamp-based 10-B integrated ADC for implantable biomedical applications," in *IEEE Transactions on Circuits and Systems-I: Regular Papers*, vol. 51, no. 1, pp. 174–178, Jan. 2004
4. N. Verma and A. P. Chandrakasan, "An ultra low energy 12-b rate-resolution scalable SAR ADC for wireless sensor nodes," *IEEE J. Solid-State Circuits*, vol. 42, no. 6, pp. 1196–1205, Jun. 2007
5. H. -C. Hong and G.-M. Lee, "A 65-fJ/conversion-step 0.9-V 200-kS/s rail-to-rail 8-bit successive approximation ADC," *IEEE J. Solid-State Circuits*, vol. 42, no. 10, pp. 2161–2168, Oct. 2007
6. S. Gambini and J. Rabaey, "Low-power successive approximation converter with 0.5 V supply in 90 nm CMOS," *IEEE J. Solid-State Circuits*, vol. 42, no. 11, pp. 2348–2356, Nov. 2007
7. M. van Elzakker, and et al., "A 1.9μW 4.4fJ/conversion-step 10b 1MS/s charge-redistribution ADC," in *Proceedings of IEEE International Solid-State Circuits Conference (ISSCC)*, pp. 244–245, Feb. 2008
8. D. C. Daly and A. P. Chandrakasan, "A 6b 0.2-to0.9V highly digital flash ADC with comparator redundancy," in *Proceedings of IEEE International Solid-State Circuits Conference (ISSCC)*, pp. 554–555, Feb. 2008
9. J. van Valburg and R. J. van de Plassche, "An 8-b 650-MHz folding ADC," *IEEE J. Solid-State Circuits*, vol. 27, no. 12, pp. 1662–1666, Dec. 1992
10. R. Y. van de Plassche and P. Baltus, "An 8-bit 100-MHz full-Nyquist analog-to-digital converter," *IEEE J. Solid-State Circuits*, vol. 23, no. 6, pp. 1334–1344, Dec. 1988
11. S. Limotyrakis, K.-Y. Nam, and B. A. Wooley, "Analysis and simulation of distortion in folding and interpolating A/D converters," in *IEEE Transactions on Circuits and Systems-II: Analog and Digital Processings*, vol. 49, no. 3, pp. 161–169, Mar. 2002
12. P. Kinget, "Device mismatch and tradeoffs in the design of analog circuits," *IEEE J. Solid-State Circuits*, vol. 40, no. 6, pp. 1212–1224, Jun. 2005
13. M. P. Flynn and D. J. Allstot, "CMOS folding A/D converters with current-mode interpolation," *IEEE J. Solid-State Circuits*, vol. 31, no. 9, pp. 1248–1257, Sep. 1996
14. M. Babaie, H. Movahedian, M. Sharif Bakhtiar, "A novel method for systematic error prediction of CMOS folding and interpolating ADC," in *Asia-Pacific Circuits and Systems Conference (APCCAS)*, pp. 1768–1771, 2006
15. A. Tajalli, Y. Leblebici, and E. J. Brauer, "Implementing ultra-high-value floating tunable CMOS resistor," in *IET Electronics Letters*, vol. 44, no. 5, pp. 349–350, Feb. 2008
16. A. Tajalli, E. J. Brauer, E. Vittoz, and Y. Leblebici, "Subthreshold source-coupled logic circuits for ultra-low-power applications," *IEEE J. Solid-State Circuits*, vol. 43, pp. 1699–1710, Jul. 2008

17. M. Beikahmadi, A. Tajalli, and Y. Leblebici, "A subthreshold SCL based pipelined encoder for ultra-low power 8-bit folding/interpolating ADC," in *Proceedings of The Nordic Microelectronics Event (NORCHIP)*, pp. 9–12, Tallin, Estonia, Nov. 2008
18. R. Roovers and M. S. J. Steyaert, "1 175 Ms/s, 6 b, 160 mW, 3.3 V CMOS A/D converter," *IEEE J. Solid-State Circuits*, vol. 31, pp. 938–944, Jul. 1996
19. Y. Li, "Design of high speed folding and interpolating analog-to-digital converter," Ph.D. Diss., Texas A&M Univ., May 2003
20. A. G. W. Venes and R. J. van de Plassche, "An 80-MHz, 8-b CMOS folding A/D converter with distributed track-and-hold preprocessing," *IEEE J. Solid-State Circuits*, vol. 31, pp. 1846–1853, Dec. 1996
21. B. Nauta and A. G. W. Venes, "A 70-MS/s 100-mW 8-b CMOS folding and interpolating A/D converter," *IEEE J. Solid-State Circuits*, vol. 30, pp. 1302–1308, Dec. 1995
22. A. Tajalli and Y. Leblebici "Nanowatt range folding-interpolating ADC using subthreshold source-coupled circuits," *J. Low-Power Electron.*, vol. 6, Apr. 2010

Chapter 9
Widely Adjustable Ring Oscillator Based $\Sigma\Delta$ ADC

9.1 Introduction

Over-sampling scheme in addition to the noise shaping property of the $\Delta\Sigma$ architectures, makes it very suitable for implementing high resolution data converters. In addition, this architecture exhibit low sensitivity to the non-ideality behavior of analog circuits, such as limited gain of amplifier, device mismatch, and offset of comparator [1, 2]. This property is specially desirable in design of low-cost and high-performance mixed-signal circuits in modern CMOS technologies.

In this chapter, some techniques for implementing power-efficient and performance scalable $\Delta\Sigma$ ADCs will be described. As will be shown, ring oscillator based $\Delta\Sigma$ ($R\Sigma\Delta$) architecture is very suitable for implementing power performance scalable ADCs. In such a topology, ring oscillator is used as quantizer block in which resolution depend on the number of delay elements inside the ring.

9.2 Background

9.2.1 *Dynamic Range*

In conventional Nyquist rate ADCs, the maximum achievable signal-to-noise ratio (SNR) depends on number of quantizer levels [3]. In other words, if the ADC uses an N bit quantizer, then the maximum achievable SNR in [dB] will be:

$$\mathrm{SNR}_{\mathrm{Max}} \approx 6.02N + 1.76 \ \ (\mathrm{dB}). \tag{9.1}$$

Here, it is assumed that the quantization noise, Δ, is a random number with uniform distribution and an average value of zero. Therefore, the quantization noise power is:

$$\overline{\epsilon_q^2} = \frac{1}{\Delta}\int_{-\Delta/2}^{\Delta/2} \epsilon_q^2 \, \mathrm{d}\epsilon_q = \frac{\Delta^2}{12} \tag{9.2}$$

A. Tajalli and Y. Leblebici, *Extreme Low-Power Mixed Signal IC Design: Subthreshold Source-Coupled Circuits*, DOI 10.1007/978-1-4419-6478-6_9,

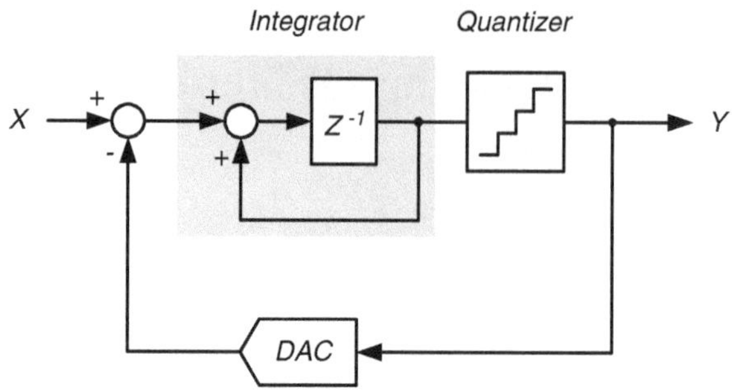

Fig. 9.1 First order ΔΣ modulator topology

while signal power is equal to $V_{\mathrm{REF}}^2/8 = 2^{2m}\Delta^2/8$ (the maximum peak-to-peak signal amplitude can not exceed V_{REF}) [3].

However, in ΔΣ ADCs, the quantizer is placed inside a feedback loop. In this way, the noise of quantizer will be shaped and moved to the frequencies higher than the bandwidth of input signal (f_{bw}). Figure 9.1 illustrates a first order discrete-time (DT) ΔΣ modulator. In this configuration, the output is:

$$y[n] = x[n] \cdot z^{-1} + q[n] \cdot (1 - z^{-1}). \tag{9.3}$$

where q stands for the quantization noise. As presented in (9.3), the input signal is transferred to the output with only a single clock phase delay. However, the quantization noise is shaped (filtered). In other words, the noise transfer function (NTF) is:

$$\mathrm{NTF}(z) = (1 - z^{-1}). \tag{9.4}$$

Therefore:

$$\|\mathrm{NTF}(z)\| = \|(1 - \mathrm{e}^{-j\omega})\| = 2\sin\frac{\omega}{2}. \tag{9.5}$$

The power spectral density of the output quantization noise, $S_Q(f)$, can be estimated with respect to the quantization noise of the internal quantizer, $S_q(f)$, as:

$$S_Q(f) = S_q(f) \cdot \|\mathrm{NTF}(\mathrm{e}^{j2\pi f/f_s})\|^2 \tag{9.6}$$

and based on this, the power of the output noise will be:

$$P = \int_0^{f_{\mathrm{bw}}} S_Q(f)\mathrm{d}f = Q_{\mathrm{rms}}^2. \tag{9.7}$$

Using (9.7), for a first order loop, the output noise power, Q_{rms}^2, versus the input noise power, q_{rms}^2 is:

$$Q_{\mathrm{rms}}^2 = \frac{\pi^2}{3 \times \mathrm{OSR}^3} \cdot q_{\mathrm{rms}}^2. \tag{9.8}$$

where over-sampling-ratio is defined by:

$$\mathrm{OSR} = \frac{f_s}{2 f_{\mathrm{bw}}}. \tag{9.9}$$

In a more general case, when there is a Lth order loop, with an ideal noise transfer function as $\mathrm{NTF}(z) = (1 - z^{-1})^L$, the in-band noise power is approximately given by:

$$Q_{\mathrm{rms}}^2 = \frac{\pi^{2L}}{(2L+1) \times \mathrm{OSR}^{2L+1}} \cdot q_{\mathrm{rms}}^2. \tag{9.10}$$

Regarding (9.10), the effective number of bits could be increased by increasing the loop order and OSR with respect to the given value or improving the resolution of the loop quantizer [1].

9.2.2 Improving the Resolution

One of the main issues with $\Delta\Sigma$ topology is that to improve the resolution, either over-sampling ratio needs to be increased or the resolution of quantizer inside the loop should be improved. Increasing OSR will make the design of high frequency ADCs very difficult. On the other side, the non-linearity of quantizers with a resolution more than one bit can degrade the circuit performance considerably. The other possibility is to increase the order of modulator [1]. Usually, the order of modulator is chosen not to be more than three due to stability issues.

Ring oscillator based $\Delta\Sigma$ ($R\Sigma\Delta$) topology has been proposed to solve these issues in high frequency and high resolution ADCs [4–6]. In this configuration, oscillator in addition to a counter act as a quantizer. As illustrated in Fig. 9.2, the

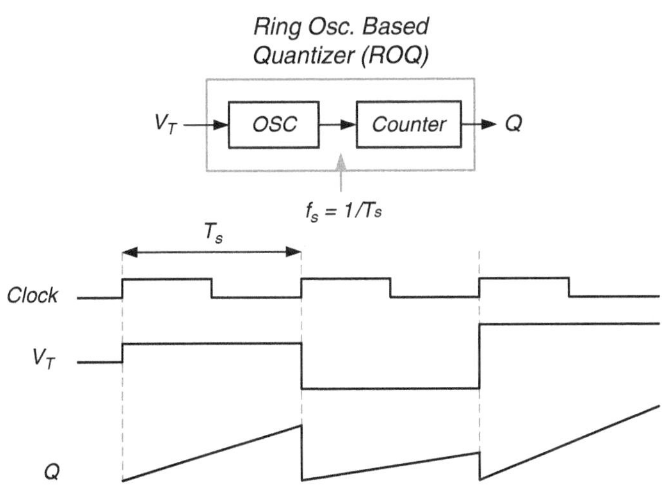

Fig. 9.2 Timing operation of a ring oscillator based quantizer (ROQ) [6]

counter is triggered by the oscillator and counts the number of transitions in a specific period of time. As shown in Fig. 9.2, the oscillation frequency of the ring oscillator, f_{osc}, is proportional to the input voltage, V_T. Therefore, the output of the counter, Q, will be also proportional to the input signal. This property helps to implement high resolution quantizers.

The ring oscillator based quantizer (ROQ), has some very interesting properties. In this configuration, the input signal is converted to a time domain parameter that means the main conversion to digital domain will be accomplished using time domain signals, i.e. clock period (T_s), and the period of oscillation of the ring oscillator ($T_{\mathrm{osc}} = 1/f_{\mathrm{osc}}$). At the end of each sampling period, there would be some time domain quantization error which is not discarded and will be considered in the next conversion step. This is because ring oscillator continues its oscillation and the output phase will be accumulated on top of the residue phase of the output signal in the last sampling time. Therefore, ROQ alone acts as a first order noise shaping quantizer with the transfer function described in (9.4) [4]. This inherent first order noise shaping characteristic will increase the noise shaping order of the final ΔΣ modulator by one.

The other interesting property of this configuration is that the delay cells in the ring oscillator that are participating in the counting process are rotating continuously. Therefore, there is a first order averaging over the delay value of these cells which reduces the sensitivity of the quantizer to the delay mismatch in the ring oscillator. Meanwhile, this continuous rotation and averaging will improve the matching (or linearity) of the DAC (digital-to-analog converter) following the ROQ stage and hence provides an intrinsic dynamic element matching (DEM) of the DAC elements [5].

In this work, we are exploiting one other interesting property of $R\Sigma\Delta$ topology for implementing scalable performance ADCs with very low power consumption.

9.3 Performance Scalability in Ring Oscillator Based ΔΣ ADCs

In this section, the possibility of scaling the sampling frequency and also dynamic range of $R\Sigma\Delta$ modulators is investigated.

9.3.1 Frequency Domain Adjustability

Assuming a linear relationship, the oscillation frequency of a voltage-controlled ring oscillator (VCO) can be represented by:

$$f_{\mathrm{osc}} = K_{\mathrm{VCO}} V_{\mathrm{in}} \tag{9.11}$$

where $K_{\rm VCO}$ is the voltage to frequency conversion gain of VCO. In $R\Sigma\Delta$ topology, the sampling frequency needs to be at least two times more than the maximum oscillation frequency of VCO, hence:

$$f_s \geq 2K_{\rm VCO}V_{\rm REF}. \tag{9.12}$$

Based on (9.12), the sampling frequency, f_s, and signal bandwidth, $f_{\rm bw} = K_{\rm VCO} \cdot V_{\rm REF}$, are constant parameters and can not be scaled. Here, $V_{\rm REF}$ is the reference voltage and hence the maximum input signal swing needs to be smaller than $V_{\rm REF}$.

On the other side, in a current-controlled ring oscillator (CCO), it is possible to scale the sampling frequency by scaling the nominal bias current of CCO as:

$$f_s \geq 4K_{\rm CCO}I_B \tag{9.13}$$

where $K_{\rm CCO}$ and I_B are the gain and the nominal bias current of CCO. Therefore, in a CCO with widely adjustable delay cells, it is possible to implement widely tunable current mode $R\Sigma\Delta$ modulators. One attractive property of CCOs is that their frequency versus input bias current characteristic is very linear while must of VCO based $R\Sigma\Delta$ modulators are suffering from poor frequency versus input voltage characteristic [6].

Using subthreshold source-coupled logic (STSCL) circuits introduced in Chap. 3 (see Fig. 9.3a), it is possible to implement very widely tunable delay elements controlled precisely by their bias current. Regarding Fig. 9.3b, the oscillation frequency of a CCO with N_d delay elements is:

$$f_{\rm osc} = \frac{I_{\rm SS}}{2\ln 2 N_d V_{\rm SW} C_L} \tag{9.14}$$

where C_L is the load capacitance at the output of each delay element, $I_{\rm SS}$ is the tail bias current, and $V_{\rm SW}$ is the voltage swing at the output of each delay cell. Regarding (9.14), the gain of a STSCL CCO is:

$$K_{\rm CCO} = \frac{1}{2\ln 2 N_d V_{\rm SW} C_L}. \tag{9.15}$$

Table 9.1 summarizes the design parameters in a CCO-based $R\Delta\Sigma$ modulator.

In the proposed STSCL ring oscillator shown in Fig. 9.3, an operational transconductance amplifier (OTA) controls the voltage swing at the output of the delay cells. The limited bandwidth of this OTA in addition to the delay in the current mirror constructed by MNB and MNR, shown in Fig. 9.3a, modifies the frequency–current relationship that has been indicated in (9.14). It can be shown that a more precise frequency-to-current relationship for the ring oscillator shown in Fig. 9.3 is:

$$f_{\rm osc} = K_{\rm CCO} I_{\rm SS} \cdot \frac{1 - s/z}{1 - s/p} \tag{9.16}$$

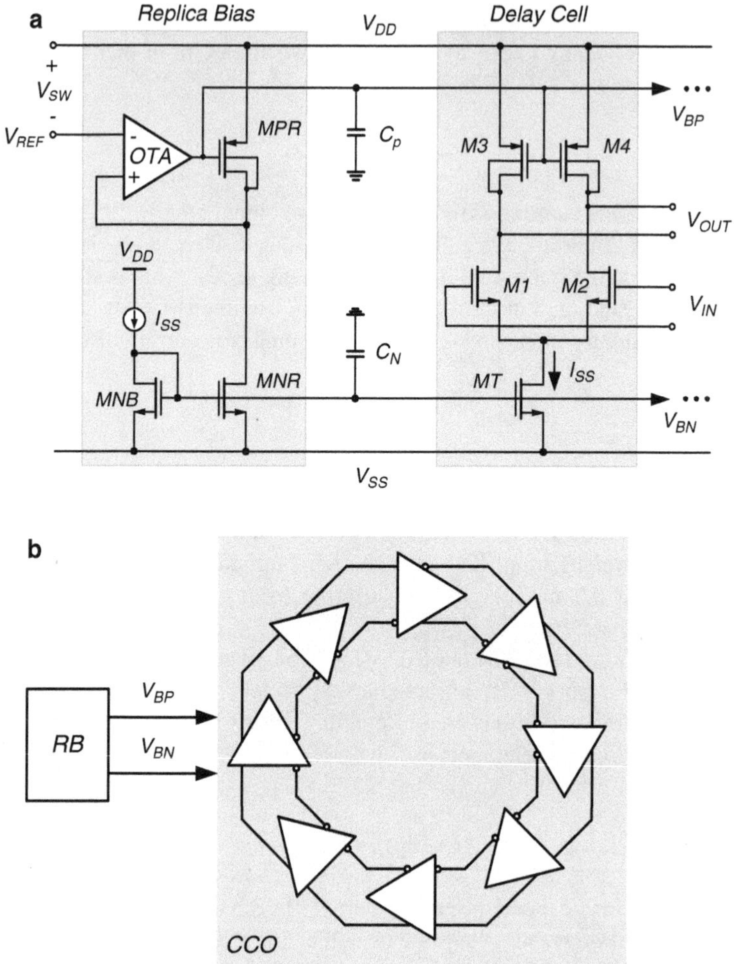

Fig. 9.3 (**a**) STSCL delay cell and replica bias circuit to generate bias voltage for PMOS and NMOS transistors. (**b**) Sample differential ring oscillator

Table 9.1 Parameter definition in CCO-based $R\Delta\Sigma$ ADC

Parameter	Value
Over-sampling ratio	$\mathrm{OSR} = f_s/(2f_{\mathrm{bw}})$
Oscillator gain	K_{CCO}
Nominal bias current of oscillator	I_B or I_{SS}
Input current	$I_{\mathrm{OSC}} = I_A \sin(2\pi f_{\mathrm{in}} t)$
Oscillation frequency	$f_{\mathrm{osc}} = K_{\mathrm{CCO}} I_{\mathrm{OSC}}$
Maximum input current	$I_{A,\mathrm{Max}} = I_{\mathrm{SS}}$
Maximum oscillation frequency	$f_{\mathrm{OSC,Max}} = K_{\mathrm{CCO}} I_{\mathrm{OSC,Max}} = 2K_{\mathrm{CCO}} I_{\mathrm{SS}}$
Sampling frequency	$f_s = 4K_{\mathrm{CCO}} I_{\mathrm{SS}}$
Signal bandwidth	$f_{\mathrm{bw}} = f_{\mathrm{in,Max}} = 2K_{\mathrm{CCO}} I_{\mathrm{SS}}/\mathrm{OSR}$

where:

$$z = -\frac{G_m + (n_p - 1)G_{\text{out}}}{(n_p - 1)C_P} \approx -\frac{G_m}{(n_p - 1)C_P} \tag{9.17}$$

and

$$p = -\frac{g_{m(n)}}{C_N} \tag{9.18}$$

Here, G_m and G_{out} are the transconductance and the output conductance the OTA, n_p is the subthreshold slope of the PMOS load devices, $g_{m(n)}$ is the transconductance of NMOS bias transistors (MNB and MNR), and C_P and C_N are the parasitic capacitances at the nodes V_{BP} and V_{BN}. As generally $C_P \gg C_N$ due to the loading at the output of OTA, therefore, usually the zero of the system, z, is closer to the origin compared to the pole, p.

Figure 9.4 shows a practical implementation of ROQ [6]. This topology avoids using counters that might affect the inherent noise shaping property of the ROQ. In contrast, the outputs of the ring oscillator are sampled in each rising edge of the sampling clock and then the number of transitions is detected by an array of exclusive OR (XOR) gates. Implementing this circuit using STSCL topology, the speed of operation of the all sections of the circuit can be scaled proportional to the bias current. This can be done by scaling the bias current of the oscillator ($I_{B,\text{OSC}}$), and the logic circuits ($I_{B,\text{LOG}}$), simultaneously.

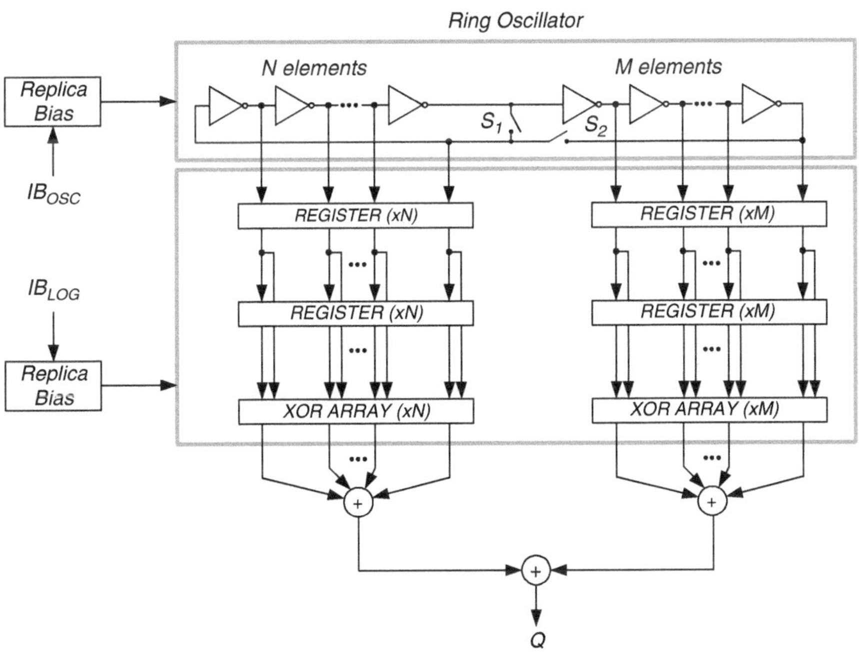

Fig. 9.4 Implementation of ring oscillator based quantizer without the need to counter as proposed in [6]. The topology is modified to make it suitable for scalable DR ADCs

9.3.2 Dynamic Range Adjustment

By changing the number of delay elements in the ring oscillator, it is possible to change the resolution of the quantizer, and hence adjust the overall SNR. Reducing the number of delay elements in the ring oscillator will reduce the overall power consumption of the modulator while the penalty is reduction in SNR. Therefore, it is possible to reduce the power consumption of the proposed data converter when the system does not require high resolution.

Assuming a constant sampling frequency, by reducing the number of delay elements (as well as the number of registers and XOR gates) from $N + M$ to N, the resolution will be reduced. This can be done using switches S_1 and S_2 in Fig. 9.4. In this new situation, the bias current of the logic part ($I_{B,\mathrm{LOG}}$) should remain unchanged while the bias current of the ring oscillator can be reduced by a factor of $(M + N)/N$. This means that the total power consumption of the quantizer can be reduced by a factor of more than $(M + N)/N$. Based on this, it is possible to implement a power-DR scalable ADC.

Figure 9.5a shows the signal-to-noise and distortion ratio (SNDR) versus input signal level for a $R\Sigma\Delta$ quantizer with 15 delay elements. Depicted by this figure, a

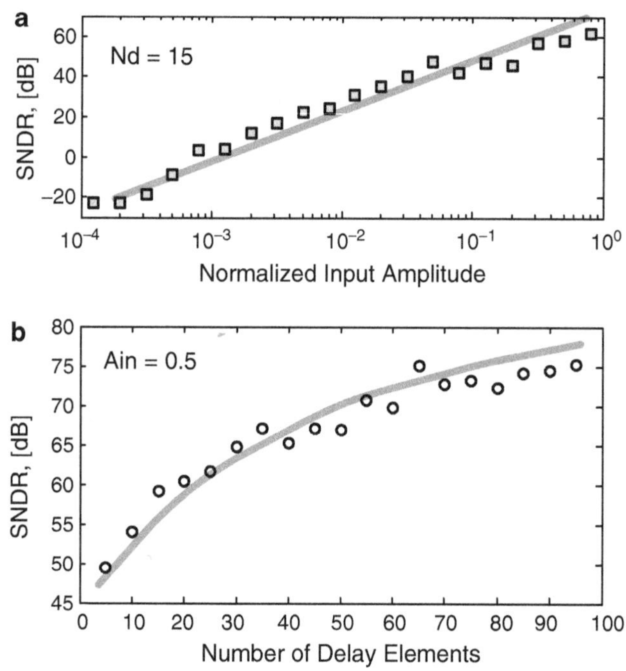

Fig. 9.5 (**a**) SNDR versus input signal amplitude based on behavioral modeling of a first order $R\Delta\Sigma$ in MATLAB (here: $N_d = 15$, and OSR $= 64$). (**b**) SNDR versus number of delay elements in the ring oscillator (here: A_{in}=0.5, and OSR $= 64$)

simple $R\Sigma\Delta$ modulator can reach to SNDR = 60 dB. The circuit DR range can be improved by using more number of delay stages as illustrated in Fig. 9.5b. Base on this figure, the resolution of the ADC can be still kept above 8 bits while the number of delay stages is only 5. This reduction in the number of delay elements will be along with reduction in power dissipation of the quantizer.

The possibility of adjusting the signal bandwidth and resolution simultaneously, make this topology very suitable for implementing reconfigurable ADCs.

9.4 Top Level Design

In this section, the main non-ideality effects in a $R\Sigma\Delta$ modulator will be studied. The results of this study could be used in circuit design step to implement a low power and high performance $R\Sigma\Delta$ modulator. A behavioral MATLAB model including different non-ideality sources is developed to investigate their effect on the system performance.

9.4.1 Sources of Non-Ideality

9.4.1.1 Delay Mismatch

In an ideal case, all the elements inside the ring oscillator exhibit the same amount of delay. Therefore, the reference sampling clock is counted by equally spaced pulses. In a real implementation, there is always some mismatch among the circuit elements, and hence among the delay values which makes the time to digital converter nonlinear. This nonlinearity can directly affect the DR at the output of quantizer.

The effect of delay mismatch in $R\Sigma\Delta$ modulators is partially similar to the effect of resistor mismatch in a resistor string based analog-to-digital converter. In this type of converters, the resistor mismatch can cause nonlinearity at the output of ADC. This effect has been studied extensively in [7]. The difference in $R\Sigma\Delta$, however, is that the delay elements are continuously changing their placement in the queue. This effect is due to this fact that the delay element that does the first transition in each conversion step depends on the oscillator phase in the previous step. This continuous change of starting point of the delay line can provide a first order averaging over the nonlinearity of the quantizer.

The delay of elements in ring oscillator can be represented by random numbers of $t_{d(i)}, i = 1, .., N_d$ with an average value of:

$$\overline{t_d} = \frac{\ln 2 V_{SW} C_L}{I_{SS}} \tag{9.19}$$

and variance of σ_{t_d}. Also, in a ring oscillator, the sum of the delay values should be equal to $T_{\text{osc}}/2$, or:

$$\sum_{i=1}^{N_d} t_{d(i)} = \frac{1}{2f_{\text{osc}}} = N_d \overline{t_d}. \tag{9.20}$$

In other words, assuming $t_{d(i)} = \overline{t_d} + \Delta t_{d(i)}$, then:

$$\sum_{i=1}^{N_d} \Delta t_{d(i)} = 0. \tag{9.21}$$

9.4.1.2 Ring Oscillator Jitter

The jitter on edges of the ring oscillator changes its instantaneous oscillation frequency. Presence of jitter in ring oscillator changes the nominal delay of a gate to:

$$t_{d(i)} = \overline{t_d} + \Delta t_{d(i)} + \partial t_{d(i)} \tag{9.22}$$

where $\Delta t_{d(i)}$ represents the delay mismatch component, and timing uncertainty has been stated by $\partial t_{d(i)}$ which has an average value of zero and variance of σ_{t_d}. Assuming that there are N complete transitions during one T_s, the timing jitter will be accumulated over N transition, and hence the last transition inside the time interval of T_s will be displaced. This displacement (d) depends on the value of $\partial t_{d(i)}$ and N. Assuming normal distribution for ring oscillator jitter [8], the variance of d will be:

$$\sigma_d \approx \partial t_{d(i)} \cdot \sqrt{N}. \tag{9.23}$$

and worst case happens when $N = N_d$. Meanwhile, as the number of delay elements increases, the oscillator jitter effect becomes more pronounced.

9.4.1.3 Sampling Clock Jitter

Sampling clock jitter is another source of error in $R\Sigma\Delta$ modulators. Sampling clock period (T_s) acts as the voltage reference in conventional ADCs. Therefore, any variation on T_s will affect the output linearity. In a real case where sampling clock contains jitter, clock period can be indicated by a random number with average value of $\overline{T_s}$ and variance of σ_{T_s}. Therefore, $T_s = \overline{T_s} + \Delta T_s$, where average value of ΔT_s is zero and its variance is σ_{T_s}.

Figure 9.6 shows the effect of clock jitter on circuit dynamic range. In this figure, the RMS value of the jitter is normalized to the clock period. As can be seen, to have a very low degradation in SNDR, the normalized RMS value of the clock jitter should be less than 0.1%.

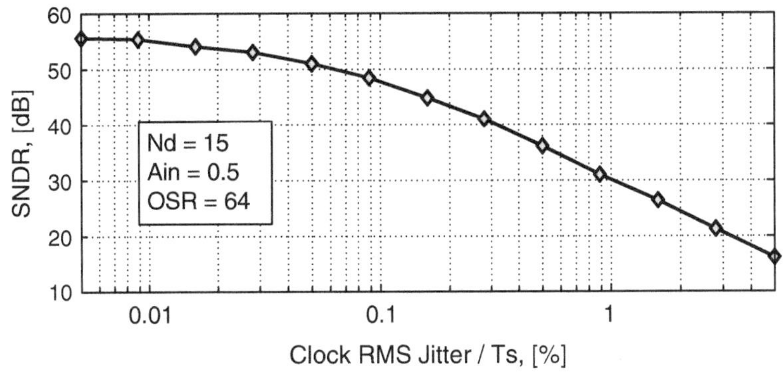

Fig. 9.6 The effect of sampling clock jitter on SNDR based on behavioral modeling in MATLAB for a first order $R\Delta\Sigma$ modulator

9.4.1.4 Comparator Meta-Stability Effect

Generally, comparators circuits are using a positive feedback to improve the speed and at the same time attain a high gain. In this case, the time needed that a comparator completes the regeneration can be approximately indicated by [3]

$$T_R = \frac{\tau_C}{A_C - 1} \cdot \ln \frac{\Delta V_O}{\Delta V_{\text{IN}}} \tag{9.24}$$

where τ_C is the characteristic time constant at the output of comparator, A_C is the small signal open loop gain of amplifier, ΔV_O is the minimum acceptable voltage swing at the output of comparator, and ΔV_{IN} is the sampled input voltage.

Regarding (9.24), ΔV_{IN} affects the regeneration time of comparator considerably and it can increase T_R indefinitely when it becomes small. Generally, in $R\Sigma\Delta$ modulator, the voltage swing at the input of comparator (or sampler stage) is large. However, there is this possibility that the sampling occurs during transition of the oscillator output. As illustrated in Fig. 9.7, when Δt gets close to zero, the ΔV_{IN} moves towards zero, and hence T_R increases.

Assuming that the comparator needs to be settled in less than half a clock period (i.e., $T_R \leq T_s/2$) and also $\Delta V_O = V_{\text{SW}} \approx 8U_T$, then

$$\Delta V_{\text{IN}} > 8U_T \cdot \exp\left(-\frac{\ln 2 \cdot (A_C - 1)N_d}{2}\right) = K_{\text{met}} \cdot U_T \tag{9.25}$$

where $A_C \approx n_p/(n_n(n_p - 1))$, as discussed in Chap. 3. This expression can be translated into a time domain equation as:

$$\Delta t_{\text{min}} \approx \Delta V_{\text{IN}} \frac{2V_{\text{SW}}}{t_r} \approx \frac{C_L}{2I_{\text{SS}}} \cdot K_{\text{met}} \cdot U_T \tag{9.26}$$

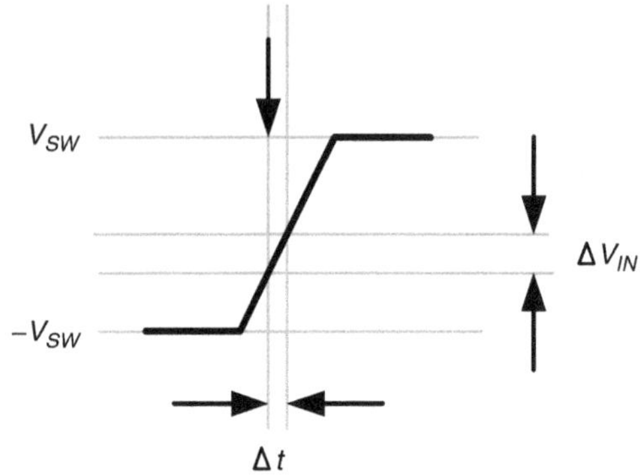

Fig. 9.7 Sampling the output of ring oscillator

Therefore, for $\Delta t < \Delta t_{\min}$, the regeneration will not be completed and the comparator output can be incorrect. Dividing this parameter to the gate delay (i.e., t_d) gives us an approximate normalized timing uncertainty as

$$\Delta t_n = \frac{\Delta t}{t_d} \approx \frac{N_d}{\ln 2} \cdot \exp\left(-\frac{\ln 2 \cdot (A_C - 1) N_d}{2}\right). \tag{9.27}$$

Regarding (9.27) and by solving $\partial \Delta t_n / \partial N_d = 0$, the worst case occurs when $N_d = 2/(\ln 2 \cdot (A_C - 1)) \approx 5$. For $N_d > 5$, the uncertainty time starts to decrease.

9.4.2 *Performance Analysis*

In a $R\Sigma\Delta$ quantizer, in each conversion step, the reference clock period is divided to an integer number of $N[n]$ and a residue $q[n] < 1$ by the ring oscillator. Indeed, the reference clock period is divided by the first N transitions of ring oscillator and there will be a residue time smaller that the delay of stage $N + 1$. Hence:

$$T_s[n] = \sum_{i=1}^{N} t_{d(i)}[n] + q[n] \times t_{d(N+1)}. \tag{9.28}$$

Replacing the different sources of non-ideality in (9.28) results in:

$$\overline{T_s}[n] = N \cdot \overline{t_d}[n] + R_t[n] \tag{9.29}$$

where R_t represents the residual time or quantization error in time domain and can be calculated by:

$$R_t[n] = \sum_{i=1}^{N} \Delta t_{d(i)} + \sum_{i=1}^{N} \partial t_{d(i)} + q[n] \cdot t_{d(N+1)} - \Delta T_s \approx Q[n] \cdot \overline{t_d}[n]. \tag{9.30}$$

Therefore, non-ideal parameters such as delay mismatch, and oscillator jitter, will be filtered similar to the quantization noise in an ideal modulator. However, the main issue here is that these non-ideality effects will increase the total quantization noise power. In addition, the total quantization noise Q depends on the input signal level through N.

The first term in right hand of (9.30) is zero when $N = N_d$ (see (9.21)) which happens when the input signal is close to its maximum value. In this special case, the effect of delay mismatch is negligible. However, as the number of transitions decreases by reducing the input signal level, the mismatch effect will become more pronounced.

The second term in (9.30), as represented in (9.23), is proportional to the time interval that the jitter will be accumulated which has an RMS value of $\partial t_d \sqrt{N}$. Therefore, this effect is more pronounced when N is larger or equivalently, input signal has larger values.

Regarding (9.30), in the presence of non-ideality effects, the quantization noise power will be increased by this factor:

$$\alpha[N] = 1 + \left(\sum_{i=1}^{(N+1)} \frac{\Delta t_d}{\overline{t_d}} + \sum_{i=1}^{N} \frac{\partial t_d}{\overline{t_d}} - \frac{\Delta T_s}{\overline{t_d}} \right) \tag{9.31}$$

Figure 9.8 shows the SNDR of a first order $R\Sigma\Delta$ quantizer versus input amplitude in presence of non-ideality effects. As can be seen, the peak SNDR value in this case drops to 52 dB.

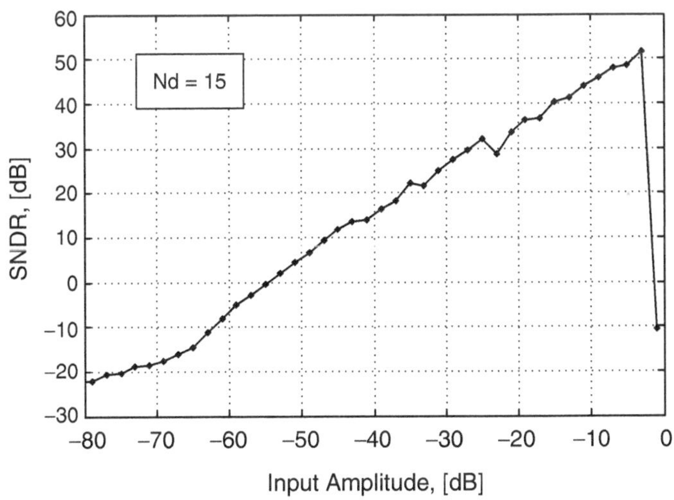

Fig. 9.8 SNDR of a first order quantizer when: $\sigma_{\text{OSC}} = 0.001t_d$, $\sigma_{\text{CK}} = 0.001T_s$, and $\Delta t_d = 0.01t_d$

9.5 Circuit Design

A current-mode $R\Sigma\Delta$ modulator has been designed in 90-nm CMOS technology. As discussed in Sect. 9.3, current-mode topology provides this possibility to have a very wide sampling frequency range with a scalable power consumption.

9.5.1 Ring Oscillator

The effect of non-ideal performance of circuit components that has been studied in Sect. 9.4 imposes some restrictions on circuit design. To keep the circuit performance on acceptable level, it is necessary to make sure that the circuit components will have the required specifications. Ring oscillator is the most important component in $R\Sigma\Delta$ topology. As explained in Sect. 9.4, oscillator jitter and delay mismatch are the main design parameters that can affect the modulator performance. In the following, the design of a ring oscillator with acceptable level of jitter and delay mismatch will be studied.

9.5.1.1 Delay Matching

The maximum acceptable mismatch on gate delay puts a lower limit on area of devices inside the delay element (see Fig. 9.3). The delay of each STSCL element can be calculated by:

$$t_d \approx \ln 2 R_L C_L = \frac{\ln 2 V_{\mathrm{SW}} C_L}{I_{\mathrm{SS}}} \tag{9.32}$$

hence:

$$\left(\frac{\Delta t_d}{t_d}\right)^2 \approx \left(\frac{\Delta V_{\mathrm{SW}}}{V_{\mathrm{SW}}}\right)^2 + \left(\frac{\Delta I_{\mathrm{SS}}}{I_{\mathrm{SS}}}\right)^2 + \left(\frac{\Delta C_L}{C_L}\right)^2. \tag{9.33}$$

Variation on V_{SW} depends on matching of PMOS load devices in delay elements ($M3$, $M4$, and MPR in Fig. 9.3a). It also depends on matching between the tail bias current of the delay elements (I_{SS}). The last term in (9.33) depends on the total capacitive load at the output of each delay element. This capacitance comes partially from interconnect parasitic capacitance, and partially from parasitic capacitance of NMOS and PMOS transistors. Therefore, a fully symmetric layout in addition to large MOS devices are required to guarantee having a good matching on load capacitance.

Defining:

$$I_0 = 2 n_p \mu_p C_{\mathrm{ox}} \frac{W}{L_{\mathrm{eff}}} U_T^2 \tag{9.34}$$

then the I/V characteristic of the PMOS load devices in delay elements depicted in Fig. 9.3a is:

$$I_{\mathrm{SD}} = I_0 \cdot \mathrm{e}^{\frac{V_{\mathrm{BG}}-V_{T0}}{n_p U_T}} \left(\mathrm{e}^{\frac{V_{\mathrm{SD}}}{U_T}-1}.\right) \tag{9.35}$$

Therefore, variation on V_{SD} of load devices due to the tail bias current and threshold voltage variation will be:

$$(\Delta V_{\mathrm{SD}})^2 \approx \left(\frac{n_p U_T}{n_p - 1} \cdot \frac{\Delta I_{\mathrm{SS}}}{I_{\mathrm{SS}}}\right)^2 + \left(\frac{\Delta V_{T0}}{n_p - 1}\right)^2. \tag{9.36}$$

For the tail bias transistors, assuming operation in subthreshold regime, then it can be shown that the current mismatch is:

$$\left(\frac{\Delta I_{\mathrm{SS}}}{I_{\mathrm{SS}}}\right)^2 \approx \left(\frac{\Delta V_{T0}}{n_n U_T}\right)^2. \tag{9.37}$$

In these calculations we have assumed that the variation due to the current gain mismatch, $\beta = \mu C_{\mathrm{ox}} W/L$, is negligible and the main source of mismatch is threshold voltage variations [9]:

$$\sigma_{V_T} = \frac{A_{V_T}}{\sqrt{W \cdot L}}. \tag{9.38}$$

Meanwhile, the variation on load capacitance can be modeled by:

$$\sigma_{C_L} = \frac{A_{C_L}}{\sqrt{W \cdot L}}. \tag{9.39}$$

Therefore, it is possible to relate the delay mismatch with the size of circuit components using (9.33) and (9.36)–(9.39).

Figure 9.9 shows the effect of delay mismatch on circuit SNDR. As can be seen, to have a drop on SNDR not more than 7 dB, the delay mismatch should

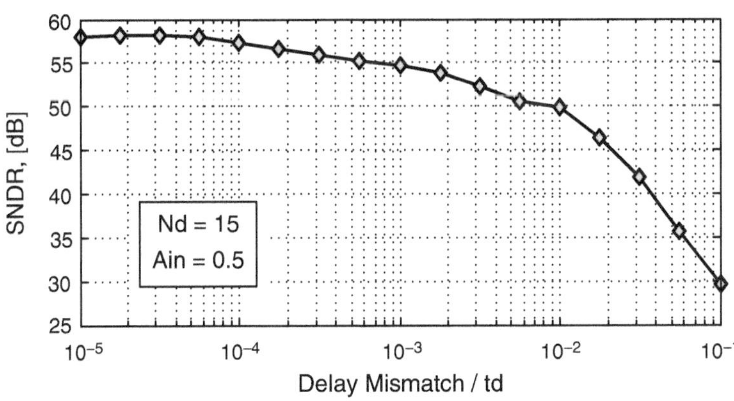

Fig. 9.9 Effect of delay mismatch on first order quantizer based on behavioral modeling in MATLAB

not exceed 1%. Considering these results and using (9.33), one can estimate the appropriate sizes for devices inside the delay elements.

9.5.1.2 Oscillator Jitter

As shown in [8], the standard deviation of jitter in an oscillator after ΔT seconds in

$$\sigma_j = \kappa\sqrt{\Delta T} \tag{9.40}$$

where κ is a proportionality constant determined by the circuit parameters. It is shown that [10]:

$$\kappa \geq \sqrt{\frac{8}{3\eta}} \cdot \sqrt{N_d \cdot \frac{kT}{P} \cdot \left(\frac{V_{\mathrm{DD}}}{V_{\mathrm{char}}} + \frac{V_{\mathrm{DD}}}{R_L I_{\mathrm{SS}}}\right)} = \sqrt{\frac{8}{3\eta}} \cdot \sqrt{\frac{kT}{I_{\mathrm{SS}}} \cdot \left(\frac{1}{V_{\mathrm{char}}} + \frac{1}{V_{\mathrm{SW}}}\right)} \tag{9.41}$$

where k is Boltzmann's constant, T is the junction temperature, N_d is the number of delay elements, P is the total oscillator power consumption, and $\eta \approx t_d/t_r$ is a function of rise time and delay in each delay element. Meanwhile, V_{char} is the characteristic voltage of the device. For long-channel devices, $V_{\mathrm{char}} = V_{\mathrm{dsar}}/\gamma$ (V_{dsat} is the gate overdrive voltage and γ is 2/3 for long-channel devices in saturation region and typically two to three times greater for short-channel devices). In short-channel devices, $V_{\mathrm{char}} = E_c L/\gamma$ (E_c is the critical electric field resulting in half carrier velocity expected from low field mobility). Assuming $V_{\mathrm{char}} \approx 4U_T$ and $V_{\mathrm{SW}} \approx 8U_T$, then

$$\kappa > \sqrt{\frac{q}{\eta I_{\mathrm{SS}}}} \tag{9.42}$$

where q is the elementary charge in Coulomb. It can be seen that the only way to reduce the jitter is to increase the tail bias current of the ring oscillator. To have a RMS jitter value not more than $\sigma_{j,\mathrm{Max}}$, tail bias current of each delay cell should be larger than:

$$I_{\mathrm{SS}} > \frac{1}{\sigma_{j,\mathrm{Max}}} \cdot \sqrt{\frac{2\ln 2q V_{\mathrm{SW}} C_L N_d}{\eta}} \tag{9.43}$$

In $R\Delta\Sigma$ topology, in each conversion step, the first transition occurs after t_d seconds with jitter variation of $\kappa \cdot \sqrt{t_d}$. The following transitions occur at $i \times t_d, i = 2, \ldots, N$ with standard variation of $\kappa \cdot \sqrt{i \cdot t_d}$. Therefore, the maximum jitter value will happen when $N = N_d$ which is equal to

$$\sigma_{j,\mathrm{Max}} \leq \kappa\sqrt{Ts} \tag{9.44}$$

Figure 9.10 shows the simulated SNDR of a first order quantizer in presence of oscillator jitter. For a quantizer with 15 delay elements, the oscillator jitter should be 10^{-3} times the delay value to have a drop on SNDR not more than 7 dB. Having

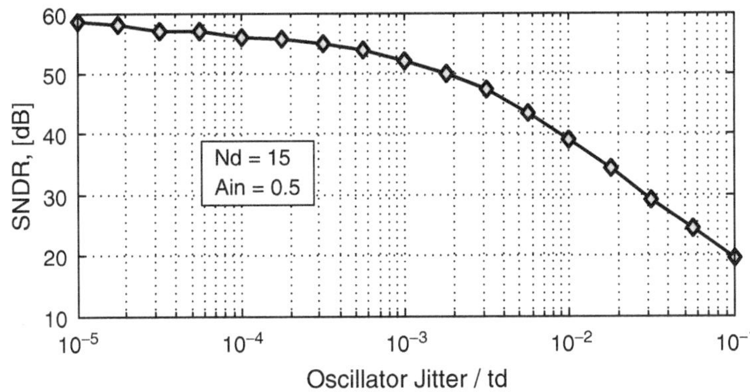

Fig. 9.10 Effect of oscillator jitter on first order quantizer based on behavioral modeling in MATLAB

extracted the maximum acceptable jitter value from system design step, (9.44) and (9.41) help to calculate the acceptable κ value, and hence the appropriate bias current of the delay cells.

9.5.2 Logic Circuit

As shown in Fig. 9.11, the logic circuit of the proposed $R\Delta\Sigma$ is constructed using DFF and XOR gates. These gates should be fast enough to make sure that the sampling and preliminary process on the sampled date will be done correctly. Meanwhile, the input referred offset of the first DFF stage is important. A reduced offset at the first stage helps to minimize the mismatch among different sampling branches, and hence reduce the modulator sensitivity to mismatch among different branches.

In this work, STSCL logic cells have been used to implement the digital part of the $R\Delta\Sigma$ system. The bias current of the digital part is proportional to the bias current of the STSCL ring oscillator. Hence, the power dissipation of the digital part is scaling with the sampling frequency.

To have a more power efficient digital part, the size of transistors in digital part are selected to be much smaller than the size of corresponding devices in the delay elements. Only, the size of devices in the first DFF has been selected to be large to suppress the offset of this stage.

9.5.3 Current-Mode Integrator

One of the main issues in design of continuous-time $R\Delta\Sigma$ is the need for implementing current mode integrators in which the output current is integration of the

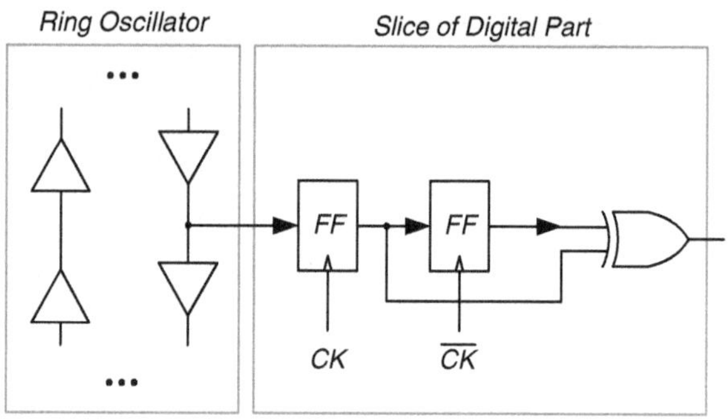

Fig. 9.11 A slice of the circuit showing part of ring oscillator and digital part

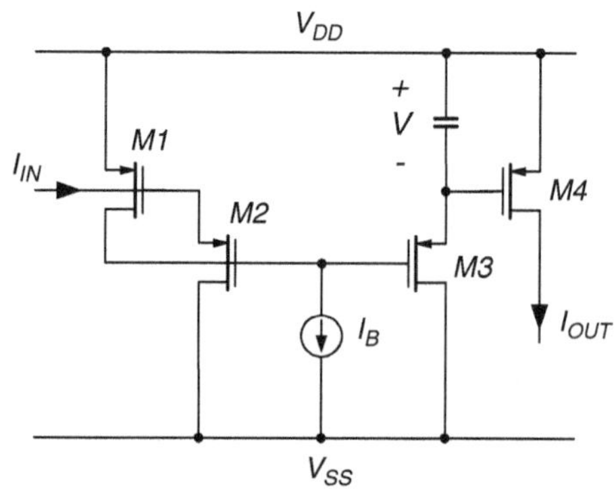

Fig. 9.12 Schematic of a companding current-mode integrator adopted from [11]

input current. Figure 9.12 shows a companding integrator uses subthreshold PMOS devices adopted from [11]. This circuit which acts as a translinear circuit can be described by:

$$I_B \cdot I_{\mathrm{IN}} = \left(C \cdot \frac{\mathrm{d}V}{\mathrm{d}T}\right) \cdot I_{\mathrm{OUT}}. \tag{9.45}$$

Assume simple exponential I/V relationship in subthreshold regime for MOS device: $I_{\mathrm{SD}} = I_b \mathrm{e}^{\frac{V_{\mathrm{SG}}}{nU_T}}$, then (9.45) can be rewritten as:

$$I_{\mathrm{OUT}} = \frac{I_B}{nU_T C} \cdot \int I_{\mathrm{IN}}(\tau) \cdot \mathrm{d}\tau = \frac{g_m}{C} \cdot \int I_{\mathrm{IN}}(\tau) \cdot \mathrm{d}\tau. \tag{9.46}$$

The DC gain of this integrator could be adjusted by proper choosing of aspect ratio of $M4$ with respect to $M1$. Also, the cutoff frequency is adjustable through I_B. A simplified circuit schematic of the current-mode integrator connected to the current steering DAC is shown in Fig. 9.13. In this circuit, signal RZ is used to construct

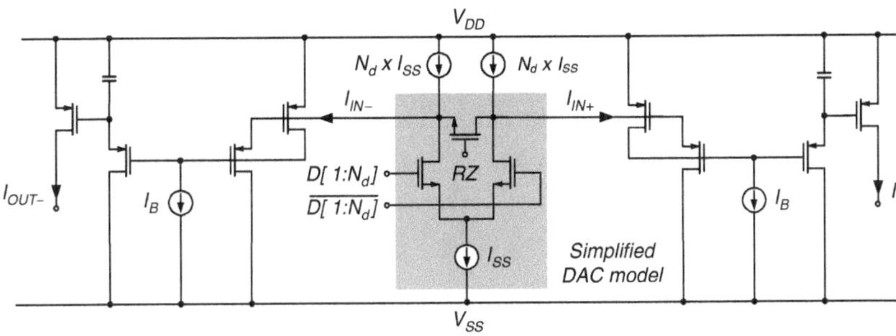

Fig. 9.13 Circuit diagram of the current steering DAC and differential current-mode integrator

a RZ DAC if necessary. To deliver the current to ring oscillator, it is necessary to convert the output differential current to single ended one, which can be done using a simple current mirror.

9.6 High Order Modulator Design

9.6.1 *Analysis and Modeling*

The $R\Delta\Sigma$ topology could be categorized as a continuous-time (CT) modulator. Design of discrete-time (DT) $\Delta\Sigma$ modulators with high order loops has been extensively studied in the literature [1]. In DT modulators, describing the desired transfer characteristics for signal and noise are relatively straightforward. The input signal appears at the output with some delay such as $\text{STF}(z) = z^{-n}$, while noise needs to be filtered output as: $\text{NTF}(z) = (1 - z^{-1})^n$.

On the other side, CT $\Delta\Sigma$ modulators consist of both continuous-time (e.g., continuous-time integrators or filters), and discrete-time parts (such as quantizer and DAC). This property makes the analysis of CT modulators more complicated [12]. A common approach for designing CT modulators is to calculate STF and NTF in CT domain and then convert them to discrete-time domain for the final design [12–14]:

$$\text{STF}(z) = \mathscr{Z}\{\mathscr{L}^{-1}\{\hat{\text{STF}}(s)\}\} \tag{9.47}$$

where $\mathscr{L}$ stands for Laplace transformation and $\mathscr{Z}$ for z-transformation. Figure 9.14 illustrates the block diagram of CT and DT $\Delta\Sigma$ modulators. In a CT modulator:

$$\text{STF}(s) = \frac{G(s)}{1 + G(s)H(s)R(s)} \tag{9.48}$$

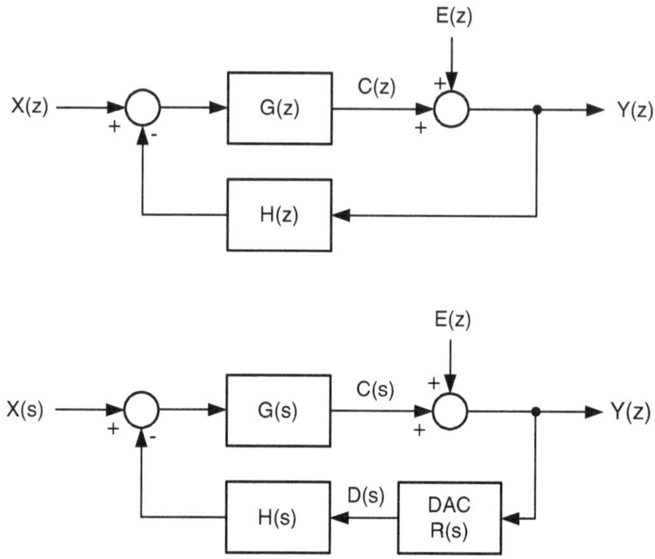

Fig. 9.14 Discrete-time and continuous-time ΔΣ modulators

Here, $R(s)$ represents the transfer function of the DAC. Depending on topology, DAC could exhibit return to zero (RZ) or non-return to zero (NRZ) specifications [14]. For an NRZ DAC:

$$R(s) = \frac{1 - e^{-sT_s}}{s} \tag{9.49}$$

where T_s is the sampling period. For a RZ DAC, the equation will change to:

$$R(s) = \frac{1 - e^{-s\tau}}{s} \tag{9.50}$$

where τ indicates the time period in which DAC is active. The NRZ DAC is a specific case of RZ DAC where $\tau = T_s$. Generally, in RZ DACs τ is selected to be equal to $T_s/2$.

While the loop gain in DT is: $F(z) = H(z)G(z)$, in CT domain it changes to: $F(s) = R(s)H(s)G(s)$ [12]. Having the CT open loop transfer function, it is possible to calculate the corresponding DT transfer function by transformation shown in (9.51):

$$F(s) = \sum_{k=1}^{N} \frac{\hat{a_k}}{s - s_k} \Leftrightarrow F(z) = \sum_{k=1}^{N} \frac{a_k}{s - z_k} \tag{9.51}$$

where:

$$z_k = e^{s_k T_s}. \tag{9.52}$$

Having $F(z)$, it is now possible to calculate the DT noise transfer function:

$$\mathrm{NTF}(z) = \frac{1}{1+F(z)}. \tag{9.53}$$

Here, a concise mathematical flow for calculating the relationship among $\hat{a_k}$ and a_k is presented [12, 15]. Assume that the impulse response of $V(s) = H(s)G(s)$ is:

$$v(t) = \mathscr{L}^{-1}\{V(s)\} = \sum_{k=1}^{N} a_k \mathrm{e}^{s_k t}. \tag{9.54}$$

Then, $f(t)$ can be calculated by:

$$\begin{aligned} f(t) &= \int_{-\infty}^{\infty} r(\theta)\cdot v(t-\theta)\cdot \mathrm{d}\theta = \int_{0}^{\tau} \sum_{k=1}^{N} a_k \cdot \mathrm{e}^{s_k(t-\theta)}\cdot \mathrm{d}\theta \\ &= \sum_{k=1}^{N} \frac{a_k}{-s_k}\cdot \mathrm{e}^{s_k t}\cdot (\mathrm{e}^{-s_k\tau}-1) \end{aligned} \tag{9.55}$$

where $r(t) = \mathscr{L}^{-1}\{R(s)\}$. Now, it is possible to calculate the discrete time value of this function by putting $t = nT_s$:

$$f(nT_s) = \sum_{k=1}^{N} \frac{a_k}{-s_k}\cdot \mathrm{e}^{s_k n T_s}(\mathrm{e}^{-s_k\tau}-1). \tag{9.56}$$

The z-domain representation of the proposed transfer function can be calculated by:

$$F(z) = \sum_{n=-\infty}^{\infty} f(n)z^{-n} = \sum_{k=1}^{N} \frac{a_k}{-s_k}\cdot (\mathrm{e}^{s_k(T_s-\tau)} - \mathrm{e}^{s_k T_s})\cdot \frac{z^{-1}}{1-z_k\cdot z^{-1}} = \sum_{k=1}^{N} \frac{\hat{a_k}}{z-z_k}. \tag{9.57}$$

Based on this, it is possible to calculate the relationship between the coefficients in s-domain and z-domain for a RZ DAC as following:

$$\hat{a_k} = \frac{a_k}{-s_k}\cdot (\mathrm{e}^{s_k(T_s-\tau)} - \mathrm{e}^{s_k T_s}) \tag{9.58}$$

For the case of an ideal integrator where the pole is placed at the origin: $s_k = 0$, L'Hôpicatl's rule can be used:

$$\hat{a_k} = a_k\cdot (T_s - \tau). \tag{9.59}$$

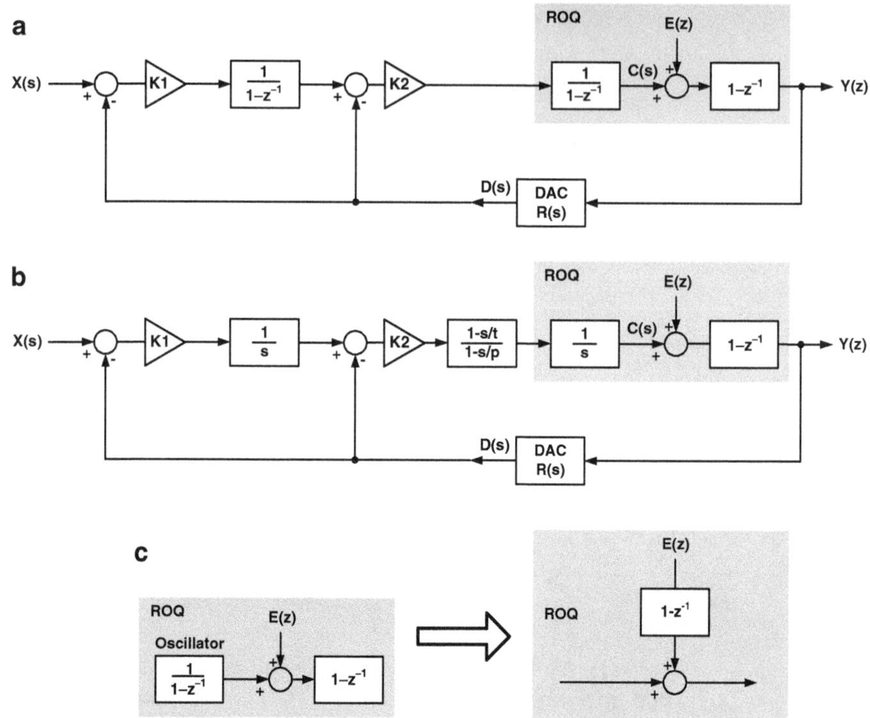

Fig. 9.15 Block diagram of a third order $R\Delta\Sigma$ modulator: (**a**) based on DT integrators, (**b**) based on CT integrators. (**c**) Model of a ROQ

Similarly, for an ideal second order z-domain integrator, the corresponding s-domain transfer function including a RZ DAC in which $\tau = T_s/2$ is [15]:

$$\mathrm{INT}(z) = \left(\frac{z^{-1}}{1-z^{-1}}\right)^2 \Leftrightarrow \hat{\mathrm{INT}}(s) = \frac{-1.5s/T_s + 2/T_s^2}{s^2}. \tag{9.60}$$

Figure 9.15a shows the discrete time model of a third order $R\Delta\Sigma$ modulator. Based on this model and using simplified model of ROC introduced in Fig. 9.15a, one can show that the ideal NTF for a discrete-time system can be achieved by setting: $K_1 = K_2 = 1$ [15].

Now, we can use the same approach to first calculate the $F(s)$ of the CT $R\Delta\Sigma$ which is shown in Fig. 9.15b, and then based on that, estimate $F(z)$ and finally NTF(z):

$$F(s) = R(s)\cdot\left(K_1 + \frac{K_2}{s}\right)\cdot\frac{1}{s}\cdot\frac{1-s/t}{1-s/p} \tag{9.61}$$

Table 9.2 Predicted SNR for different sets of parameters ($OSR = 128$)

K_1	K_2	Normalized p	Normalized t	SNR, [dB]
$4/T_s$	$0.5/T_s$	−0.3	−0.3	105

Here, pole, p, and zero , t,[1] of the replica bias circuit have been also included in the model. Rearranging (9.61), $F(s)$ can be written as [16]:

$$F(s) = R(s) \cdot \left(\frac{A}{s} + \frac{B}{s^2} + \frac{C}{s-p} \right) \tag{9.62}$$

where:

$$A = K_2 \cdot \left(\frac{p}{t} - 1 \right) - K_1 \cdot p \tag{9.63}$$

$$B = K_2 \tag{9.64}$$

$$C = K_1 \cdot \frac{p}{t} \tag{9.65}$$

Using the regulations have been developed in [12]:

$$H(z) = \frac{\frac{A}{2}}{z-1} + \frac{\frac{3B}{8} \cdot z + \frac{1}{8}}{(z-1)^2} + \frac{C}{p} \cdot \frac{e^p - e^{\frac{p}{2}}}{z - e^p} \tag{9.66}$$

The set of parameters in (9.66) should be selected somehow to give the desired noise transfer function which is ideally $\mathrm{NTF}(s) = (1-z^{-1})^2$ for a third order loop.[2] Table 9.2 shows the predicted signal to noise ratio for the proposed modulator based on (9.66). In this table, values of pole and zero are normalized to the sampling frequency.

9.6.2 Behavioral Modeling

Following the analysis made for a third order noise shaping loop, a model for the proposed modulator has been developed. The goal has been to include all the sources of nonidealities and study the effect of each one on system performance more precisely. Figure 9.16 shows the detailed results of the behavioral modeling made in MATLAB/Simulink.

[1] Notation for zero has been changed from z to t in order to avoid mixing it with the notation of z in z-transform.

[2] The extra factor of $(1-z^{-1})$ for having a third order noise shaping comes from the ring oscillator (see Fig. 9.15).

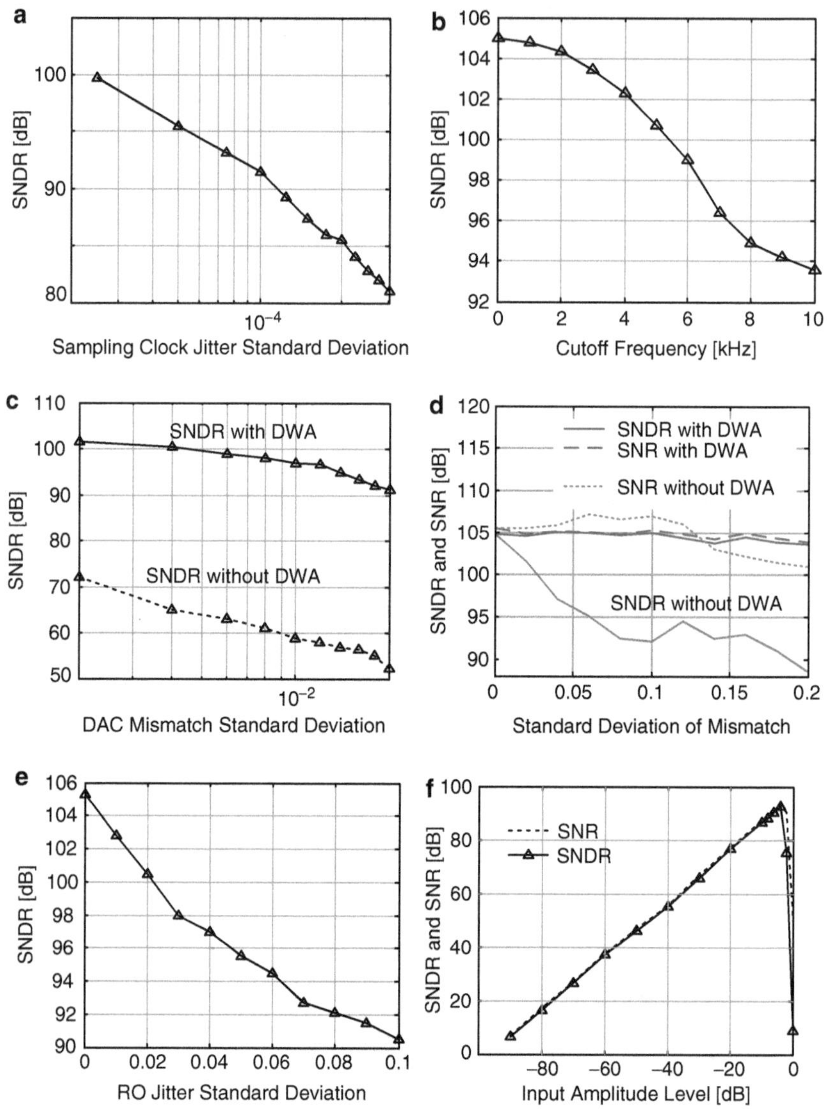

Fig. 9.16 Performance of a third order $R\Delta\Sigma$ based on behavioral modeling in MATLAB: (**a**) Effect of sampling clock jitter on SNDR. (**b**) Effect of leaky integrator on SNDR. (**c**) Effect of DAC component mismatch on SNDR, with and without DWA. (**d**) Effect of delay element mismatch on SNR and SNDR. (**e**) Effect of ring oscillator jitter on system performance. (**f**) SNR and SNDR of the system including all nonideal effects

There are two features of the modulator that are influenced by the sampling clock jitter. First one is the decision point of the quantizer. Variable clock cycle period will add uncertainty when it comes to counting transitions inside the cycle. This can be easily modeled by adding this uncertainty to the model of the quantizer when using the duration of the clock period to compare it to the sum of overall delays [15]. The second feature is the influence of clock jitter on the DAC performance. Since DAC is switched by the clock, clock jitter affects the shaping of the DAC pulses making them either wider or narrower in term of time duration. To remove this effect from the time domain in order to improve the simulation speed, instead of changing the width of the pulses by adding uncertainty, we have changed the amplitude. Since the pulses will be shaped further by an integrator, this will have completely the same effect [15].

Based on Fig. 9.16a, to keep the performance above 90 dB, it is necessary to reduce the standard deviation of the clock jitter below 10^{-4}. In this simulation, the variation on delay mismatch is set to 0.1, and oscillator jitter is 0.01 with a DAC mismatch of 0.01.

Figure 9.16b shows the effect of a leaky integrator in which transfer characteristics is $1/(1 + s/p)$ instead of $1/s$. Here, it is assumed that the signal bandwidth is 4 kHz. Regarding this figure, the cutoff frequency of the nonideal filter should be less than 1 kHz to have negligible degradation in performance.

Typical curve of the measured SNDR versus standard deviation of DAC mismatch is given in Fig. 9.16c. In these measurements, filter cutoff frequency was set to 1 kHz, delay mismatch was set to 0.1 standard deviation, and ring oscillator jitter was set to 0.01 standard deviation. It can be observed from the figure that improvement due to DWA is significant. Already for standard deviation of DAC mismatch of 0.002, there is a difference of almost 30 dB in SNDR which means that improvement of DAC linearity due to DWA is crucial in this topology. Still, as expected, it can be observed that sensitivity to this mismatch is quite high even including DWA. For the level of standard deviation of 0.02, the decrease in achieved SNDR is approximately 10 dB.

To measure how much inherent data weighted averaging in ring oscillator helps to reduce the effect of DAC mismatch and also delay mismatch, some simulations with and without this effect has been performed. The influence of delay mismatch is important since it determines how relaxed final design of ring oscillator stages can be. In other words, less sensitivity to mismatch means higher level of tolerable mismatch which allows using smaller transistors when designing inverter (delay) cells. It is expected that DWA performed by the ROQ improves the resistance of system to mismatch. A typical curve displaying measured SNDR and SNR with and without DWA is shown in Fig. 9.16d. As it can be observed from the figure, SNR does not show significant fall when increasing standard deviation of the mismatch which is expectable since DWA improves linearity.

Ring oscillator jitter is derived as a group influence of all the sources of noise within one delay cell. The level of acceptable oscillator jitter also relates to individual delay cell design, as well as biasing, since sources of the noise are devices inside the cell, but also both supply voltage and tail currents. A typical curve of oscillator jitter influence on SNDR of ADC is given in Fig. 9.16e. For these measurements, filter cutoff frequency was also set to 1 kHz, with delay mismatch set to 0.1 standard deviation. As it can be observed from the figure, deviation of the delay introduced by the oscillator jitter should generally be kept below 0.02 value of standard deviation. In other words, compared to nominal delay, this is what is the tolerable delay variance (0.022), introduced by the oscillator jitter.

Finally, Fig. 9.16f shows the dynamic range of the proposed third order modulator in presence of all different sources of nonidealities. In this plot, the pole of the nonidea integrator is set to be 1 kHz, delay mismatch is set to be 0.1, ring oscillator jitter is 0.01, DAC mismatch is 0.01, and sampling clock jitter is set to be 0.0001. Based on behavioral modeling, the peak SNDR value is 93dB with an overload level of around −4 dB.

9.7 Simulations and Experimental Results

A first order noise shaped quantizer has been designed and implemented in a conventional CMOS 90 nm technology. Figure 9.17 shows the mask layout of the proposed circuit. This prototype contains a test structure to study the matching properties of STSCL circuits by putting several ring oscillators with different distances to a common replica bias circuit. The output of each ring oscillator can be probed and measured separately.

Two different versions of ROQ have been also implemented. The type A circuit is a simple first order modulator while type B circuit is a second order ROQ. The type B modulator uses a current-steering type DAC to close the loop. The area of this modulator is 250 μm × 400 μm.

Since the current levels are very low, a programmable current scaler circuit has been employed to scale down the input DC and sinusoidal currents. The input current of the modulator can be adjusted between 20 pA and 100 nA. Based on simulation results, the linearity of the current scaler circuit is better than 80 dB.

Figure 9.18 shows the supply current of the proposed $R\Delta\Sigma$ modulator when biased at $I_{SS(nom)} = 1$ nA. Simulation results show that the variation on supply current at clock transitions is only 15% of the total circuit current consumption.

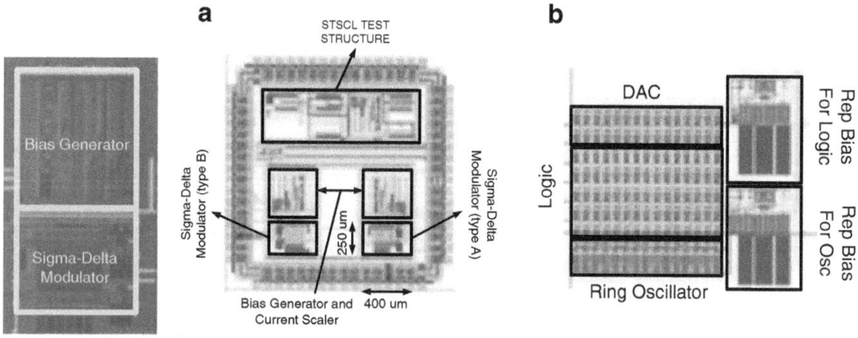

Fig. 9.17 (**a**) Chip phot and mask layout of the test chip fabricated in 90-nm CMOS technology. (**b**) Mask layout of the quantizer circuit

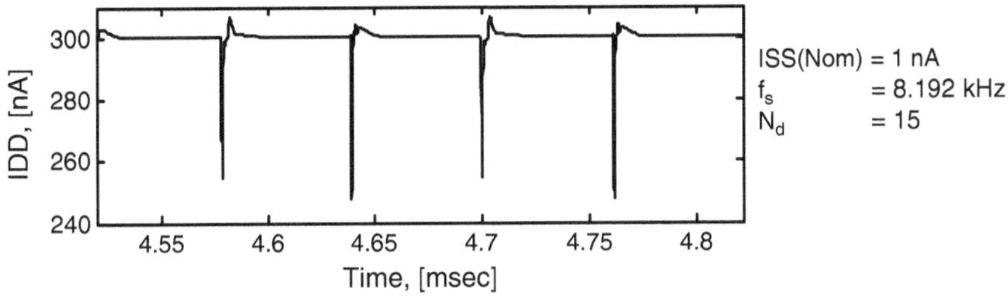

Fig. 9.18 Simulated supply current consumption of the $R\Delta\Sigma$ modulator for $I_{SS(nom)} = 1$ nA. The variation on supply current is about 15% of the total circuit current consumption

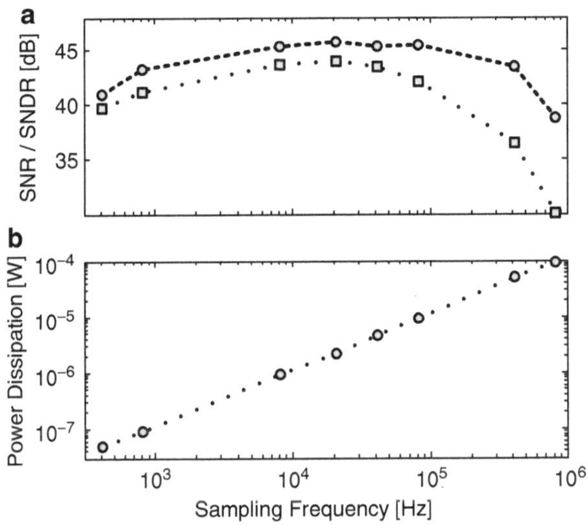

Fig. 9.19 Measurement results in different sampling frequencies: (**a**) SNR and SNDR values and (**b**) Power dissipation of the modulator. Here: OSR = 64, $A_{IN} = -20$ dB, $V_{DD} = 1.2$ V

Figure 9.19 summarizes the measurement results for the proposed ROQ. In these simulations, the input amplitude is $A_{IN} = -20$ dB and OSR = 64. While the power dissipation of the modulator scales linearly with the sampling frequency, the SNR value stays above 40 dB over more than three decades of variation on sampling frequency. The power dissipation of modulator is 37 pW/Hz.

9.8 Conclusion and Discussion

In this chapter, the possibility of implementing ultra-low-power ADCs with scalable performance has been studied. To achieve the desired flexibility in performance, ring oscillator based $\Delta\Sigma$ topology has been selected. Using STSCL (subthreshold SCL) building blocks, the proposed ADC can achieve three decades of operating range.

Meanwhile, the effect of non-ideal circuit behavior on the system performance has been studied. This study helps to optimize the circuit parameters with respect to the system requirements.

A test chip has been implemented in 90-nm CMOS technology occupying $250\,\mu m \times 400\,\mu m$. The power consumption of the proposed ROQ is about 37 pW/Hz for sampling frequencies ranging from 160 Hz to 820 kHz with a peak SNDR value of 65 dB.

References

1. S. R. Norsworthy, R. Schreier, G. C. Temes, *Delta-Sigma Data Converters: Theory, Design, and Simulation*, IEEE, 1997
2. B. E. Boser and B. A. Wooley, "The design of Sigma-Delta modulation analog-to-digital converters," *IEEE J. Solid-State Circuits*, vol. 23, no. 6, pp. 1298–1308, Dec. 1988
3. B. Razavi, *Principles of Data Conversion System Design*, IEEE, 1995
4. A. Iwata, N. Skimura, M. Nagata, and T. Morie, "An architecure of delta sigma A-to-D converter using a voltage controlled oscillator as a multi-bit quantizer," in *Proceedings of IEEE International Symposium on Circuits and Systems*, pp. 445–448, May 1998
5. R. Naiknaware, H. Tang, and T. Fiez, "Time-referenced single-path multi-bit ΔΣ ADC using VCO-based quantizer," in *IEEE Transactions Circuits Systems-II: Analog Digital Signal Processings*, vol. 47, no.6, pp. 596–602, Jun. 2000
6. M. Z. Staayer and M. H. Perrot, "A 12-Bit, 10-MHz bandwidth, continuous-time ΔΣ ADC with a 5-bit, 950-MS/s VCO-based quantizer," *IEEE J. Solid-State Circuits*, vol. 43, no. 4, pp. 805–814, Apr. 2008
7. S. Kuboki, K. Kato, N. Miyakawa, and K. Matsubara, "Nonliearity analysis of resistor string A/D converters," in *IEEE Transactions on Circuits and Systems*, vol. 29, no. 6, pp. 383–390, Jun. 1982
8. J. McNeill, "Jitter in ring oscillators," *IEEE J. Solid-State Circuits*, vol. 32, pp. 870879, Jun. 1997
9. P. Kinget, "Device mismatch and tradeoffs in the design of analog circuits," *IEEE J. Solid-State Circuits*, vol. 40, no. 6, pp. 1212–1224, Jun. 2005
10. A. Hajimiri, S. Limotyrakis, and T. H. Lee, "Jitter and phase noise in ring oscillators," *IEEE J. Solid-State Circuits*, vol. 34, no. 6, pp. 790-804, Jun. 1999
11. E. Seevinck, "Companding current-mode integrator: a new circuit principle for continuous-time monolithic filters," in *IEE Electronics Letters*, no. 24, vol. 26, pp. 2046–2047, Nov. 1990
12. O. Shoaei, "Continuous-time Delta-Sigma A/D converters for high speed applications" *Ph.D. Dissertation*, Carleton University, 1995
13. O. Bajdechi and J. H. Huijsing, *Systematic Design of Sigma-Delta analog-to-digital converters*, Kluwer, 2004
14. J. A. Cherry and W. M. Snelgrove, "Excess loop delay in continuous-time Delta-Sigma modulators," in *IEEE Transactions on Circuits and Systems-II*, vol. 46, no. 4, pp. 376–389, Apr. 1999
15. N. Kotic, "Design of a ring oscillator based delta-sigma modulators," *Master Thesis*, Ecole Polytechnique Fédérale de Lausanne (EPFL), Switzerland, 2010
16. D. San Martin Molina, "Design of a very low power delta-sigma analog to digital converter," *Master Thesis*, Ecole Polytechnique Fédérale de Lausanne (EPFL), Switzerland, 2009

Chapter 10
Wide Tuning Range PLL

10.1 Introduction

Many applications require wide tuning range phase-locked loops (PLLs) to generate pure and well controlled periodic signals [1–3]. PLLs might be used for controlling the operation condition of specific parts of a system such as in continuous-time filters [4].

Clock generation for digital systems is an example where the system clock frequency needs to be scaled in a very wide range for power saving purpose [5]. In this approach, the clock frequency is adjusted through a controlling unit with respect to the work load of the system [6]. Therefore, the operating frequency should be adjustable over a very wide range, and hence it requires a very wide tuning range clock generator [7, 8].

In this work, several wide tuning range analog and digital integrated circuit have been implemented. To adjust the operating condition of these circuits during the operation, a precise controlling unit is required. This adjustment can be done using a wide tuning range PLL circuit which is the main concern of this chapter. The PLL circuit that will be developed here is needed to be compatible with the circuit building blocks that have already been developed using subthreshold source-coupled topology.

In the rest of this chapter, the main design issues associated with wide tuning range PLLs will be studied and finally the implementation of the proposed PLL in 0.13- μm CMOS technology will be explained.

10.2 Wide Tuning Range PLLs

PLL circuits are widely used in telecommunication and digital processing systems for different purposes [1]. In telecommunication systems, PLLs have been widely used to generate very low-jitter reference frequencies for modulating or demodulating the RF and baseband signals [2, 9]. In these applications, stability, settling time, and phase noise of the output oscillating signal in addition to the circuit power

A. Tajalli and Y. Leblebici, *Extreme Low-Power Mixed Signal IC Design: Subthreshold Source-Coupled Circuits*, DOI 10.1007/978-1-4419-6478-6_10,

consumption and cost are the most important design parameters. Some recently developed applications such as multi-band or multi-standard transceivers have made the design of wide tuning range PLLs or frequency synthesizers very demanding [10, 11].

The design of wide tuning range PLLs have been for a longtime interesting in digital integrated systems [7, 12]. Having an arbitrary clock generator makes it possible to adjust the operating condition of the digital system in a wide range, and have a close to optimum power-performance compromise point [6, 13].

There are some special issues with design of wide tuning range PLLs which are not a concern in design of conventional PLLs. The design of PLLs with a limited tuning range have been widely studied in literature [1, 14]. The developed design methodologies for conventional PLLs can be extended for deigning wide tuning range PLLs with some modifications specially in the topology of the circuit [3].

In this section, after a very short introduction on charge-pump based PLLs, the main issues with the design of wide tuning range PLLs will be studied.

10.2.1 Background

The performance of a PLL circuit highly depends on the specifications of the components inside the loop. Figure 10.1 shows a charge-pump based PLL (CPLL) [1]. Based on this topology, the phase of the input signal, f_P, and the output of frequency divider circuit, f_D, are compared with a phase-frequency detector (PFD) circuit. The error signal at the output of PFD which depends on the phase and frequency difference between the two inputs is filtered to remove its high frequency components, and finally is applied to the ring oscillator in order to adjust its oscillation frequency.

The main specifications of a PLL, such as settling time (t_{ss}) and phase noise, depend on the loop characteristics such as loop bandwidth (ω_C), phase margin (PM), and loop gain ($| T(j0) |$). Changing the operating frequency of the loop will definitely change some of the parameters such as ω_C and settling time which is not unexpected. However, to ensure the stability of the circuit, some parameters such as PM and $| T(j0) |$ should to be kept almost unchanged.

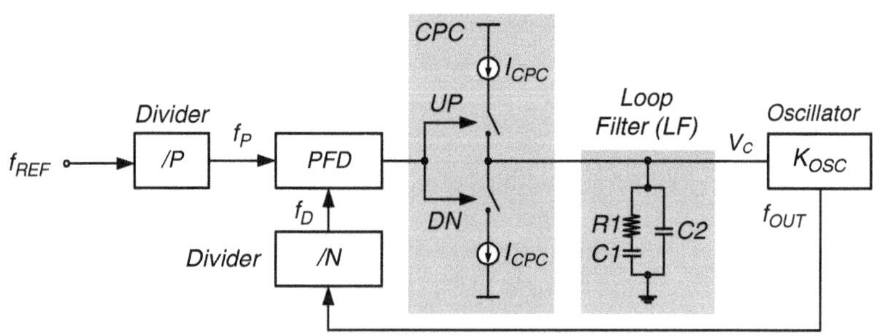

Fig. 10.1 Conventional charge-pump PLL (CPLL) topology

Here, we will first study the main constraint in design of conventional PLLs and then the analysis will be extended for wide tuning range PLLs. The goal is to derive a methodology for implementing a stable PLL circuit for a given reference frequency range.

Using continuous-time approximation [1] for the PLL shown in Fig. 10.1, the open loop gain can be calculated as

$$T(s) = \frac{I_{\text{CPC}}}{2\pi} \cdot LF(s) \cdot \frac{K_{\text{OSC}}}{s} \cdot \frac{1}{N} \tag{10.1}$$

where K_{OSC} is the oscillator sensitivity factor and is defined as the variation at the output oscillation frequency divided by the input controlling signal.

The loop filter should be designed based on jitter and dynamic performance requirements of the system. In Fig. 10.1, R_1 and C_1 create a zero to make the loop stable. The noise associated with R_1 can degrade the phase noise at the output of oscillator, hence it is recommended to choose a small enough value for R_1 [9]. Meanwhile, C_2 is used to reduce the ripples on controlling signal, V_C, and hence reduce the pattern jitter [8]. However, the extra phase lag associated with by the extra pole created by C_2 will cause some stability issues. As will be shown later, the ratio of C_1/C_2 should be selected very carefully to avoid instability. To reduce the pattern jitter which is mainly due to the variations on the controlling signal, V_C, the order of the loop filter can be increased even more [4, 9].

In the simplified circuit diagram depicted in Fig. 10.1 for the loop filter:

$$\text{LF}(s) = \frac{C_1}{C_1 + C_2} \cdot \frac{1}{sC_1} \cdot \frac{1 - s/z}{1 - s/p} \tag{10.2}$$

where

$$p = -\frac{C_1 + C_2}{R_1 C_1 C_2} \tag{10.3}$$

and

$$z = -\frac{1}{R_1 C_1} = -\frac{1}{\tau}. \tag{10.4}$$

To study the stability of the circuit, phase margin (PM) or damping factor (ζ) of this system can be estimated. The phase margin of the proposed third order PLL is

$$\text{PM} = \tan^{-1}(\tau\omega_C) - \tan^{-1}\left(\frac{\tau\omega_C}{b+1}\right) \tag{10.5}$$

where $b = C_1/C_2$, $\tau = R_1 C_1$, and ω_C is the loop crossover frequency. It can be shown that the phase margin can be maximized if [9]

$$\omega_C = \frac{\sqrt{b+1}}{\tau}. \tag{10.6}$$

The value of the phase margin in this case will be

$$\mathrm{PM}_{\mathrm{Max}} = \tan^{-1}\left(\sqrt{b+1}\right) - \tan^{-1}\left(\frac{1}{\sqrt{b+1}}\right) \tag{10.7}$$

which depends only on b. To impose the constraint indicated in (10.6), ω_C can be calculated from (10.2), and hence:

$$\frac{I_{\mathrm{CPC}}}{2\pi} \cdot \frac{K_{\mathrm{OSC}}}{N} \cdot \frac{b}{b+1} = \frac{C_1}{\tau^2}\sqrt{b+1}. \tag{10.8}$$

The design can be completed by selecting a proper value for $\omega_C = 2\pi f_C$ and b [9]. The crossover frequency is generally selected at least 10–20 times smaller than the input clock frequency, f_P, to make sure that the continuous-time approximation remains valid [1]. Therefore

$$f_C < \frac{f_P}{M_F} = \frac{f_{\mathrm{REF}}}{M_F P} \tag{10.9}$$

where $M_C = 10$ to 20. In addition, b can be selected to have the desired phase margin as indicated in (10.7). Having b and ω_C, the value of τ can be calculated from (10.6). The next step is to calculate the charge-pump bias current from (10.8).

The design also can be proceeded based on estimating of damping factor (ζ) instead of PM [3, 8]:

$$\zeta = \frac{1}{2} \cdot \sqrt{\frac{1}{N} \cdot I_{\mathrm{CPC}} K_{\mathrm{OSC}} R_1^2 C_1} \tag{10.10}$$

where

$$\omega_C = \frac{2\zeta}{R_1 C_1}. \tag{10.11}$$

After choosing a proper value for ζ, then the proper value for different elements can be derived.

10.2.2 Wide Tuning Range CPLL

To implement a scalable output frequency PLL, it is possible to change the input frequency (f_{REF}), or the division ratio of the frequency dividers (N and P) shown in Fig. 10.1. Therefore, the effect of changing these three parameters on the loop dynamic behavior needs to be studied.

To have a stable PLL, based on the analysis results presented in Sect. 10.2.1, it is necessary to properly set the values of ω_C and $|\ z\ |$ with respect to the reference frequency. In addition, the ratio of capacitances in loop filter (b) should be selected carefully. This parameter is independent of the input reference frequency. Finally the bias current of CPC circuit needs to be selected with respect to the input frequency and division ratio.

This discussion implies that by scaling the input frequency or division ratio, there are some parameters that can be kept constant (such as b), while some other parameters need to be scaled (such as I_{CPC}, ω_C, and τ). Therefore, it is necessary to determine the requirements on each design parameter and make sure that by scaling the operating condition the system remains stable while the system performance (phase noise, settling time, and etc.) will be maintained.

The design process can be started by estimating the value of τ with respect to the input reference frequency:

$$\tau = R_1 C_1 = \frac{\sqrt{b+1}}{\omega_C} = \left(\frac{\sqrt{b+1}}{2\pi} \cdot M_F\right) \cdot \frac{1}{f_P} \tag{10.12}$$

where M_F and b are constant numbers could be selected from (10.9). Therefore, τ depends only on f_P, and not on N. The next step is to calculate the charge-pump bias current from (10.8) using τ calculated in (10.12)

$$I_{CPC} = 8\pi^3 \cdot C_1 \cdot \frac{\sqrt{b+1}}{b} \cdot \frac{1}{M_F^2} \cdot \frac{1}{K_{OSC}} \cdot N \cdot \left(\frac{f_{REF}}{P}\right)^2 \tag{10.13}$$

which indicates that for constant values of C_1, f_{REF}, and M_F, the charge pump bias current needs to be changed proportional to N and inversely proportional to the square value of P. Therefore, a CPC with programmable or adjustable bias current is required. Design of a charge pump circuit with a bias current proportional to N/P^2 will be complicated and requires a complex current switching network. A remedy for simplifying the circuit topology is to use a current-controlled oscillator (CCO) instead of a voltage controlled oscillator in which:

$$K_{OSC} = \frac{\partial I_C}{\partial V_C} \cdot \frac{\partial f_{OSC}}{\partial I_C} = G_m \cdot K_{CCO}. \tag{10.14}$$

Based on (10.14), a transconductance, G_m, is required to convert the controlling voltage to controlling current. In this case, the controlling current is equal to the oscillator current: $I_C = I_{OSC} = \frac{N}{P} \times \frac{f_{REF}}{K_{CCO}}$. Therefore, the controlling current is always proportional to N/P. Based on this, if we make G_m value proportional to its current, i.e.: $G_m = I_C/V_{char}$, then using (10.13):

$$I_{CPC} = \frac{I_{OSC}}{N} \times \left(\frac{V_{char}}{V_{SW}} \cdot \frac{C_1}{C_L} \cdot \frac{\sqrt{b+1}}{b} \cdot \frac{4\pi^3}{\ln 2 M_F^2 N_d}\right) \tag{10.15}$$

As a conclusion, it is sufficient to make the bias current of the charge pump circuit proportional to I_C/N which can be simply done as shown in Fig. 10.2.

In transconductor circuit, V_{char} depends on the circuit topology used to produce G_m. For example, for a single MOS transistor biased in subthreshold regime: $V_{char} = U_T$ (thermal voltage) and in strong inversion: $V_{char} = V_{DSsat}$ (gate overdrive voltage).

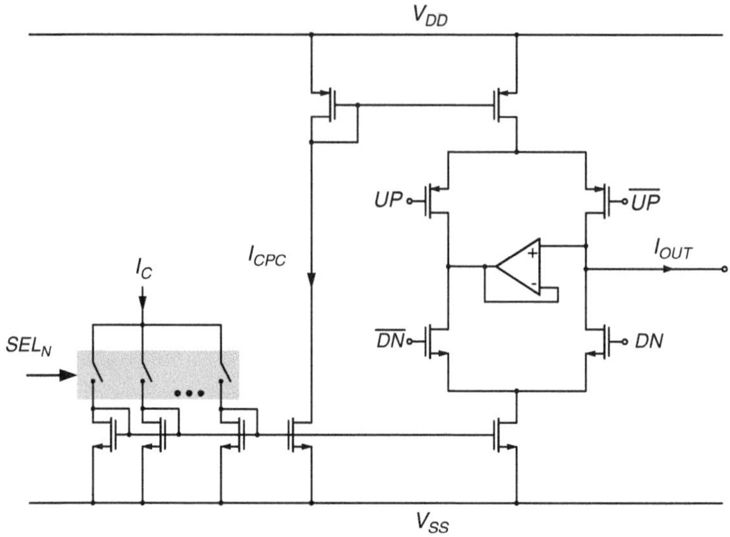

Fig. 10.2 Charge pump circuit with programmable bias current

Table 10.1 Summary of the main design parameters of wide tuning range CPLL

Parameter	Value
Reference frequency	f_{REF}
Oscillation frequency	$f_{OSC} = \frac{N}{P} \cdot f_{REF}$
Oscillator bias current	$I_{OSC} = \frac{f_{OSC}}{K_{CCO}} = \frac{N}{P} \cdot \frac{f_{REF}}{K_{CCO}}$
Number of delay stages in ring oscillator	N_d
Capacitive load at the output of each delay stage	C_L
Voltage swing at the output of each STSCL gate	V_{SW}
Oscillator gain	$K_{CCO} = \frac{1}{2 \ln 2 N_d V_{SW} C_L}$
Transconductance	$G_m = \frac{I_C}{V_{char}} = \frac{I_{OSC}}{V_{char}}$
Charge pump current	$I_{CPC} = \frac{I_{OSC}}{N} \times \left(\frac{V_{char}}{V_{SW}} \cdot \frac{C_1}{C_L} \cdot \frac{\sqrt{b+1}}{b} \cdot \frac{4\pi^3}{\ln 2 M_F^2 N_d} \right)$
Loop filter resistance	$R_1 = \left(\frac{\sqrt{b+1}}{2\pi} \cdot \frac{M_F}{C_1} \right) \cdot \frac{P}{f_{REF}}$

In addition to that, based on (10.12), τ is inversely proportional to f_P or directly proportional to P. Since C_1 is constant, R_1 needs to be proportional to P. Therefore, to complete the circuit design, a resistance proportional to P is also required to make sure that the system will remain stable:

$$R_1 = \left(\frac{\sqrt{b+1}}{2\pi} \cdot \frac{M_F}{C_1} \right) \cdot \frac{P}{f_{REF}}. \qquad (10.16)$$

Table 10.1 summarizes the results of this discussion.

10.2.3 Design Issues with Wide Tune PLLs

There are several concerns with the design of self-biased wide tuning range PLL circuits. The first issue is implementing scalable CPC and G_m circuits whose bias current needs to be controlled precisely with respect to the values of P and N. In addition to these two circuits, R_1 also needs to have a scalable value proportional to P. The other very important issue here is maintaining the constraint indicated in (10.9) during the transitions when P or N are suddenly changing from one value to a new value.

Consider a PLL where its reference bias current, I_R, is generated with respect to the desired operation frequency. This types of PLLs whose bias current is generated automatically with respect to their operation condition are called self-biased PLLs [3, 15]. In the presence of any change at f_P, the system starts to track this modification and consequently I_R will be adjusted, correspondingly.

If f_P reduces suddenly from f_{P0} to $f_{P0}-\Delta f_P$, then it takes a specific amount of time for the self-biased circuit to track this change (see Fig. 10.3a). During this time interval, ω_C of the loop slowly starts to reduce from f_{P0}/M_F to $(f_{P0}-\Delta f_P)/M_F$. However, during this time, the ratio of f_P to ω_C will be smaller than M_F. If this ratio gets smaller than a specific value, then PLL could become unstable. If $M_F = 20$ and suppose for stability considerations, it is required to ensure $f_P/\omega_C/(2\pi)$ be always larger than 10. Then, it implies that Δf_P should be always less than $f_{P0}/2$ and it is very likely that PLL becomes unstable for $\Delta f_P > f_P/2$.

On the other hand, if f_P increases suddenly from f_{P0} to $f_{P0}+\Delta f_P$, there would be no stability issue. However, in this case the loop bandwidth becomes much smaller than $(f_{P0}+\Delta f_{P0})/M_F$ during the transition and hence loop may not be able to track the input frequency properly. This phenomena happens because the signal at the output of PFD circuit will be suppressed by the loop filter, and hence the useful information at the output of PFD will be partially discarded. Therefore, it is important to make sure that the bandwidth of loop filter is higher than $|f_{DIV}-f_P|$. This constraint ensures that the low frequency component at the output of the PFD will lie at the pass-band of the loop filter and hence will be applied to the oscillator to adjust its frequency. This effect is shown in Fig. 10.3b. As described

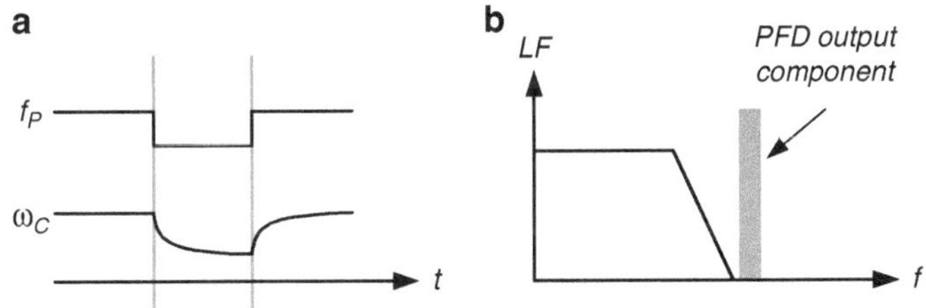

Fig. 10.3 (**a**) Transient loop response to the variation at the input frequency of the PLL. (**b**) The effect of small loop filter bandwidth with discarding the desirable component at the output of PFD

in [16], loop filter generally does not completely suppress the Δf_P which means the oscillation frequency of ring oscillator will be modulated by this even very small signal:

$$v_{\mathrm{OSC}}(t) = \cos\left(\omega_0 t + K_{\mathrm{OSC}} \int A_{\Delta f_P} \cos(2\pi \Delta f_P \mathrm{d}t)\right) \tag{10.17}$$

where $A_{\Delta f_P}$ is the amplitude of signal at the output of filter and in $f = \Delta f_P$. The modulated output of ring oscillator in combination with PFD circuit will result in a DC component that will adjust the oscillator frequency until it locks. If a large frequency jump (Δf_P) occurs, then it might take several beat cycles (or cycle slips) before lock is achieved.

These two phenomena will limit the maximum speed for changing the output frequency of a wide tuning range self-biased PLL.

10.3 Circuit Design

10.3.1 Proposed PLL Topology

One of the main goals in this design is to implement a wide tuning range PLL with a scalable power consumption. Having a scalable power consumption will improve the power efficiency of the circuit. To achieve this requirement, the power consumption of all the sub-blocks of the PLL needs to be proportional to the operating frequency. Referring to Fig. 10.1, using subthreshold source-coupled circuit for different sub-blocks such as PFD, frequency divider, and ring oscillator, can help to satisfy this requirement.

Figure 10.4 illustrates the proposed self-biased adaptive bandwidth PLL. In this topology, two frequency dividers have been used to program the output oscillation frequency. A CCO has been used to achieve a very wide tuning range.

The controlling current, I_C, is adjusted automatically by changing the value of P. By changing P, the input frequency of the PFD can be programmed as $f_P = f_{\mathrm{REF}}/P$. The PFD circuit compares f_P and f_{OSC} and generates proper controlling signals for CPC to adjust the oscillation frequency of CCO. A SCL-to-CMOS converter is used to convert the SCL signal levels at the output of PFD to full swing CMOS levels to be applied to the CPC circuit [8].

The transconductor (G_m) shown in Fig. 10.4 is used to convert the controlling voltage (V_C) to controlling current (I_C) in which

$$I_C = G_m \cdot V_C. \tag{10.18}$$

A copy of the controlling current is applied to different parts of the circuit to scale their bias current proportional to I_C. In this configuration, when P changes, the controlling current and hence the bias current of different parts of the circuit such

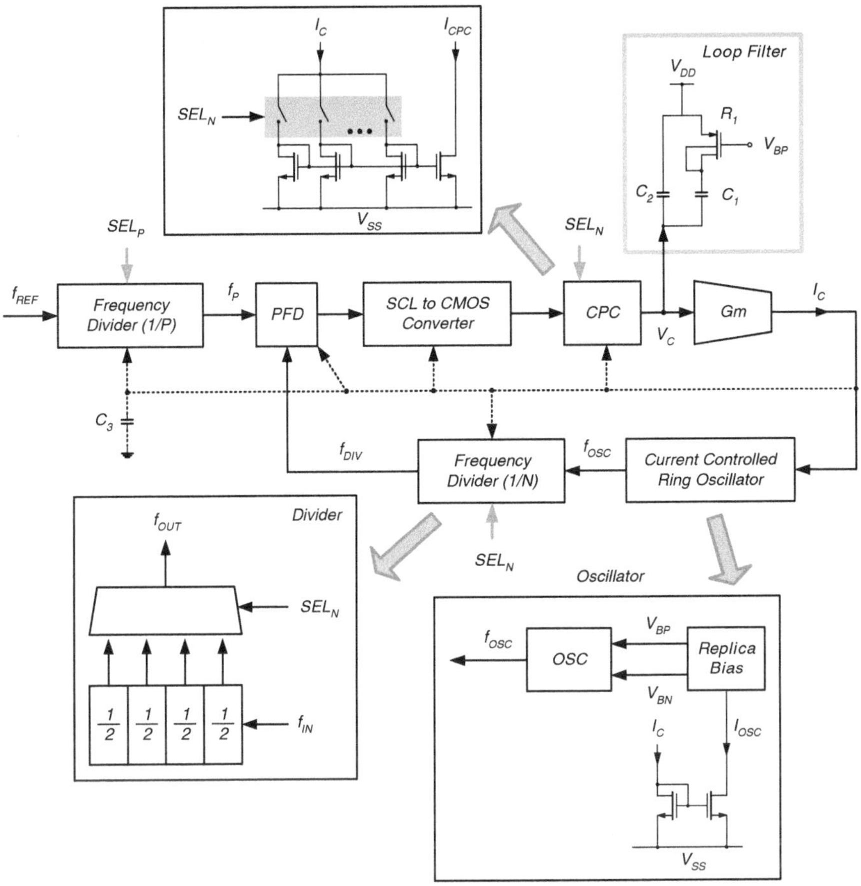

Fig. 10.4 Topology of the proposed self-biased adaptive bandwidth PLL

as PFD, dividers, CPC, and SCL-to-CMOS converter, will be scaled proportional to the variation on I_C. This property helps to have a power scalable PLL.

Meanwhile, I_C is applied to a replica bias circuit to generate the proper bias voltages for NMOS and PMOS devices in each STSCL gate. The bias voltage for PMOS devices, V_{BP}, is also applied to the PMOS resistance inside the loop filter (R_1). Therefore, the resistance $R1$ would have a resistivity proportional to the operating condition as explained in Sect. 10.2.2. Adjusting the bias current of CPC circuit and resistance in the loop filter will provide a adjustable frequency response for the loop. Meanwhile, the operating frequency of CCO can be adjusted through changing the frequency division ratio inside the loop, N. As illustrated in Fig. 10.4, N can be controlled by signal SEL_N. A multiplexer (MUX) is used to select one of the outputs of a chain of divide-by-two circuits and hence select a proper value for N as shown in this figure.

To keep the frequency characteristics of the loop unchanged by varying N, the bias current of CPC circuit should be adjusted with respect to the value of N. This adjustment can be done by scaling controlling current in the CPC as shown in Fig. 10.4.

To filter the ripples on controlling signal, it is possible to put a capacitance at the output of G_m cell. The G_m cell and this extra parasitic capacitance creates a low frequency pole at the loop filter with a cutoff frequency of

$$\omega_p = \frac{G_m}{C_3}. \tag{10.19}$$

To make sure the PLL will stay stable in presence of new pole, ω_p should be selected carefully [4].

With these characteristics, the topology shown in Fig. 10.4 provides a self-biased and adaptive bandwidth PLL can be used in a very wide range of operation.

10.3.2 Ring Oscillator

The heart of the proposed wide tuning range PLL circuit is a current-controlled ring oscillator based on STSCL topology. It is possible to design this oscillator compatible to the critical path in the digital circuit and hence adjust the system clock with respect to the delay of the critical path.

In this work, we are using STSCL buffers as delay stages as illustrated in Fig. 10.5. Based on this topology, the replica bias circuit generates the proper bias voltages for NMOS biasing transistor (MT) and PMOS load devices ($M3$ and $M4$). To have a balanced capacitive loading at the output of all delay elements, an interchange placement for delay elements has been used as depicted in Fig. 10.5.

Figure 10.6 shows the tuning range of STSCL based ring oscillators with 8 and 24 number of delay stages versus tail bias current of each delay cell (I_{SS}). As it can be seen, the oscillation frequency ranges from below 1 kHz to about 100 MHz (about six decades) by adjusting the bias current. It is also noticeable that the tail bias current can be reduced down to only 10 pA per cell for very low oscillation frequencies.

To achieve more tuning range, the number of delay elements inside the oscillator can be changed. Based on the simulation results shown in Fig. 10.6, the oscillation frequency will scale inversely proportional to the number of delay elements inside the loop.

An output buffer is used to connect the oscillator output to the frequency divider circuit without disturbing the performance of the oscillator.

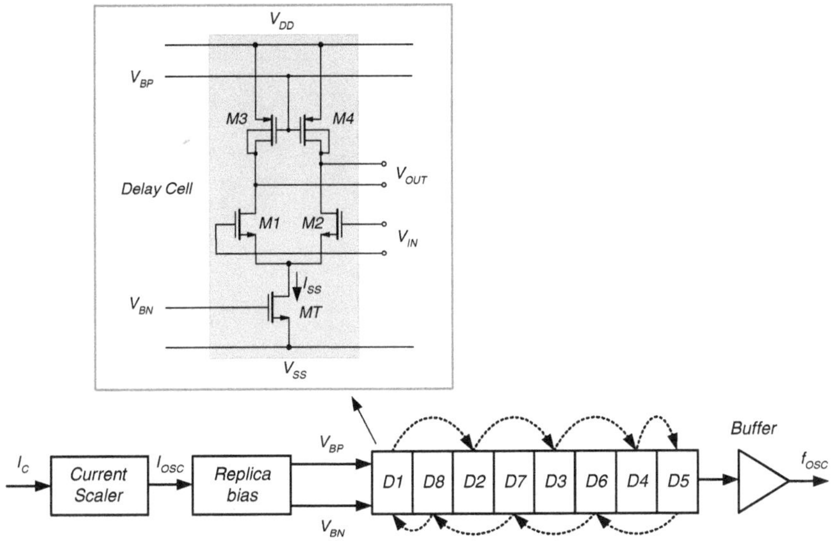

Fig. 10.5 Current-controlled ring oscillator structure uses STSCL cells as delay stages

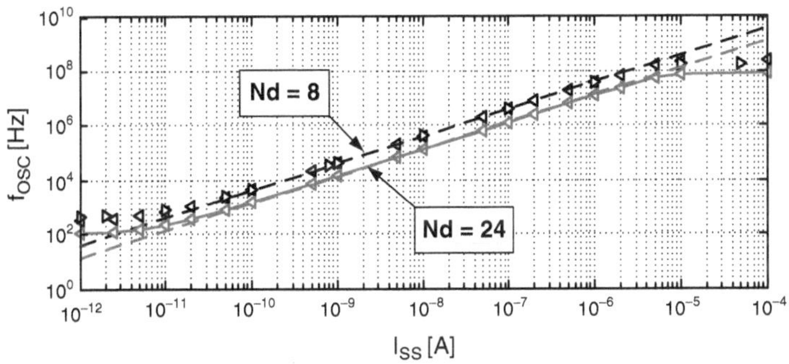

Fig. 10.6 Simulated tuning range of STSCL ring oscillator with 8 and 24 delay elements designed in 0.13-μm CMOS technology

10.3.3 Frequency Divider and Phase-Frequency Detector (PFD)

Frequency divider and PFD circuits, both are designed based on STSCL topology. This topology allows to have a power consumption proportional to the operating frequency and hence make a more power efficient circuit.

A programmable divide by 1/2/4/8/16 circuit is used for this PLL. Figure 10.7 shows the proposed frequency divider and also its building blocks. The divider is based on D flip-flop (DFF) circuits constructed based on STSCL latches (Fig. 10.7a). Using a MUX, as illustrated in Fig. 10.4, the division ratio of the circuit can be programmed.

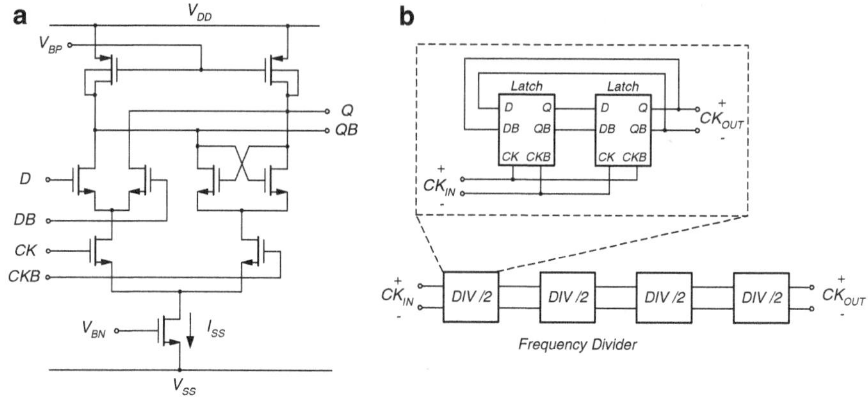

Fig. 10.7 Frequency divider circuit: (**a**) STSCL latch circuit schematic and (**b**) Frequency divider

As the operation frequency of the circuit decreases through the divider chain, the bias current of the divider-by-two stages can be scaled down to improve the power efficiency of the circuit. The PFD used in this work is based on the conventional topology explained in [9], while using STSCL building blocks.

10.3.4 Transconductor

A key component in design of the proposed wide tuning range PLL is the transconductor used to convert the controlling voltage to current. As illustrated in Fig. 10.6, the tail bias current or gate delay in STSCL topology can be adjusted over a very wide range. To cover such a wide range, the output current swing of the transconductor used inside the loop needs to be very wide and at the same time the transconductance should satisfy the loop stability requirements. As illustrated in Fig. 10.8, the core of the proposed transconductor is a PMOS based resistance with shorted bulk to drain similar to the load resistance of STSCL gates. $M1$ in this configuration acts as a current buffer. The resistivity of $M2$ is controlled by a local loop and is equal to $R_{M2} = V_{SW}/I_C$ ($V_{SW} = V_{DD} - V_{REF}$). Based on simulation results, this circuit can provide an output current between 40 pA and 800 nA with a transconductance proportional to current.

10.4 Simulation and Experimental Results

Figure 10.9 shows the simulated transient response of the PLL in different operating frequencies. The time scale of the graph is normalized to the oscillation period. As can be seen, the transient response of the PLL remains invariant with the frequency scaling and hence the ratio of the settling time respect to the oscillation period remains almost unchanged.

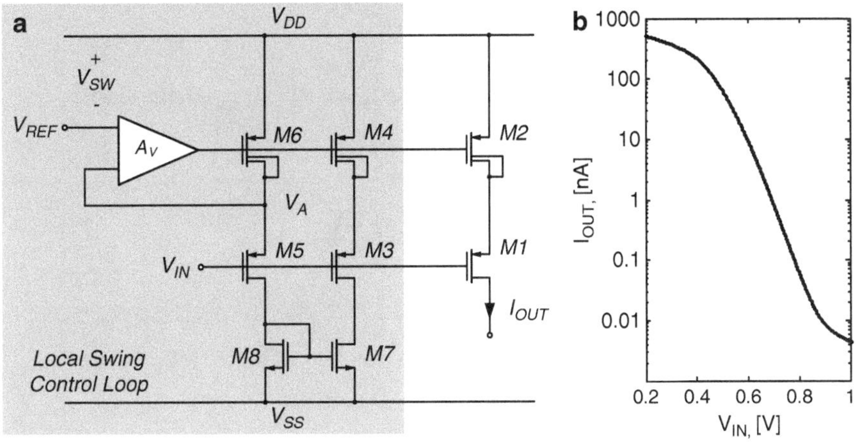

Fig. 10.8 (**a**) Wide swing transconductor. (**b**) I–V characteristics of the transconductor

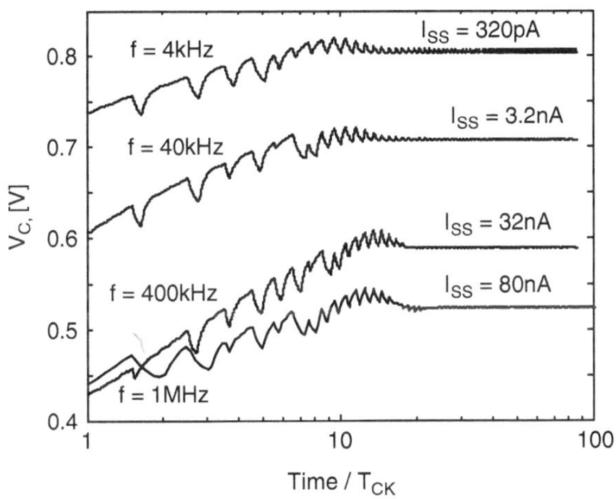

Fig. 10.9 Simulated transient response of the PLL in different frequencies

It is important to observe the performance of circuit in presence of large frequency jumps at the input. Figure 10.10 shows the simulated controlling voltage (top) and the controlling current (bottom) when there is a very big (×200) at the input. As the cutoff frequency of the loop filter is proportional to the controlling current, the loop responses very fast to the jumps in frequency and it remains stable. In this simulation, frequency changes on both directions (increasing by a factor of 200 and then reducing by the same factor) has been considered to make sure that in both cases the stability will be maintained. As mentioned before, the transconductor used for converting the controlling voltage to controlling current is biased in subthreshold regime with an exponential I–V characteristics. This property helps to maintain a

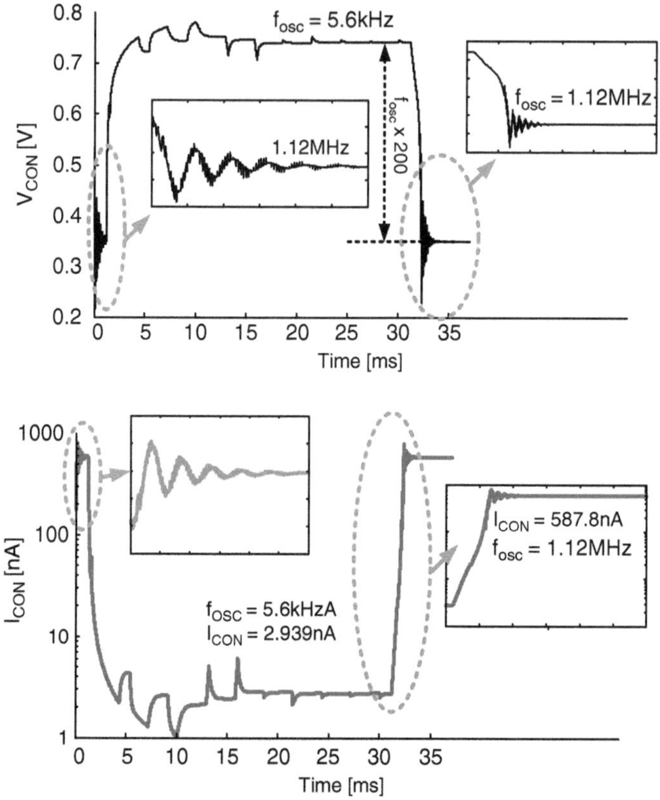

Fig. 10.10 Simulated transient response of the PLL when there is a jump at the input frequency. In this simulation, the initial input frequency is $f_1 = 1.12\,\text{MHz}$ and then there is a jump to $f_2 = f_1/200 = 5.6\,\text{kHz}$. At the end of simulation, again there is a jump back to f_1

very large tuning range. As illustrated in Fig. 10.10, the controlling current (which in plotted in logarithmic scale), is changing by a factor of 200, while controlling voltage is changing only between 0.35 and 0.75 V, approximately.

The proposed PLL has been implemented in a conventional 0.13-μm CMOS technology with an active area of $300\,\mu\text{m} \times 200\,\mu\text{m}$, whose photomicrograph is shown in Fig. 10.11. In this design, division ratio inside the loop can be set to $N = 2^n$ where n =0–4. As depicted in Fig. 10.12, measurement results show that the frequency of the 8-stage ring oscillator used in the PLL can be adjusted from 1 kHz to 3 MHz. Selecting other outputs of the loop divider, as shown in Fig. 10.4, makes the tuning range wider by the division factor. The circuit power consumption is 9 pW/Hz while in low frequencies power efficiency degrades due to the overhead of the biasing circuitry. At 3 MHz oscillation frequency, PLL consumes 20 μW, while it consumes 300 nW at 1 kHz. Measurements show that the supply voltage can be reduced down to 0.9 V with some reduction on the tuning range due to limitation at the output current of the transconductor. In the measurements shown in Fig. 10.12,

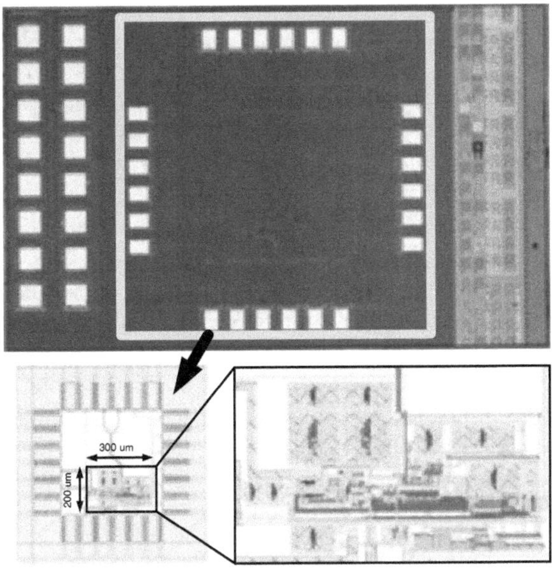

Fig. 10.11 Mask layout of the proposed wide tuning range PLL implemented in 0.13- μm CMOS technology and occupying 300 μm× 200 μm area

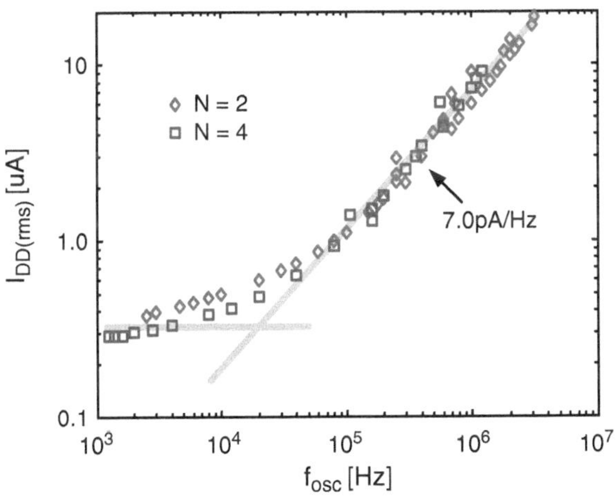

Fig. 10.12 Measured rms supply current consumption versus oscillation frequency for two different loop-divider values

voltage swing at the output of STSCL gates are set to be $V_{SW} = 200\,\text{mV}$. By reducing V_{SW}, noise margin and gain of STSCL gates reduces and hence oscillator stops operating. Measurements show that V_{SW} could be reduced down to 170 mV without degrading the performance and improve proportionally the operating frequency.

10.5 Conclusions

Design of wide tuning range and stable PLLs is a challenging task. To maintain the stability in different output frequencies, special techniques are required to be used. One common approach is implementing adaptive bandwidth loop filters.

In this chapter, an adaptive bandwidth and self-biased PLL compatible to the subthreshold source-coupled circuits is developed. The bandwidth of the proposed PLL is scaled with respect to the operating frequency using a self-biased technique. Based on the proposed approach, the bias current of the charge-pump circuit and also the zero in the loop filter both are scaling with respect to the operating conditions.

A test chip has been implemented in 0.13-μm CMOS technology occupying $300\,\mu\text{m} \times 200\,\mu\text{m}$. Simulation results show that the output frequency can be adjusted over three decades ranging from 300 Hz to 3 MHz. The power consumption of the PLL also scales with respect to the output frequency. This circuit can be used to tune the specifications of analog or digital circuit designed based on subthreshold source-couple topology.

References

1. F. Gardner, "Charge-pump phase-locked loops," in *IEEE Transactions on Communication*, vol. 28, pp. 1849–1858, Nov. 1980
2. T. H. Lee, *The Design of CMOS Radio-Frequency Integrated Circuits*, Cambridge University Press, Second Ed., 2004
3. J. G. Maneatis, "Low-jitter process-independent DLL and PLL based on self-biased techniques," *IEEE J. Solid-State Circuits*, vol. 11, pp. 1723–1732, Nov. 1996
4. A. Tajalli, P. Muller, and Y. Leblebici, "A power-efficient clock and data recovery circuit in 0.18 μm CMOS technology for multi-channel short-haul optical data communication," *IEEE J. Solid-State Circuits*, vol. 42, pp. 2235–2244, Oct. 2007
5. T. Ebuchi, Y. Komatsu, T. Okamoto, Y. Arima, Y. Yamada, K. Sogawa, K. Okamoto, T. Morie, T. Hirata, S. Dosho, and T. Yoshikawa, "A 125-1250 MHz process-independent adaptive bandwidth spread spectrum clock generator with digital controlled self-callibration," *IEEE J. Solid-State Circuits*, vol. 44, no. 3, pp. 763–772, Mar. 2009
6. A. P. Chandrakasan and R. W. Brodersen, *Low Power Digital CMOS Design*, Kluwer, 1995
7. M. Mansuri and C. -K. K. Yang, "A low-power adaptive bandwidth PLL and clock buffer with supply-noise compensation," *IEEE J. Solid-State Circuits*, vol. 38, no. 11, pp. 1804–1812, Nov. 2003
8. J. G. Maneatis, J. Kim, I. McClatchie, J. Maxey, and M. Shankaradas, "Self-biased high-bandwidth low-jitter 1-to-4096 multiplier clock generator PLL," *IEEE J. Solid-State Circuits*, vol. 38, no. 11, pp. 1795–1803, Nov. 2003
9. H. Rategh and T. H. Lee, *Multi-GHz Frequency Synthesis & Division*, Kluwer, 2001
10. T. Wu, P. K. Hanumolu, K. Mayaram, U.-K. Moon, "Method for a constant loop bandwidth in LC-VCO PLL frequency synthesizers," *IEEE J. Solid-State Circuits*, vol. 44, no. 2, pp. 427–435, Feb. 2009
11. A. Tajalli, P. Torkzadeh, and M. Atarodi, "A wide tuning range, fractional multiplying delay-locked loop topology for frequency hopping applications," in *Analog Integrated Circuits and Signal Processings*, vol. 46, no. 3, pp. 203–214, Mar. 2006

12. M. Brownlee, P. K. Hanumolu, K. Mayaram, U. -K. Moon, "A 0.5-GHz to 2.5-GHz PLL with fully differential supply regulated tuning," *IEEE J. Solid-State Circuits*, vol. 41, no. 12, pp. 2720–2728, Dec. 2006
13. G. Yan, C. Ren, Z. Gzo, Q. Ouyang, and Z. Chang, "A self-biased PLL with current-mode filter for clock generation," in *IEEE International Solid-State Circuits Conference (ISSCC)*, pp. 420–421, Feb. 2005
14. P. K. Hanumolu, M. Brownlee, K. Mayaram, and U. -K. Moon, "Analysis of charge-pump phase-locked loops," in *IEEE Transactions on Circuits and Systems-I: Regular Papers*, vol. 51, no. 9, pp. 1665–1674, Sep. 2004
15. S. Sidiropoulos, D. Liu, J. Kim, G. Wei, and M. Horowitz, "Adaptive bandwidth DLLs and PLLs using regulated supply CMOS buffers," in *Symposium on VLSI Circuits Digest of Technical Papers*, pp. 124–127, Jun. 2000
16. B. Razavi, "Monolithic phase-locked loops and clock recovery circuits: theory and design," Wiley, 1996

Chapter 11
Conclusions

In this work, the potentials of subthreshold MOS devices for implementing power-performance scalable integrated systems have been studied. It has been shown that the exponential I–V characteristics of this type of devices can help to implement very widely tunable analog and digital integrated circuits. Meanwhile, low current density of transistors in this region of operation makes them very suitable for implementing ultra-low-power circuits.

The book starts with a brief overview on power-performance scalable systems and their applications. A short study on previous art on widely tunable digital and analog integrated circuits have been provided. This type of circuits can be used in multi-standard or multi-purpose flexible integrated systems. In addition, many modern complicated integrated systems are employing power management systems in which power-performance scalable circuits are the essential building blocks. In the rest of this report, different techniques for implementing widely tunable digital and analog integrated circuits have been proposed. These techniques are categorized in two parts.

In the first part, implementing ultra-low-power source-coupled logic circuits have been extensively studied and explored. This part includes some novel techniques for implementing subthreshold SCL circuits [1–5]. In addition, the performance of this type of logic circuits is compared analytically with conventional CMOS topology for ultra-low-power applications [6–8]. Several techniques for improving the performance of subthreshold SCL (STSCL) circuits have been proposed that make this logic family a suitable topology for implementing ultra-low-power systems [9–13].

In the second part of this work, widely tunable analog integrated circuits such as continuous-time filters, analog-to-digital data converters, and phase-locked loop systems have been considered. Employing an approach compatible with the design of subthreshold SCL digital circuits for implementing analog circuits provides this opportunity to implement widely adjustable complex mixed-mode integrated systems.

Two different approaches have been proposed for implementing continuous-time filters. The filters are based on MOSFET-C and g_m-C topologies showing few decades of tuning range [14–16]. Meanwhile, two analog-to-digital data converters based on folding and interpolating and ring oscillator $\Sigma\Delta$ topologies have been introduced. Both converters can be operated over a very wide frequency range [9, 10].

A. Tajalli and Y. Leblebici, *Extreme Low-Power Mixed Signal IC Design: Subthreshold Source-Coupled Circuits*, DOI 10.1007/978-1-4419-6478-6_11,

The wide tuning range phase-locked loop (PLL) circuit introduced in this work can be employed for controlling the operating condition of digital or analog integrated circuits. This is especially demanding for power management purpose in modern integrated systems. In this context, the stability issue in widely tunable PLL circuits has been studied.

11.1 Main Contributions

Ultra-low-power and configurable circuits are becoming essential parts of modern integrated systems. Circuits operating in different conditions and with different standards can help to reduce the product cost, and power consumption, while at the same time can keep the performance very high. Power-performance scalability of circuits is also very important for implementing energy efficient circuits which are very demanding in many different industrial or biological applications.

The main focus of this research has been developing some novel techniques for implementing configurable integrated circuits with the possibility of embedding them in ultra-low-power systems. As it has been shown, MOS devices biased in weak-inversion could be used for this purpose. By moving toward smaller device future sizes in MOS technology, there is an evident improvement in speed of operation while area can be shrunk considerably. However, in deep submicron technologies, the leakage current increases and many other side effects become more pronounced limiting the performance or usability of the MOS devices for specific applications. This problem can be seen more evidently in ultra-low-power systems used in many modern applications such as biological systems, sensor networks, data acquisition systems, battery operating systems, etc.

Standard digital CMOS circuits implemented in deep submicron technologies suffer from subthreshold leakage current. The extra power consumption due to the leakage current is becoming a significant part of the total power dissipation of digital systems. Subthreshold source-couple logic (STSCL) topology introduced in this work is a new approach to use more efficiently the available energy [1, 2]. In this topology, there is a good control on the static current of each cell even in advanced technologies such as in CMOS 65 nm [7]. This property can help to reduce the static power consumption of each cell well below the subthreshold leakage current exists in static CMOS topology [6, 7].

The main problem in design of STSCL circuits with ultra-low-power consumption is implementing very high-valued load resistances with a very small area occupation and good control on their resistivity. In this work, bulk-drain shorted PMOS devices with close to minimum sizes are introduced to implement the desired load devices [1,5]. Using this approach, it is possible to reduce the static power consumption of each gate down to few pico-Amperes. Several test structures have been designed and implemented using the proposed topology. Meanwhile, two standard cell libraries have been implemented in 0.18- μm and 90- nm CMOS technologies.

An analytical approach has been proposed to compare the performance of STSCL circuits with conventional static CMOS digital circuit in different operating frequencies [6]. Explored by this analysis, the operating range that STSCL topology exhibits a better power efficiency compared with the conventional CMOS depends on average activity rate of a system [7]. It has also been shown that in very low activity rate systems, where traditionally static CMOS is widely used, STSCL topology can offer a better compromise. Using STSCL topology, a novel static random-access memory (SRAM) with very low stand-by current has been developed [8]. This circuit has been used successfully as a test vehicle for showing the performance of very low activity rate STSCL circuit.

To improve the power-delay performance of STSCL circuits, several techniques have been proposed in this work. Pipelining is a powerful approach for increasing the activity rate of a circuit, and hence considerably improve the efficiency of source-coupled logic circuits [5]. To reduce the overhead associated with the extra circuitry required for using pipelining technique in STSCL topology, some new power and area efficient techniques have been developed [11]. This technique has been first used to implement an encoder as a part of an ultra-low-power analog-to-digital converter circuit [9].

Using simple source follower buffers in each STSCL gate, as it is introduced in this research, can improve the power efficiency of this type of circuits by a factor of about two [3, 12]. In addition, this technique can simplify the development of standard cell library of STSCL circuits and reduce the required area for each cell [13].

The second part of this work concerns with the design of ultra-low-power and scalable analog circuits. The scope of this part of research has been developing a high performance analog front end compatible with the logic circuitry developed in the first part in order to construct the fundamentals of implementing scalable performance mixed-mode circuits.

Based on the high-valued resistance topology used in STSCL configuration, a floating high-valued resistance has been introduced and used for implementing a widely tunable MOSFET-C filter [14]. The proposed floating resistance in addition to a scalable power operational transconductor amplifier are the main building blocks for constructing this filter.

In addition, a transconductor-C filter with five decades of tunability has been proposed using very basic differential transconductors. To improve the linearity performance of the filter, a new modified biquadratic topology has been used [15]. Based on the proposed topology, instead of converting differential voltages to current and then summing or subtracting them in current domain, the summation or subtraction is done in voltage domain somehow to reduce the required linearity range of each transconductance, and then the signal is converted to current. It has been shown that this technique helps to improve the total harmonic distortion by 30 dB.

To construct scalable performance analog-to-digital converters (ADCs) which are essential building blocks in all modern mixed-mode circuit, two different topologies have been developed. The first ADC is based on folding and interpolating topology which is constructed based on subthreshold source-coupled circuits to be able to reduce the power dissipation below 1 μW [17]. The PMOS load device has been modified to improve the bandwidth of subthreshold source-couple circuits. A pipelined STSCL encoder has been employed to have a very low power consumption in digital part [9, 10]. A novel resistor ladder with very high resistivity has been used in the proposed folding and interpolating ADC. The second ADC is constructed using ring oscillator based ΣΔ topology. The delay elements and logic circuit of this ADC are based on STSCL topology. This ADC can be used for quantizing input current levels in the range of few pico-Amperes.

Finally, a widely tunable phase-locked look has been developed. The importance of this block is that it can be used for controlling the operating condition of miscellaneous digital and analog circuits with respect to the operating frequency. To cover a very wide tuning range, it is necessary to overcome different issues such as covering a large operating frequency and also stability. Several techniques for keeping the system stable over its wide tuning range has been developed. Self-biased charge pump and loop filter circuits have been designed for this purpose.

The collection of circuit techniques introduced in this work provides the basics for implementing complicated ultra-low-power mixed-mode integrated circuits.

11.2 Perspectives

This research introduces a set of new design styles for implementing ultra-low-power and scalable circuits. The proposed techniques have been utilized for designing some different circuit families successfully. There is a considerable potential for using these techniques for many other circuits and applications.

Reduce Leakage: One of the main concerns in this work has been developing some techniques for reducing the logic cell leakage or static power consumption. The main limiting factor for reducing the static power consumption of STSCL circuits below tens of pico-Amperes is the forward bias current of source-bulk PN junction of PMOS load devices. Simple circuit techniques for overcoming this problem will be a step forward for implementing STSCL circuits. A possible remedy can be putting a resistance in series to the bulk-drain connection such as the one used in implementing pre-amplifiers in Chap. 8.

Area Efficient Standard Cell Library: In this work, few techniques for implementing efficient standard cell libraries have been proposed. These techniques can help to improve the area and power efficiency of the library very much. However, still there is this possibility to improve especially the area efficiency of logic cells. Making smaller standard cells will result in more power and speed efficient designs. Design of small area logic cells can be specially important for memory cells. The area of the first memory cell developed in this work is relatively large. By a careful design, the size of devices and hence area of a cell can be reduced.

Linear and High-Valued Resistance: On analog part, implementing a high-valued resistance with high linearity performance is very important. There are many emerging applications that need high performance filters and analog signal processing units and in this type of circuits linear high-valued resistances are essential components. The floating high-valued resistance developed in this work exhibits a medium level linearity performance.

Ultra-Low Power Data Converters: In addition to the techniques introduced in this work for implementing ADCs, it would be interesting to study the efficiency of the other topologies for designing ADCs. For reducing the power consumption even more, successive approximation (SAR) topology can be used. Meanwhile, using offset cancelation techniques can help to improve the power efficiency of the ADC.

Reference Voltage and Current Generator: To complete the design, a very low power and precise integrated voltage reference biased in subthreshold regime is required. This reference voltage can be used for controlling the voltage swing in STSCL gates and also generate reference currents for biasing purposes.

FPGA and FPAA: The versatility of the proposed topologies in this work are very suitable for implementing digital and mixed-signal programmable gate arrays. This type of programmable gate arrays may include different essential blocks such as data converters, phase-locked loop and clock generators, continuous-time filters, and a rich array of digital blocks. It is also possible to change the driving strength or operating frequency of the digital building blocks very simply by changing bias voltage as explained in Chap. 4. Such a topology can be a good match for ultra-low power applications.

References

1. A. Tajalli, E. Vittoz, Y. Leblebici, and E.J. Brauer, "Ultra low power subthreshold current mode logic utilizing a novel PMOS load device," in *IEE Electronics Letters*, vol. 43, no. 17, pp. 911–913, Aug. 2007
2. A. Tajalli, E. Vittoz, Y. Leblebici, and E.J. Brauer, "Ultra low power subthreshold current mode logic utilizing a novel PMOS load device concept," in *Proceedings of European Solid-State Circuits Conference (ESSCIRC)*, pp. 304–307, Munich, Germany, Sep. 2007
3. A. Tajalli, M. Alioto. E.J. Brauer, and Y. Leblebici, "Design of high performance subthreshold source-coupled logic circuits," in *Proceedings of International Workshop on Power and Timing Modeling, Optimization and Simulation (PATMOS)*, Lisbon, Portugal, Oct. 2008
4. A. Tajalli, Y. Leblebici, and E.J. Brauer, "Pico-watt source-coupled logic circuits," in *Proceedings of International Conference on Very Large Scale Integration (VLSI-SoC)*, Rhode Island, Greece, Oct. 2008
5. A. Tajalli, E.J. Brauer, Y. Leblebici, and E. Vittoz, "Subthreshold source-coupled logic circuits for ultra low power applications," *IEEE J. Solid-State Circuits*, vol. 43, no. 7, pp. 1699–1710, Jul. 2008
6. A. Tajalli and Y. Leblebici, "Subthreshold leakage reduction: A comparative study of SCL and CMOS design," to appear in *IEEE International Symposium on Circuits and Systems (ISCAS)*, Taipei, Taiwan, May 2009

7. A. Tajalli and Y. Leblebici, "Leakage current reduction using subthreshold source-coupled logic," in *IEEE Transactions on Circuits and Systems-II: Express Briefs (Special Issue on Nanocircuits)*, vol. 56, no. 5, pp. 347–351, May 2009
8. A. Tajalli and Y. Leblebici, "Subthreshold SCL for ultra-low-power SRAM and low-activity-rate digital systems," to appear in *European Solid-State Circuits Conference (ESSCIRC)*, Athen, Greece, Sep. 2009
9. M. Beikahmadi, A. Tajalli, and Y. Leblebici, "A subthreshold SCL based pipelined encoder for ultra-low power 8-bit folding/interpolating ADC," in *Proceedings of The Nordic Microelectronics Event (NORCHIP)*, Tallin, Estonia, pp. 9–12, Nov. 2008
10. A. Tajall and Y. Leblebici, "Ultra-low power mixed-signal design platform using subthreshold source-coupled circuits," to appear in *Design And Test in Europe (DATE)*, Dreseden, Germany, Mar. 2010
11. A. Tajalli, E.J. Brauer, and Y. Leblebici, "Ultra low power 32-bit pipelined adder using subthreshold source-coupled logic with 5fJ/stage PDP," *Elsevier Microelectron. J.*, vol. 40, no. 6, pp. 973–978, Jun. 2009
12. A. Tajalli, F. Gurkaynak, M. Alioto, Y. Leblebici, and E.J. Brauer, "Improving the power-delay product in SCL circuits using source follower output stage" in *Proceedings of IEEE International Symposium on Circuits and Systems (ISCAS)*, Seattle, pp. 145–148, USA, May 2008
13. A. Tajalli, M. Alioto, and Y. Leblebici, "Power-delay performance improvement of subthreshold SCL circuits," in *IEEE Transactions on Circuits and Systems-II: Express Briefs*, vol. 56, no. 2, pp. 127–131, Feb. 2009
14. A. Tajalli, Y. Leblebici, and E.J. Brauer, "Implementing ultra high value tunable CMOS resistors," in *IEE Electronics Letters*, vol. 44, no. 5, pp. 349–350, Feb. 2008
15. A. Tajalli, and Y. Leblebici, "Linearity improvement in biquadratic transconductor-C filters," in *IEE Electronics Letters*, vol. 43, no. 24, Dec. 2007
16. A. Tajalli and Y. Leblebici, "A widely-tunable and power-scalable MOSFET-C filter operating in subthreshold," in *Custom Integrated Circuits Conference (CICC)*, San Jose, USA, pp. 593–596, Sep. 2009
17. A. Tajalli and Y. Leblebici "Nanowatt range folding-interpolating ADC using subthreshold source-coupled circuits," *J. Low-Power Electron.*, vol. 6, Apr. 2010
18. A. Tajalli and Y. Leblebici, "A slew controlled LVDS output driver circuit in 0.18m CMOS technology," *IEEE J. Solid-State Circuits*, vol. 44, no. 2, pp. 538–548, Feb. 2009

Index

FSC
www.fsc.org
MIX
Papier | Fördert gute Waldnutzung
FSC® C083411

Kolibri 360 GmbH
Ferdinand-Jühlke-Straße 7
99095 Erfurt, Deutschland
produktsicherheit@kolibri360.de